ICSOT Korea: Safety of Offshore and Subsea Structures in Extreme and Accidental Conditions 2015

Busan, South Korea
15 - 16 September 2015

ISBN: 978-1-5108-9199-9

Printed from e-media with permission by:

Curran Associates, Inc.
57 Morehouse Lane
Red Hook, NY 12571

Some format issues inherent in the e-media version may also appear in this print version.

Printed by Curran Associates, Inc. (2019)

The Royal Institution of Naval Architects
8-9 Northumberland Street
London, WC2N 5DA
United Kingdom

Phone: 44 207 235 4622
Fax: 44 207 259 5912

publications@rina.org.uk

Additional copies of this publication are available from:

Curran Associates, Inc.
57 Morehouse Lane
Red Hook, NY 12571 USA
Phone: 845-758-0400
Fax: 845-758-2633
Email: curran@proceedings.com
Web: www.proceedings.com

CONTENTS

AUTHORS' CONTACT DETAILS

Alan Loynd
Branscombe Marine Consultants
branscombe@netvigator.com

Ana Ivanović
University of Aberdeen
a.ivanovic@abdn.ac.uk

Bongsik Choo
KAIST
bschoo@kaist.ac.kr

Dae Kyeom Park
Pusan National University
greatddalgi@pusan.ac.kr

Ho-Gyun Park
Korean Register
hgpark@krs.co.kr

Igor A. Guz
University of Aberdeen
i.guz@abdn.ac.uk

Jeom Kee PAIK
Pusan National University
jeompaik@pusan.ac.kr

Jin Hyung PARK
Yokogawa Electric Korea
jinhyung.park@kr.yokogawa.com

John Waltham-Sajdak
Alion Offshore
jsajdak@alionscience.com

JooShin PARK
Samsung Heavy Industries Co., Ltd.
jooshin.park@samsung.com

José Manuel CABRERA MIRANDA
Pusan National University
cabrera@pusan.ac.kr

Jung Kwan Seo
Pusan National University
seojk@pusan.ac.kr

Jung Min Sohn
Pukyeong National University
jminz@pknu.ac.kr

KIM Hyungju
Norwegian University of Science and Technology
hyung-ju.kim@ntnu.no

Kyung Won Yang
DNV GL Oil & Gas
kyung.won.yang@dnvgl.com

Petter Ellingsen
DNV GL
hans.petter.ellingsen@dnvgl.com

Prof. Sang Gap Lee
Korea Maritime and Ocean University
sglee@kmou.ac.kr

Saara Keränen
DNV GL
saara.keraenen@dnvgl.com

Sang Eui Lee
Pusan National University
euui2000@pusan.ac.kr

Shuai Yuan
Zhejiang University
yuanshuaizheda@163.com

Soo Young Bae
Korea Marine Equipment Research Institute
bsy7000@komeri.re.kr

Sung In Park
DSME
parksungin@DSME.co.kr

Tae-Min Cho
Samsung Heavy Industries Co., Ltd.
tm1.cho@samsung.com

Takashi MIWA
Kobe University
miwa@maritime.kobe-u.ac.jp

Yanbin Wu
Kobe University
128w201w@stu.kobe-u.ac.jp

Yonghwan Kim
Seoul National University
yhwankim@snu.ac.kr

NOVEL MATERIALS FOR OFFSHORE AND SUBSEA STRUCTURES: SAFETY IMPLICATIONS

I A Guz, University of Aberdeen, Scotland, UK
M Menshykova, University of Aberdeen, Scotland, UK
J K Paik, Pusan National University, South Korea

SUMMARY

The paper reviews practical applications of composite materials in oil and gas industry. A special consideration has been paid to possible future opportunities in offshore and onshore usage; in particular, to long fibre-reinforced composite pipes. The problem of thick-walled filament-wound multi-layered composite pipes subjected to outer pressure is considered as an example. An analytical method is used for failure analysis for different lay-ups and loading conditions. The stress distributions through the pipe thickness for various lay-ups are computed for fibre-reinforced pipes under different outer pressure magnitudes.

1. INTRODUCTION

Having advantages over conventional materials in weight, strength, stiffness, corrosion resistance composite materials are becoming more and more attractive for structural and non-structural applications in the oil and gas industry. However, in spite of their undoubted advantages, the main of which may be the ability to be tailored for specific purpose, their introduction in the industry is a very slow process. Lack of appropriate performance information, regulatory requirements, efficient design procedures and reparability issues are the main obstacles.

1.1 USE OF COMPOSITES IN OIL AND GAS INDUSTRY

Fibre-reinforced composites are the alternatives of steel, metal and wood in many applications. This may be due to their high specific tensile and compressive strength, good fatigue and corrosion resistance and their suitability for the production of complex-shape components with reduced manufacture times compared to conventional metallic materials. Ease of handling, high adaptability, economy of fabrication and damage tolerance are the driving force of the application of the fibre-reinforced composite materials.

High specific strength as well as overall weight-saving and resistance to a wide range of fluids (including seawater, aerated water, H_2S, CO_2 and hydrocarbons) which can attack metals results in growing interest in their use for applications within the oil and gas sector. The areas within the oil industry, where the composites are used include rigid pipework, coiled tubing, rigid and flexible risers, structural repairs and others. Gibson [1] described composite pipework as one of the most established uses of composites within the oil and gas industry. One of the more recent papers on the topic was published by Taheri [2], where fibre-reinforced composites for pipes and tanks manufacture and rehabilitation in oil and gas industry were considered. The author also outlined various applications of the FRPs in relation to pipes and risers.

The exploitation of GFRP offshore was inhibited by the knowledge that the strength of most GFRP systems is reduced at elevated temperature and by the absorption of water. The results of the tests conducted to characterise the strength reduction of three GFRP composite materials as a function of temperature and testing environment can be found in [3]. In [4, 5] authors considered the behaviour of glass-fibre-reinforced unsaturated polyester laminates fabricated by using the vacuum-assisted resin infusion technique after immersing in the artificial seawater for different durations of time. The biaxial failure envelope and creep testing of GFRP in high temperature (up to 160°C) liquids were investigated in [6]. The tests were made on dry and after exposure to saturation in various working environments (sea water, crude oil, water with dissolved CO_2, and water with dissolved H_2S) specimens, highlighting the sensitivity to loading direction with respect to fibre orientation.

The corrosive behaviour of PMC (polymer matrix composite) structure is much better than the steel structure in the marine environment. In view of the wide application of polymeric composites in offshore structures, submersibles and highway infrastructure, the effect of seawater on deformation, strength and durability of materials were considered in [7]. The effect of moisture absorption on the compressive, tensile and shear behaviours of PMC has been investigated by many researchers [8-12]. It is known from their research results that moisture absorption causes plasticisation and swelling of the epoxy, weakening the interfacial strength between fibres and epoxy, which results in the reduction of tensile and shear strength. However, the effect of moisture absorption on fracture behaviour is different depending on the studies conducted.

In [13] the effect of seawater on failure pressure and impact behaviours of glass-epoxy pipes was experimentally investigated. The effect of seawater absorption on the interlaminar fracture behaviour of carbon/epoxy composites was investigated in [14]. The effect of hydrostatic pressure on the compressive and shear behaviour of carbon/epoxy composites was considered in [15-18]. The tensile behaviour of carbon/epoxy composites under a hydrostatic pressure environment was investigated in [19]. It is generally accepted that the tensile and compressive strengths of polymeric composites increase as the applied hydrostatic pressure increases.

The combined effect of seawater absorption and hydrostatic pressure on the mechanical behaviour of PMC materials was considered in [20, 21]. For this purpose, fracture tests were performed using seawater-absorbed carbon/epoxy composite, whose seawater absorption capability was at a maximum absorption level, in a hydrostatic pressure environment. Compliance, fracture load and fracture toughness were determined as a function of applied hydrostatic pressure. After the fracture tests were performed, optical microscope examination was made on the fracture surfaces to determine the fracture mechanism of carbon/epoxy composites in a deep-sea environment. Comparative study of dry- and seawater-absorbed specimens showed that seawater-absorbed specimen has a lot more epoxy fracture than the dried one. There is a clear need in more precise modelling of underwater compressive behaviour of composites. It can be achieved, in particular, by incorporating into modelling the exact analytical solutions developed by Guz [22-27].

Composite cylinders are widely used in the oil industry, designed to resist either internal pressure (e.g. fluid transport and storage, cooling systems) or external pressure (e.g. underwater or downhole applications). One of the main concerns with these structures is the propagation of interlaminar defects, produced during manufacture or service. Understanding the factors affecting the delamination resistance of the laminate used in the cylinders is essential if their safe working envelope is to be defined. In [28], the influence of winding angle was examined in order to optimize the cylinder's performance. In the paper, the delamination resistance of filament wound composite cylinders and the influence of delamination damage on the failure of externally pressurized composite cylinders were considered. Pressure tests on cylinders without implanted defects and with impact damage and implanted defects to simulate fabrication and in-service damage were performed. A series of external pressure tests was performed on the glass/epoxy, carbon/epoxy and carbon/PEEK cylinders to examine the influence of different types of delaminations.

Full-scale application of composites in rigid risers is expected within the next decade. Since risers operate in seawater, effects of seawater are examined by aging carbon- and glass-reinforced epoxy laminates of different thickness and lay-up orientation at temperature experimentally [29, 30]. The analytical scaling methodology that depends on the geometry, constituent material properties as well as location, orientation hybridization of the reinforcing fibres was developed but not verified for full-scale composite tubes [31, 32]. This information is particularly important for another class of risers hybrid flexible tubulars which are in various stages of maturity in the oil industry. These risers are either fully filament wound composites with thermoplastic matrix or have stainless steel internal carcass with an extruded polymer fluid barrier, a carbon steel interlocked hoop strength layer and helically wound composite tensile armour (e-glass, aramid, carbon fibres). Flexible risers are the tubes which achieve their flexibility by virtue of the fact that the load-bearing components are free to move relative to one another. There are significant opportunities for the use of unidirectional carbon fibre composite elements, which may be either thermoplastic or thermoset-based, in the armour for weight-saving, and because of corrosion problems with steel. This could represent a significant future application for composites [1].

The other use of composite materials in the offshore industry is the insulation coating of the production infrastructure at great depths. To limit heat losses and so to avoid the formation of hydrate and wax plugs inside subsea production flowlines or risers under high pressure and low external temperature conditions, even during production shutdowns, the pipelines need to be thermally insulated. One of the most efficient types of thermal insulation systems is the multilayered structure made of several materials of different thicknesses directly applied to the external surface of the steel pipe. Currently used materials in thermal insulating multilayered systems for deep sea applications include massive polymers and syntactic foams, composed of hollow glass microspheres embedded in a polymer matrix. These composites must combine thermal insulation function, low buoyancy while providing good compressive strength. Syntactic foam coatings are expected to fulfil today's challenges as pipelines are required to operate in more and more severe conditions. Syntactic foam is a lightweight composite material made from tiny hollow microspheres in a polymeric resin binder, along with other fillers and additives. Because the micro-spheres are filled with air, syntactic foam is low in both density and thermal conductivity. Typically, the choice of syntactic materials includes rigid binders and flexible binders. Rigid binders add support and reinforcement to the micro-spheres, giving greater strength. The most common rigid binder resin choice is epoxy, because of its superior strength and resistance to hot and wet conditions. However, excessive rigidity may result in cracking under thermal shock and cycling. Flexible binders reduce this prospect. Flexibility is desirable in many applications, and can offer some protection against cracking. However, flexible binders must be carefully designed to avoid degradation under hot, wet conditions, and they do not strongly reinforce micro-sphere fillers, detracting from their effectiveness in density and thermal conductivity [33-38].

Composite spoolable or coiled tubing is also intensively utilised technology in the oil industry (composite coiled or spoolable tubing refers specifically to Fiberspar and similar products). Spoolable tube is a thick-walled pipe which must withstand number of loads, such as bending, thermal, axial loads, internal and external pressure, etc. Tubing of this type can only be coiled by allowing the matrix to crack extensively, i.e. a brittle matrix composite is used with the matrix in an almost completely cracked condition. Of course, such a tube would leak unless it is equipped with a thermoplastic liner. With the increasing length of the boreholes, the use of steel coiled tubing is limited by tensile load capacity, combined allowed axial force and pressure and tube dimension. Composite coiled tubing allows overcoming the steel coiled tubing limited working depth problem as they have such advantages as being lightweight, possessing high strength and low density. The coiled tube itself is the thick-walled pipe of several kilometres length. Consequently, the continuous manufacturer processes such as pultrusion, pull winding or pull braiding technique is used for fibre-reinforced composite coiled tubing fabrication. One of the applications of coiled tubing in the oil industry is the well intervention operation. To use composite coiled tubing for high-pressure down-hole applications it must be capable to withstand a high working pressure, high temperature, bending and axial load.

The humid ageing of syntactic foams under hydrostatic pressure, up to 100 MPa, has been widely studied [39, 40]. But the combination of high pressure/high temperature conditions (up to 30 MPa/130 ^{0}C) is a recent concern: a pioneering research programme including ageing in hot-wet conditions has shown that traditional industrial syntactic foams were undergoing rapid and severe degradation in seawater at temperatures as low as 60 ^{0}C [41]. There is experimental evidence that a key factor may be the chemical degradation of glass microspheres due to the "entrance" of water [42].

Functionally graded materials are becoming increasingly viable thanks to polymer composite materials. Syntactic foams have become attractive for aerospace and marine applications owing to low density, high water resistance and high strength offering a variety of options for the designer for creating property gradients. Three-dimensional elasticity solution for a functionally graded simply supported plate subjected to transverse loading was obtained by Kashtalyan [43]. Several analytical and numerical studies are available on statically loading, impact loading, free vibration and thermoelasticity analysis of functionally graded core sandwich panels. paper [44] presents a three-dimensional elasticity analysis of sandwich panels with a functionally graded core subjected to transverse loading.

The application of composites in the subsea industry is based on thermoset matrix composites, mainly epoxies. Potentially thermoplastic composites can replace these. In [45], the results from preliminary screening tests on the selection of thermoplastic composite materials were obtained. Based on these results, glass/PEI and carbon/PEEK materials were selected for further evaluation.

Composites are routinely used to repair metal structures as damaged or corroded pipelines. These repairs have been a common practice in the oil industry and are generally easier and faster to carry out as compared to the traditional metallic method. At the moment the research is focused on the development of effective subsea composite repair systems and was investigated in [46-48]. The effectiveness of FRP composites for the subsea repair of steel pipeline was analysed in [46]. The performance of a carbon/epoxy repair installed and cured underwater has been tested [47].

1.2 FIBRE-REINFORCED COMPOSITE PIPES

The composite pipes from glass-fibre-reinforced epoxy (GFRE) have been used onshore for both low- and high-pressure applications with a wide variety of fluids, including hydrocarbons. Offshore glass-fibre-reinforced resin matrix composites are used for relatively low-pressure aqueous services. The effect of water absorption of the mechanical behaviour of fibreglass pipes used for offshore service waters was considered in [49]. Considering the fact that GFRE has the high specific strength and corrosion resistance, which make it ideal for the petrochemical industry, the use it offshore likely to change in the future [1]. However, the wall thickness of high-pressure GFRE pipe tends to be inconvenient for manufacture and handling, because of the high effective safety factor on the GFRE pipe. The development of steel strip laminate overcomes the problem [50]. The effect of artificial seawater on the impact behaviour of filament-wound GRE pipes has been experimentally studied in [51].

Since the properties of the fibre composite are different along the fibre and in transverse direction, the properties of the filament wound composite pipe can be easily changed by changing the number of winding angles and the fibre direction in each layer. The studies of thin-walled fibre-reinforced composite pipes were made by a number of researchers. Works [52-57] show that the winding angle plays an important role in the filament-wounded pipes design. The failure processes of thin-walled glass-reinforced polyester pipes wounded at "ideal angle" of $55^0 44'$ were studied by Hull, Legg and Spencer [52]. The series of closed-end and unrestrained end internal pressure tests to failure of the pipes were conducted. In their next work Spencer and Hull [53] extended their initial investigation and considered pipes wound at 35^0, 45^0, 65^0 and 75^0. Their investigation showed a pronounced effect of the winding angle on the micro-mechanisms associated with deformation, weepage and fracture. In [55], Rosenow considered filament-wound pipes wounded at six different angles under different types of loading: biaxial pressure, hoop pressure and tensile loading. Comparing the results with the classical lamination theory, he got good agreement between theory and experiment. His investigation showed that the optimal winding angle depends primarily on state of loading and should be 54.75^0 for biaxial pressure loading, 75^0 for hoop pressure loading and the lowest possible angle for tensile loading. Soden, Kitching and Tse [56] considered ±55^0 thin-walled filament wounded tubes under combinations of internal pressure and axial tensile and compressive loads to get the failure envelope. The buckling behaviour of ±55^0 filament wound pipes under combinations of external pressure and axial loading were considered in [57]. The analytical procedure for the orthotropic cylindrical shells was developed by Wild and Vickers [58]. The procedure allows to assess the stresses and deformations of filament-wound structures and the effect of wind angle. Their procedure is based on the Leikhnitskii [59] solution for the problem of plane stresses in the multi-layered thin circular cylindrically orthotropic shell.

The studies mentioned above were focused on thin-walled cylindrical fibre-reinforced structures. The number of studies on the behaviour of thick-walled fibre-reinforced tubular structures is more limited.

Thermal stress analysis of thick fibre-reinforced laminated rings was made by Ajit [60]. Rodriguez and Ochoa [32] have presented an evaluation of the behaviour of filament-wound spoolable composites. A combination of four-point bending testing and FEA was used to determine the minimum spool radii. An analytical solution for design and analysis of laminated composite tubes is given in [61]. Xia et al. [62] considered the thick-walled sandwich pipe with nonreinforced core layer and alternate-ply skin layers. In this paper, the method for stress-strain deformation under pure bending analysis is presented. The theoretical approach for the strength and failure mode prediction of filament-wound composite pipes under bending load was considered in [63]. Stress distribution in spoolable fibre-reinforced composite pipes was investigated in [64]. The cyclic internal pressure and temperature loading was considered in [65]. The stress analysis of multi-layered filament-wound composite pipes was carried out obtaining the time-dependent stress, strain and deformation distribution. The analytical method to study the mechanical response of thick-walled composite pipe subjected to internal pressure was proposed in [66]. The effect of random design variables on the pipe behaviour was analysed in the paper. Xia et al. [67] presented a simplified elastic solution to analyse the stress and deformation of multi-layered filament-wound composite pipes under internal pressure. The thermomechanical loading of thick-walled composite pipes was considered in [68, 69]. In [68] the analysis of filament-wound sandwich pipe was carried out and in [69] the multi-layered filament-wound composite pipe was considered under internal pressure and thermomechanical loading with thermal variation. Guedes [70] presented an analytical approach to time-dependent stress–strain state calculation in nonlinear viscoelastic multi-layered composite cylinders.

The following subsection will show how the usage of composite materials can contribute to tackle one of the challenging problems in the oil and gas industry.

2. STRESS AND FAILURE ANALYSIS FOR THIK-WALLED FIBRE COMPOSITE PIPES

In the oil and gas industry composite coiled tubing is used for different well intervention operations. In this case the main loads the pipe is subjected are bending load (when it is spooled on the drum), and the outer pressure and thermal loading in the well bore. Coiled tube is a thick-walled pipe of several kilometres length. Pultrusion, pull winding and pull braiding processes can be used to manufacture the fibre-reinforced pipe of such length. Pultrusion process is the cheapest and easiest one, however the resulting product is highly anisotropic. Pull braiding is the most expensive process and also the most difficult from manufacturer point of view. In this paper, we will focus on a pipe, produced by the winding process, subjected to outer pressure loading, which is important for well intervention operations and subsea applications. However, in the applications of water depth over 1000 m, normally, the composite pipes include metal layers (such as subsea flexible pipe) to resist the high tensile load and external buckling load. The research on mechanical response of non-metal composite pipe in the offshore applications was published in [71, 72]. The method presented in this paper can be modified to solve the problem of multi-layered thick-walled fibre-reinforced pipes with metal layer under pressure loading.

2.1 STRESSES IN THE PIPE CALCULATION

Let us consider a multilayered fibre-reinforced filament-wound composite pipe under outer pressure. The number of layers in the pipe is N, inner radius of the pipe is denoted by r_0 and outer radius by r_a. The pipe is subjected to the outer pressure p_a. The radial, hoop and axial coordinates are denoted by r, θ and z respectively (see Fig. 1).

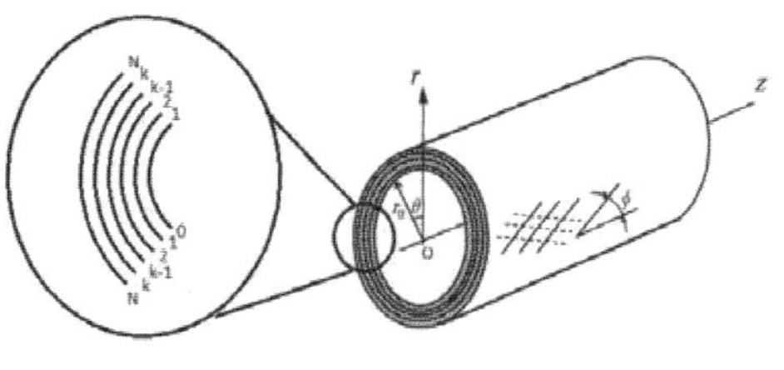

Figure 1. Multi-layered filament-wound pipe in cylindrical coordinates (adapted from [64]).

As we consider the axisymmetric tube, all displacements, strains and stresses are independent of θ. The radial displacements, r, are independent on axial displacements, z, as well as axial displacements are not depended on the radial ones. Thus, the displacements are

$$u_r = u_r(r), \quad u_\theta = u_\theta(r,z), \quad u_z = u_z(z), \quad (1)$$

where u_r, u_θ and u_z are radial, hoop and axial displacements, respectively.

The strain-displacement relations can be written as [67]

$$\varepsilon_r^{(k)} = \frac{du_r^{(k)}}{dr}, \quad \varepsilon_\theta^{(k)} = \frac{u_r^{(k)}}{r}, \quad \varepsilon_z^{(k)} = \varepsilon_0,$$
$$\gamma_{zr}^{(k)} = 0, \quad \gamma_{\theta r}^{(k)} = 0, \quad \gamma_{z\theta}^{(k)} = \gamma_0 r, \quad (2)$$

where γ_0 is a twist of the pipe per unit length and ε_0 is constant.

The elastic constitutive equations in the global cylindrical coordinates (z, θ, r) for the orthotropic layer k can be written as [67]

$$\begin{bmatrix} \sigma_z \\ \sigma_\theta \\ \sigma_r \\ \tau_{\theta r} \\ \tau_{zr} \\ \tau_{z\theta} \end{bmatrix}^{(k)} = \begin{bmatrix} \overline{C}_{11} & \overline{C}_{12} & \overline{C}_{13} & 0 & 0 & \overline{C}_{16} \\ \overline{C}_{12} & \overline{C}_{22} & \overline{C}_{23} & 0 & 0 & \overline{C}_{26} \\ \overline{C}_{13} & \overline{C}_{23} & \overline{C}_{33} & 0 & 0 & \overline{C}_{36} \\ 0 & 0 & 0 & \overline{C}_{44} & \overline{C}_{45} & 0 \\ 0 & 0 & 0 & \overline{C}_{45} & \overline{C}_{55} & 0 \\ \overline{C}_{16} & \overline{C}_{26} & \overline{C}_{36} & 0 & 0 & \overline{C}_{66} \end{bmatrix}^{(k)} \begin{bmatrix} \varepsilon_z \\ \varepsilon_\theta \\ \varepsilon_r \\ \gamma_{\theta r} \\ \gamma_{zr} \\ \gamma_{z\theta} \end{bmatrix}^{(k)} \quad (3)$$

Each layer of the pipe is a unidirectional long fibre composite. If we define 1 as the principal axis along the fibre, and 2 and 3 as transverse directions, the principal stiffness matrix can be written in terms of engineering constants as

$$\begin{bmatrix} C_{11} & C_{12} & C_{13} \\ C_{12} & C_{22} & C_{23} \\ C_{13} & C_{23} & C_{33} \end{bmatrix} = \begin{bmatrix} \frac{1}{E_1} & -\frac{\nu_{12}}{E_1} & -\frac{\nu_{12}}{E_1} \\ -\frac{\nu_{12}}{E_1} & \frac{1}{E_2} & -\frac{\nu_{23}}{E_2} \\ -\frac{\nu_{12}}{E_1} & -\frac{\nu_{23}}{E_2} & \frac{1}{E_2} \end{bmatrix}^{-1} \quad (4)$$

Using the stiffness transformation matrix the off-axis stiffness constants can be calculated from stiffness constants along principal material directions [67].

The equilibrium equations in cylindrical coordinates for a long axisymmetric tube under prescribed loading have the following form [67, 73]:

$$\frac{d\sigma_r^{(k)}}{dr} + \frac{\sigma_r^{(k)} - \sigma_\theta^{(k)}}{r} = 0, \quad (5)$$

$$\frac{d\tau_{\theta r}^{(k)}}{dr} + \frac{2\tau_{\theta r}^{(k)}}{r} = 0, \quad (6)$$

$$\frac{d\tau_{zr}^{(k)}}{dr} + \frac{\tau_{zr}^{(k)}}{r} = 0. \quad (7)$$

Integration of the last two equations gives the shear stresses as [67, 73]

$$\tau_{\theta r}^{(k)} = \frac{U^{(k)}}{r^2}, \quad \tau_{zr}^{(k)} = \frac{W^{(k)}}{r}, \quad (8)$$

where $U^{(k)}$ and $W^{(k)}$ are unknown constants of integration.

After combining the first equilibrium equation (5) with the elastic constitutive equations (3), the strain-displacement relations (2) and the displacement field (1) we obtain the second-order ordinary differential equation [67] the solution of which can be written as follows:

$$u_r^{(k)} = A^{(k)} r^{\beta(k)} + B^{(k)} r^{-\beta(k)} + \alpha_1^{(k)} \varepsilon_0 r + \alpha_2^{(k)} \gamma_0 r^2 \quad (9)$$

where

$$\beta^{(k)} = \sqrt{\frac{\overline{C}_{22}^{(k)}}{\overline{C}_{33}^{(k)}}}, \quad \alpha_1^{(k)} = \frac{\overline{C}_{12}^{(k)} - \overline{C}_{13}^{(k)}}{\overline{C}_{33}^{(k)} - \overline{C}_{22}^{(k)}},$$
$$\alpha_2^{(k)} = \frac{\overline{C}_{26}^{(k)} - 2\overline{C}_{36}^{(k)}}{4\overline{C}_{33}^{(k)} - \overline{C}_{22}^{(k)}}. \quad (10)$$

For the isotropic or transversely isotropic layers the solution is expressed as [73]

$$u_r^{(k)} = A^{(k)} r + B^{(k)} r^{-1}. \quad (11)$$

The boundary conditions for the laminated tube subjected to outer pressure are [67]:

$$\sigma_r^{(1)}(r_0) = 0, \; \sigma_r^{(N)}(r_a) = -p_a,$$
$$\tau_{\theta r}^{(1)}(r_0) = \tau_{zr}^{(1)}(r_0) = 0, \; \tau_{\theta r}^{(N)}(r_0) = \tau_{zr}^{(N)}(r_0) = 0. \quad (12)$$

From the traction continuity between the layers follows that the two interlaminate shear stresses ($\tau_{\theta r}$ and τ_{zr}) were continuous from layer to layer. For a long laminated tube subjected to axisymmetric uniform loading along its length they are zeros. As the layers are perfectly bonded, all displacements must also be continuous from layer to layer. Consequently, the continuity conditions for the considered problem are [67]

$$U^{(k)} = W^{(k)} = 0, \quad k = \overline{1, N}, \quad (13)$$

$$\tau_{\theta r}^{(k)}(r_0) = \tau_{zr}^{(k)}(r_0) = 0, \quad k = \overline{1, N}, \quad (14)$$

$$u_r^{(k)}(r_k) = u_r^{(k+1)}(r_k), \; \sigma_r^{(k)}(r_k) = \sigma_r^{(k+1)}(r_k), \; k = \overline{1, N-1}. \quad (15)$$

The axial force at the end of the tube is determined by integrating the axial stress over the area of cross section.

For the N-layered tube with axial force equal to zero the expression has the form:

$$2\pi \sum_{k=1}^{N} \int_{r_{k-1}}^{r_k} \sigma_z^{(k)}(r) r dr = 0 . \quad (16)$$

Similarly one can determine the torque, by integrating the moment of the shear stress $\tau_{z\theta}$ over the cross-sectional area of the tube. In the considered problem the torque is equal to zero, consequently the expression is written as follows [67]:

$$2\pi \sum_{k=1}^{N} \int_{r_{k-1}}^{r_k} \tau_{z\theta}^{(k)}(r) r^2 dr = 0 . \quad (17)$$

From the strain-displacement relations (2) and the expression for displacements (9) the layer strains can be determined. The expressions for layer stresses are obtained from constitutive equation (3) and layer strains. Substituting the expression for displacement (9) and layer stresses into boundary and continuity conditions and expressions for axial force and torque we obtain the system of equations, solving which the stress in the pipe can be found:

2.2 FAILURE ANALYSIS

A comprehensive analysis of the failure criteria for composite materials was given in the World Wide Failure Exercise [74], where their advantages, drawbacks and the applicability issues were considered in detail. In this study, for the failure prediction we use the Tsai–Hill failure criterion introducing some appropriate modifications to this criterion found in the literature.

The Tsai–Hill failure coefficient (FC) has the form

$$FC = \frac{\sigma_1^2}{X_T^2} - \frac{\sigma_1 \sigma_2}{X_T^2} + \frac{\sigma_2^2}{Y_T^2} + \frac{\tau_{12}^2}{S^2} \quad (18)$$

where X_T and Y_T are longitudinal and transverse tensile strength, respectively, and S is the shear strength.

It is generally recognised that the Tsai–Hill failure criterion underestimates the failure stress because the transverse tensile strength of a unidirectional lamina is generally much less than its transverse compressive strength and the compressive strengths are not used in the Tsai–Hill failure theory. The modified form of the Tsai–Hill failure criterion was formulated by Kawai and Saito [75]. For the off-axis tensile loading it agrees with the original formulae, however it takes a modified form for the off-axis compressive loading:

$$FC = \frac{\sigma_1^2}{X_T^2} - \frac{\sigma_1 \sigma_2}{X_T^2} + \frac{\sigma_2^2}{Y_T^2} + \frac{\tau_{12}^2}{S^2}, \quad \text{if } \sigma_2 > 0 ;$$

$$FC = \frac{\sigma_1^2}{X_C^2} - \frac{\sigma_1 \sigma_2}{X_C^2} + \frac{\sigma_2^2}{Y_C^2} + \frac{\tau_{12}^2}{(S - \mu_L \sigma_2)^2}, \quad \text{if } \sigma_2 < 0 , \quad (19)$$

where μ_L is a material constant that characterises the in-plane shear strength differential effect.

According to the criterion, the failure occurs when the failure coefficient becomes greater than 1. The criterion can be expressed as follows:

$$\begin{aligned} FC < 1 &\Rightarrow \text{no failure} \\ FC \geq 1 &\Rightarrow \text{failure} \end{aligned} \quad (20)$$

3. RESULTS AND DISCUSSION

As a numerical example the pipes of different designs made of carbon/epoxy fibre composite (T300/LY5052) subjected to outer pressure are considered. Taking the outer pressure into account is particularly important for well-intervention operations and subsea applications. The properties of the composite are given in Table 1.

Table 1. Properties of carbon/epoxy unidirectional fibre composite (T300/LY5052) [69]

E_1 (GPa)	135
E_2 (GPa)	8
G_{12} (GPa)	3.8
ν_{12}	0.27
ν_{23}	0.49
Longitudinal tensile strength (MPa)	1860
Transverse tensile strength (MPa)	76
Longitudinal compressive strength (MPa)	1470
Transverse compressive strength (MPa)	85
Shear strength (MPa)	98

The pipe of 3 cm outer diameter and 2 cm inner diameter under 60 MPa outer pressure was considered.

Figures 2-4 give detailed through the thickness variation of failure criterion coefficient as a function of winding angle. The suitable winding angles for pipe with inner 0^0 fibre layer and outer layers of opposite fibre orientation vary from around 45^0 to 90^0 for 60MPa outer pressure (Figure 2). A small jump in failure coefficient distribution occurs on the interface of 0^0 and α fibre layers.

Figure 3 shows that if we add outer layer of 90^0 fibre orientation, the pipe will be able to withstand the outer pressure of 60MPa. However the failure coefficient distribution will not be as smooth, having more jumps. In Figure 4 the distribution for $[\alpha,-\alpha]$ lay-up is presented. This pipe design gives a very smooth distribution of failure criterion coefficient, however not all winding angles are suitable to withstand outer pressure. The winding angles which allow the pipe withstand 60MPa outer pressure are vary from around 45^0 to 90^0.

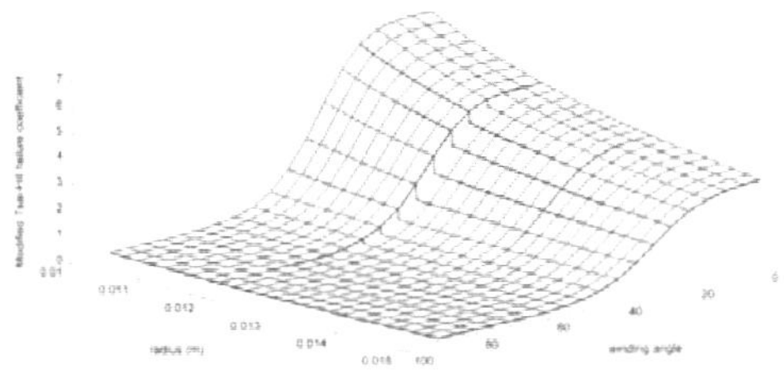

Figure 2. The thickness variations of the failure criterion coefficient for $[0,\alpha,-\alpha]$ lay-up and 60 MPa outer pressure.

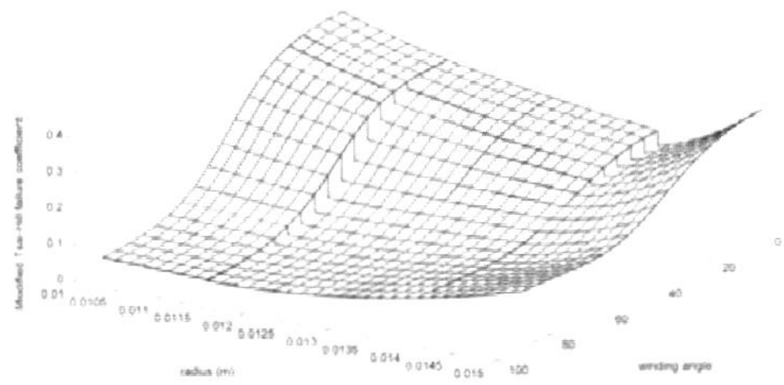

Figure 3. The thickness variations of the failure criterion coefficient for $[0,\alpha,-\alpha,90]$ lay-up and 60 MPa outer pressure.

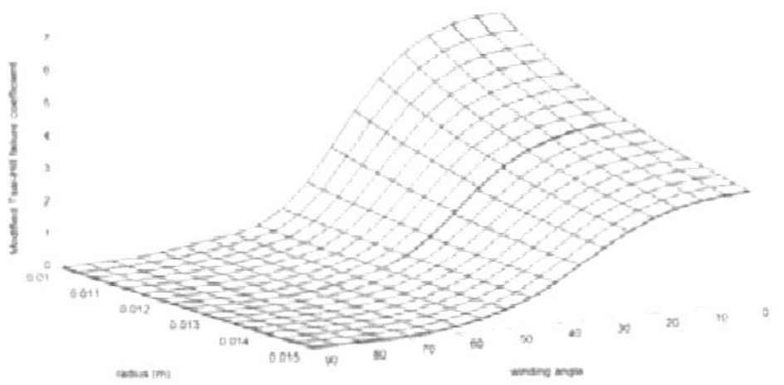

Figure 4. The thickness variations of the failure criterion coefficient for $[\alpha,-\alpha]$ lay-up and 60 MPa outer pressure.

4. CONCLUSIONS

This paper presents a review of the application of composite materials in the oil and gas industry. In particular the potential use of different types of composite materials (such as GFRP, PMC, syntactic foams etc.) in the industry is considered. Special attention is paid to long fibre-reinforced pipes, focusing on thin- and thick-walled filament-wound pipes under different types of loading.

As an example of modelling composite pipes, the problem of thick-walled multi-layered filament-wound pipe subjected to outer pressure was solved analytically. Taking the outer pressure into account is particularly important for well-intervention operations and subsea applications.

The method presented in this paper can be adopted in the future for modelling multi-layered thick-walled pipes with metal layers. The latter would be important for the deep-water offshore applications (such as subsea flexible pipes) since for the installations of water depth over 1000m, the composite pipes usually include metal layers to resist the high tensile load and external buckling load.

5. REFERENCES

1. GIBSON, A.G., 'Composite materials in the offshore industry'. Comprehensive Composite Materials, 2000.

2. TAHERI F., 'Advanced fiber-reinforced polymer (FRP) composites for the manufacture and rehabilitation of pipes and tanks in the oil and gas industry', In: Bai J, editor. Advanced fiber-reinforced polymer (FRP) composites for structural applications. Woodhead Publishing Limited, 2013.

3. HALE, J.M., GIBSON, A.G.: 'Coupon tests of fibre reinforced plastics at elevated temperatures in offshore processing environments'. Journal of Composite Materials, 1998.

4. GU, H., 'Behaviours of glass/fiber unsaturated polyester composites under seawater environment', Materials & Design, 2009.

5. GU, H., 'Dynamic mechanical analysis of the seawater treated glass/polyester composites', Materials & Design, 2009.

6. HALE, J.M., GIBSON, A.G., SPEAKE, S.D., 'Biaxial failure envelope and creep testing of fibre reinforced plastic pipes in high temperature aqueous environments', Journal of Composite Materials, 2002.

7. WEITSMAN, Y.J., ELAHI, M., 'Effects of fluids on the deformation, strength and durability of polymeric composites - An overview', Mechanics of Time-Dependent Materials, 2000.

8. RUSSELL, A.J., STREET, K.N., 'Moisture and temperature effects on the Mode I and Mode II interlaminar fracture of graphite/epoxy composites', Key Engineering Materials, 1989.

9. OGI, K., TAKEDA, N., 'Effect of moisture content on nonlinear deformation behavior of CF/epoxy composites', Journal of Composite Materials, 1997.

10. TOGO, M., NAKAMURA, T., TAKAHASHI, K., 'Effects of moisture absorption on the dynamic interlaminar fracture toughness of

carbon/epoxy composites', Journal of Composite Materials, 2000.

11. WEITSMAN, Y.J., 'Effects of Fluids on Polymeric Composites – A Review', University of Tennessee Report MAES98-5.0-CM, August, 1998.

12. DAVIES, P., CARLSSON, L.A., 'Delamination of composite cylinders'. Major Accomplishments in Composite Materials and Sandwich Structures, 2009.

13. DENIZ, M., OZDEMIR, O., OZEN, M., KARAKUZU, R., 'Failure pressure and impact response of glass-epoxy pipes exposed to seawater', Composites: Part B, 2013.

14. SLOAN, F.E., SEYMOUR, R.J., 'The effect of seawater exposure on mode I interlaminar fracture and crack growth in graphite epoxy', Journal of Composite Materials, 1992.

15. SHIN, E.S., PAE, K.D., 'Effects of hydrostatic pressure on the torsional shear behavior of graphite/epoxy composites', Journal of Composite Materials, 1992.

16. SHIN, E.S., PAE, K.D., 'Effects of hydrostatic pressure on in-plane shear properties of graphite/epoxy composites', Journal of Composite Materials, 1992.

17. PAE, K.D., RHEE, K.Y., 'Effects of hydrostatic pressure on the compressive behavior of thick laminated 45^0 and 90^0 unidirectional graphite-fiber/epoxy-matrix composites', Composites Science and Technology, 1995.

18. RHEE, K.Y., PAE, K.D., 'Effects of hydrostatic pressure on the compressive properties of laminated, 0^0 unidirectional, graphite fiber/epoxy matrix thick-composite', Journal of Composite Materials, 1995.

19. ZINOVIEV, P.A., TSVETKOV, S.V., KULISH, G.G., VAN DEN BERG, R.W., VAN SCHEPDAEL, L.J.M.M., 'The behavior of high-strength unidirectional composites under tension with superposed hydrostatic pressure'. Composites Science and Technology, 2001.

20. RHEE, K.Y., HA, S.R., PARK, S.J., KIM, H.J., JUNG, D.H., 'Fracture behavior of seawater-absorbed carbon/epoxy laminated composites in the hydrostatic pressure condition', Materials Science and Engineering A, 2006.

21. RHEE, K.Y., LEE, S.M., PARK, S.J., 'Effect of hydrostatic pressure on the mechanical behavior of seawater-absorbed carbon/epoxy composite', Materials Science and Engineering A, 2004.

22. GUZ, IA., 'Composites with interlaminar imperfections: Substantiation of the bounds for failure parameters in compression', Composites Part B, 1998.

23. SOUTIS, C., GUZ I.A., 'Predicting fracture of layered composites caused by internal instability', Composites Part A, 2001.

24. GUZ, I.A., HERRMANN, K.P., 'On the lower bounds for critical loads under large deformations in non-linear hyperelastic composites with imperfect interlaminar adhesion', European Journal of Mechanics, A/Solids, 2003.

25. GUZ, I.A., RUSHCHITSKY, J.J., 'Theoretical description of a debonding mechanism in fibrous micro- and nano-composites', International Applied Mechanics, 2004

26. GUZ, I.A., RODGER, A.A., GUZ, A.N., RUSHCHITSKY, J.J., 'Predicting the properties of micro and nanocomposites: From the microwhiskers to bristled nano-centipedes', The Philosophical Transactions of the Royal Society A, 2008.

27. MENSHYKOVA, M.V., GUZ, I.A., MENSHYKOV, O.V., 'A unified computational approach to instability of periodic laminated materials', CMES, 2009.

28. DAVIES, P., CARLSSON, L.A., 'Delamination of composite cylinders', Major Accomplishments in Composite Materials and Sandwich Structures, 2009.

29. OCHOA, O.O., ROSS, G.R., 'Hybrid composites: models and tests for environmental aging', Journal of Reinforced Plastics and Composites, 1998.

30. GRANT, T.S., BRADLEY, W.L., 'In situ observations in SEM of degradation of graphite/epoxy composite materials due to seawater immersion', Journal of Composite Materials, 1995.

31. CHOUCHAOUI, C.S., OCHOA, O.O., 'Similitude study for laminated cylindrical tube under tension, torsion, bending, internal & external pressure. Part I: Governing equations and Part II: Scale models', Composite Structures, 1999.

32. RODRIGUEZ, D.E., OCHOA, O.O., 'Flexural response of spoolable composite tubulars: an integrated experimental and computational

assessment', Composite Science and Technology, 2004.

33. SONG, B., CHEN, W., YANAGITA, T., FREW, D.J., 'Temperature effects on dynamic compressive behavior of an epoxy syntactic foam', Composite Structures, 2005.

34. SAUVANT-MOYNOT, V., DUVAL, S., GIMENEZ, N., KITTEL, J., 'Hot wet aging of glass syntactic foam coating monitored by impedance spectroscopy', Progress in Organic Coatings, 2007.

35. BOUCHONNEAU, N., SAUVANT-MOYNOT, V., CHOQUEUSE, D., GROSJEAN, F., PONCET, E., PERREUX, D., 'Experimental testing and modelling of an industrial insulated pipeline for deep sea application', Journal of Petroleum Science and Engineering, 2010.

36. BOUCHONNEAU, N., SAUVANT-MOYNOT, V., GROSJEAN, F., CHOQUEUSE, D., PONCET, E., PERREUX, D., 'Thermal insulation material for subsea pipelines: benefits of instrumented full scale testing to predict the long term thermo-mechanical behavior', Proceedings of the Offshore Technology Conference — OTC 18679, Houston, Texas, USA, 2007.

37. SAUVANT-MOYNOT, V., GIMENEZ, N., SAUTEREAU, H., 'Hydrolytic ageing of syntactic foams for thermal insulation in deep water: degradation mechanisms and water uptake model', Journal of Materials Science, 2006.

38. GROSJEAN, F., BOUCHONNEAU, N., CHOQUEUSE, D., SAUVANT-MOYNOT, V., 'Comprehensive analyses of syntactic foam behaviour in deepwater environment', Journal of Materials Science, 2009.

39. WATKINS, L., 7th International Conference on Offshore Mechanics and Arctic Engineering, Houston, Texas, USA, 1988.

40. FINE, T., SAUTEREAU, H., SAUVANT-MOYNOT, V., 'Innovative processing and mechanical properties of high temperature syntactic foams based on a thermoplastic/thermoset matrix', Journal of Materials Science, 2003.

41. CHOQUEUSE, D., CHOMARD, A., BUCHERIE, C., 'Insulation materials for ultra deep seaflow assurance: evaluation of the material properties', Proceedings of the Offshore Technology Conference — OTC 14115, Houston, Texas, USA, 2002.

42. KOCHETKOV, V.A., MAKSIMOV, R.D., 'Water absorption and swelling of glass/epoxy syntactic foams', Mechanics of Composite Materials, 1996.

43. KASHTALYAN, M., 'Three-dimensional elasticity solution for bending of functionally graded rectangular plates', European Journal of Mechanics A/Solids, 2004.

44. KASHTALYAN, M., MENSHYKOVA, M., 'Three-dimensional elasticity solution for sandwich panels with a functionally graded core', Composite Structures, 2009.

45. DAVIES, P., RIOU, L., MAZEAS, F., WAENIER, P., 'Thermoplastic composite cylinders for underwater applications', Journal of Thermoplastic Composite Materials, 2005.

46. SHAMSUDDOHA, M., ISLAM, M.M., ARAVINTHAN, T., MANALO, A., LAU, K., 'Effectiveness of using fibre-reinforced polymer composites for underwater steel pipeline repairs', Composite Structures, 2013.

47. MALLY, T.S., JOHNSTON, A.L., CHANN, M., WALKER R.H., KELLER, M.W., 'Performance of a carbon-fibre/epoxy composite for the underwater repair of pressure equipment', Composite Structures, 2013.

48. ALEXANDER, C., 'Development of a composite repair system for reinforcing offshore risers', Ph.D, Texas A&M University, 2007.

49. D'ALMEIDA, J.R.M., DE ALMEIDA, R.C., DE LIMA, W.R., 'Effect of water absorption of the machanical behavior of fiberglass pipes used for offshore service water', Composite Structures, 2008.

50. FRIEDRICH, R., 'Composite Materials for Offshore Operations-2 (CMOO-2)', In: Wang, S.S., Williams, J.G., Lo, K.H. (eds.) pp. 183–196. American Bureau of Shipping, Houston, TX, 1999.

51. DENIZ, M., KARAKUZU, R., 'Seawater effect on impact behaviour of glass-epoxy composite pipes', Composites: Part B, 2012.

52. HULL, D., LEGG, M.J., SPENCER, B., 'Failure of glass/polyester filament wound pipe', Composites, 1978.

53. SPENCER, B., HULL, D., 'Effect of winding angle on the failure of filament wound pipe', Composites, 1978.

54. UEMURA, M., FUKUNAGA, H., 'Probabilistic burst strength of filament wound cylinders under internal pressure', Journal of Composite Materials, 1981.

55. ROSENOW, M.W.K., 'Wind angle effects in glass fiber-reinforced polyester filament wound pipes', Composites, 1984.

56. SODEN, P.D., KITCHING, R., TSE, P.C., 'Experimental failure stresses for $\pm 55^0$ filament wound glass fiber reinforced plastic tubes under biaxial loads', Composites, 1989.

57. MISTRY, J., GIBSON, A.G., WU, Y-S., 'Failure of composite cylinders under combined external pressure and axial loading', Composite Structures, 1992.

58. WILD, P.M., VICKERS, G.W., 'Analysis of filament-wound cylindrical shells under combined centrifugal, pressure and axial loading', Composites Part A: Applied Science and Manufacturing, 1997.

59. LEKHNITSKII, S.G., 'Theory of elasticity of an anisotropic body', Moskow: MIR Publishers, 1981.

60. AJIT, K.R., 'Response of thick laminated composite rings to thermal stresses', Composite Structures, 1991.

61. STARBUCK, J.M, EBERLE, C., 'Analytical solution for the design of spoolable composite tubing', Third International Conference on Composite Materials for Offshore Operations, Houston, TX, October 31 – November 2, 2000.

62. XIA, M., TAKAYANAGI, H., KEMMOCHI, K., 'Bending behaviour of filament-wound fiber-reinforced sandwich pipes', Composite Structures, 2002.

63. NATSUKI, T., TAKAYANAGI, H., TSUDA, H., 'Prediction of bending strength for filament-wound composite pipes', Journal of Reinforced Plastics and Composites, 2003.

64. MENSHYKOVA, M., GUZ, I.A., 'Stresses analysis of layered thick-walled composite pipes subjected to bending loading', International Journal of Mechanical Sciences, 2014.

65. ANSARI, R., ALISAFAEI, F., GHAEDI, P., 'Dynamic analysis of multi-layered filament-wound composite pipes subjected to cyclic internal pressure and cyclic temperature', Composite Structures, 2010.

66. BOUHAFS, M., SEREIR, Z., CHATEAUNEUF, A., 'Probabilistic analysis of the mechanical responce of thick composite pipes under internal pressure', International Journal of Pressure Vessels and Piping, 2012.

67. XIA, M., TAKAYANAGI, H., KEMMOCHI, K., 'Analysis of multi-layered filament-wound composite pipes under internal pressure', Composite Structures, 2001.

68. XIA, M., KEMMOCHI, K., TAKAYANAGI, H., 'Analysis of filament-wound fibre-reinforced sandwich pipe under combined internal pressure and thermomechanical loading', Composite Structures, 2001.

69. BAKAIYAN, H., HOSSEINI, H., AMERI, E., 'Analysis of multi-layered filament-wound composite pipes under combined internal pressure and thermomechanical loading with thermal variations', Composite Structures, 2009.

70. GUEDES, R.M., 'Nonlinear viscoelastic analysis of thick-walled cylindrical composite pipes', International Journal of Mechanical Sciences, 2010.

71. BAI Y, XU W, CHENG P, WANG N, RUAN W., 'Behaviour of reinforced thermoplastic pipe (RTP) under combined external pressure and tension', Ships and Offshore Structures, 2014.

72. BAI Y, RUAN W, CHENG P, YU B, XU W., 'Buckling of reinforced thermoplastic pipe (RTP) under combined bending and tension', Ships and Offshore Structures, 2014.

73. HERAKOVICH, C.T., 'Mechanics of fibrous composites', John Wiley and Sons, Inc., 1998.

74. HINTON, M.J., KADDOUR, A.S., SODEN, P.D., editors. 'Failure criteria in fibre reinforced polymer composites: The World-Wide Failure Exercise', Elsevier, 2004.

75. KAWAI, M., SAITO, S., 'Off-axis strength

differential effects in unidirectional carbon/epoxy laminates at different strain rates and predictions of associated failure envelopes', Composites: Part A, 2009.

6. AUTHORS' BIOGRAPHIES

Professor Igor Guz is Head of School of Engineering at the University of Aberdeen in Scotland. He joined Aberdeen 2002 as a Lecturer becoming personal Chair in Engineering in 2007 and established Sixth Century Chair in Solid Mechanics in 2010. Prior to joining Aberdeen, he worked at various positions in the Timoshenko Institute of Mechanics (Kiev), Vienna University of Technology, University of Cambridge, University of Stuttgart, Imperial College London and University of Paderborn.

Professor Guz received the Doctor of Sciences (DSc) degree in 1994. In 1995, he was awarded the Prize of the Academia Europaea, and in 2009, The State Prize of Ukraine in Science and Technology. He was also a recipient of the Research Fellowships of the Alexander von Humboldt Foundation, the Research Fellowships of the Royal Society, Lise Meitner Research Fellowship of the Austrian Science Foundation, and organiser and Co-chairman of the Euromech Colloquium 400 in London. In 2004, he founded the inter-disciplinary Centre for Micro- and Nanomechanics (CEMINACS) in Aberdeen.

Professor Guz's main research interests are in solid mechanics; advanced anisotropic materials; three-dimensional theory of stability; multi-scale 3-D modelling of composite materials and structures; fracture mechanics; numerical methods; rock mechanics; nanomaterials and nanotechnologies. His research work has appeared in more than 300 publications including 4 books and over 130 journal papers. He made over 40 invited talks at leading universities in the UK, USA, Australia, Germany, Denmark, Austria, Slovakia, Chile Korea and Ukraine.

Dr Marina Menshykova is a teaching fellow at the University of Aberdeen, Scotland. She started to work at the university in 2009 and she worked as a KTP (Knowledge Transfer Partnership) Associate in the partnership between Ziebel UK and the University of Aberdeen, lecturer, research and teaching fellow.

Dr Menshykova's main research interests are in solid mechanics; non-linear and linear fracture mechanics; computational mechanics (BEM and FEM); iterative optimization algorithms and modelling of composite materials and structures. She has more than 40 scientific publications including a chapter of the book and 24 papers in international peer reviewed journals.

Professor Jeom Kee Paik is a Professor of the Department of Naval Architecture and Ocean Engineering and serves as president of the Korea Ship and Offshore Research Institute (KOSORI) and as director of the Lloyd's Register Foundation Research Centre of Excellence at Pusan National University (PNU) in Korea. In 2015, he also joined University College London's Department of Mechanical Engineering as a professor and faculty member. He is a Fellow and Vice President of SNAME, USA and a Fellow and Council Member of RINA, UK. He was awarded William Froude Medal from RINA in 2015, the David W. Taylor Medal from SNAME in 2013, and a Doctor Honoris Causa from the University of Liege in Belgium in 2012. His research interests are safety engineering and design of ships and offshore structures in extreme and accidental conditions associated with risk assessment and management.

REDUCING PROJECT RISK AND DELAYS IN INTERNATIONAL OFFSHORE EPC PROJECTS

Hans Petter Ellingsen, DNV GL Oil & Gas, Korea
Hyun Chang Shin, DNV GL Oil & Gas, Korea

SUMMARY

This paper discusses DNV GL's key observations with regards to critical success factors and pitfalls in offshore oil & gas projects built in Korea. DNV GL has in close dialogue with the industry and people involved in ongoing projects looked into causes for delays and prepared initiatives and suggestions to improve efficiency and reduce project risk in offshore Oil & Gas projects. As a result, a JIP was initiated to establish a new international industry standard for offshore oil and gas projects. The importance of standardization and the initial results of the ongoing Offshore Standardization JIP will be presented and discussed.

NOMENCLATURE

JIP	Joint industry Project
CMC	Certification of Material and Components
VMC	Verification of Materials and Components
EPC	Engineering Procurement and Construction
FPSO	Floating Production Storage Offloading vessel
TLP	Tension Leg Platform
FLNG	Floating Liquefaction Natural Gas Unit
MOU	Mobile Offshore Unit
ISO	International Organization for Standardization
NMA	Norwegian Maritime Authority
ANSI	American National Standards Institute
CAPEX	Capital Expenditures
OPEX	Operational Expenditures

1. BACKGROUND

Korea has longstanding traditions within the shipbuilding industry with the main yards Hyundai Heavy Industries (HHI), Daewoo Shipbuilding & Marine Engineering (DSME) and Samsung Heavy Industries (SHI) delivering commercial ship traffic vessels of all types such as bulk carriers, container ships, oil tankers and LNG carriers. The main difference between Korean and other Asian and European yards is the size and the scale of simultaneous projects being constructed. European yards are generally smaller handling a limited number of concurrent projects whereas the Korean yards are a lot bigger with typically around 50 projects being constructed in parallel. Commercial ship traffic vessels and offshore projects are often being built in the same construction halls, applying more or less the same construction methods side by side. According to data from Clarksons 2015 as shown in Figure 1, South Korea and China have the largest market shares of the offshore yards when measured in monetary value of projects.

The following two trends have been taking place over the last two decades in Korea:

- The main yards are increasingly taking on engineering scope in addition to construction scope. This development happened first for commercial vessels and subsequently for Mobile Offshore Units (MOUs) such as drilling units.
- The main yards have built up capacity to also handle detailed engineering and construction for large and complex offshore projects, e.g. FPSOs, TLPs, FLNGs, jacket platforms, semi-submersible vessels, etc..

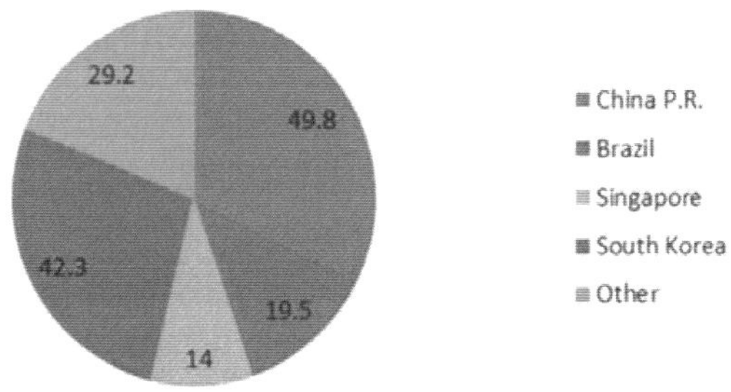

Figure 1: Order book contract volume in $bn with offshore yards in different countries from 2015
(Source: Clarksons 2015)

The typical constellations for large and complex offshore projects also include engineering companies. The interfaces between yards and engineering companies, including a timely handover of drawings issued for construction, have to be managed efficiently and are typically one of the main challenges in the EPC projects.

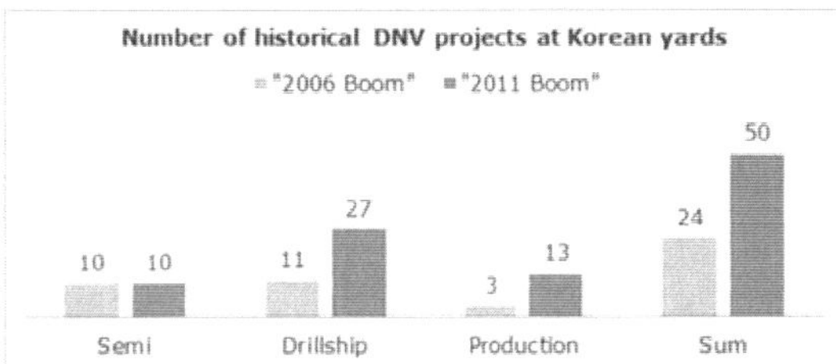

Figure 2: DNV GL involvement for different type of units for the year's 2006 and 2011

DNV GL has had a strong presence in Korea since 1977 and has been able to follow closely the rapid development of the Korean yards. Figure 1 shows DNV GL involvement for different type of units for the year's 2006 and 2011. The yards are currently (third quarter 2015) working at a high activity level. However, due to the current low oil price and rig market situation; new contracts for both Mobile Offshore Units (MOUs) and large EPC production projects has come to a sudden halt.

Many yards are experiencing significant delivery delays for their offshore projects. This can among other things be a result of less familiar specifications, requirements and processes resulting in misunderstandings, rework and delays during design and construction. Operators address this challenge through onsite teams of up to 300 people. Even with a high workforce on site the operators sometimes decide to move units from the yards for final completion closer to the final destination or offshore. Delays are also partly a result of resource and capacity limitations during offshore boom periods such as the most recent activity peak in 2014.

2. PROJECT RISKS AND DELAYS

There are multiple and complex causes for delays in offshore oil & gas projects. Some observed challenges based upon DNV GL's experience are as follows:

- Large variations and unique requirements for every project.
- Projects experience challenges related to cultural and organisational differences.
- The Asian yards' organizational set up and construction processes represent a new and unfamiliar setting for many western companies.
- Inconsistent quality in engineering, construction and equipment manufacturing throughout the project lead to non-compliance, rework and delays.
- Complex interface management and misalignment between owner, engineering, yards and equipment subcontractors leads to inefficiency in project execution.

Tools, processes and solutions to address the causes described above exist to some extent today such as processes for verification and certification of components and equipment, vendor pre-qualification and project quality plans. There is however significant room for improvement, even within areas where well established processes are implemented.

Critical success factors and criteria for different project stages collated through DNV GL's experience with large and complex offshore EPC projects in Korea is shown in Figure 3.

There are significant differences between shelf state requirements and it is therefore very important to select EPC partners with sufficient experience on the specific shelf state for the project. Recognized engineering companies often come with a price. If companies without sufficient experience within the specific shelf state requirements and application area are selected, then this should also result in a more comprehensive follow up through audits and controls throughout the project.

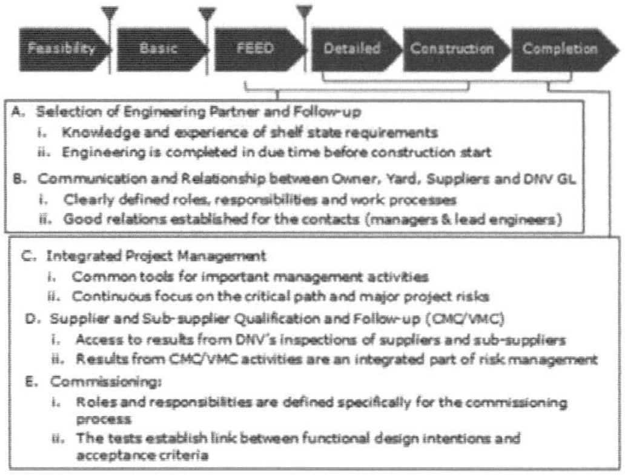

Figure 3: Key aspects and criteria for different project stages collated through DNV GL's experience with large and complex offshore EPC projects in Korea

Once competent partners have been selected, it is very common that engineering is delayed during detailed engineering and late changes occur in the phase between detailed design and construction. This violates the good practice principle that engineering should be completed before construction is started and often leads to significant rework and delays. Subsequently, an established process for management of change becomes critical for addressing late changes in engineering.

Pre-qualification of vendors is a critical process often underestimated in EPC projects. Sub-standard delivery and quality issues can often be linked back to lack of pre-qualification. It is therefore essential to have sufficient competent resources to follow up in reviewing and inspecting information related to vendor pre-qualification. Subsequently, certification and verification of material and components as well as second part inspections are critical activities to ensure that vendor packages are delivered according to the project specifications.

Quality remains a major challenge and concern in all EPC projects. A typical example is related to under reporting of welding faults or poor quality in the execution of NDT such as for ultra-sonic inspections. The main recommendation to mitigate this risk is a close follow up through training and qualification of UT technicians. Welding and material faults discovered in late stages are difficult and expensive to rectify also involving additional risk for delays.

Cultural and organizational differences pose a key challenge for many owners executing projects in South East Asia and the Far East. In peak construction periods (e.g. 2006, 2011 and 2014) the yards are resource constrained and owners often do not have sufficient knowledge of the yards' organizational set up and construction processes to rectify or address concerns about delays. Studies and research to address cultural differences such as between Norway and Korea are available through INTSOK Global Procurement Processes

– Phase I [1] and Master Thesis Managing the Efficiency of Foreign Engineering Contracts: A Study of a Norwegian and South Korean Project Interface [2].

Overall, different challenges in projects require different tools, processes and solutions. Large variations in requirements can be addressed by standardization as well as consistent use of existing standards. The focus for knowledge and experience should be on the yard's construction practices and processes (e.g. quality assurance of welded connections) and standards and regulations (e.g. applying NORSOK for the Norwegian Continental shelf). Common tools (e.g. 3D models and drawing specifications), communication protocols and clearly defined roles and responsibilities are important to manage the interface between engineering, construction and other stakeholders. Finally, comprehensive processes for vendor pre-qualification, verification and certification are crucial for ensuring quality during construction.

3. STANDARDIZATION

DNV GL has in close dialogue with the industry and people involved in ongoing projects looked into causes for delays and prepared initiatives and suggestions to improve efficiency and reduce project risk in offshore Oil & Gas projects. As a result, a JIP was initiated to establish a new international industry standard for offshore oil and gas projects between DNV GL, Hyundai Heavy Industries (HHI), Samsung Heavy Industries (SHI), Daewoo Shipbuilding and Marine Engineering Company (DSME), Korea Offshore and Shipbuilding Association and the Korea Marine Equipment Research Institute.

The cost of offshore projects has been on the rise and the large variations and unique requirements for every project is a cost driving factor which should be addressed. Unfamiliar specifications and processes are today resulting in re-work, delays, misunderstandings and are adding thousands of engineering hours to the projects. Suppliers are struggling to deliver materials and equipment of specified quality without delays or large additional costs. Oil companies are mobilizing big teams to follow-up yards and suppliers. Owners are typically mobilizing 100 -300 people at the yards for large offshore projects compared to typically 25 for mobile offshore units (MOUs).

The yards construction process is geared towards maritime work processes and standards. This is different from the additional requirements specified for most Oil & Gas projects. Offshore projects are developed according to different set of standards depending on applicable regulations, owner's requirements and operational context.

The general trend is that regulators are moving away from including prescriptive requirements in their regulations introducing what is often referred to as a goal or performance based regulatory regime. In the goal based regime safe operation is achieved by setting and achieving goals defined in performance standards, rather than by following prescriptive rules. Standards such as ISO, NORSOK, API and Class rules are however still applied for offshore projects with large variations in prescriptive requirements for design and construction.

The current market situation will force the industry to find new ways to reduce cost and standardization will be a key success factor for the offshore oil and gas industry to reduce the general cost level in the projects without compromising on quality or safety. The main benefits can be summarized as follows:

- Predictable component/equipment requirements and capabilities
- Less time spent on systems engineering due to increased predictability in EPC
- Adequate and risk based requirements for audits, inspections and verification
- Less time spent on discussions and delays related to pre-qualification of subcontractors
- Easier comparison of EPC tenders
- Less time spent on procurement
- Reduced EPC delivery time
- Reduced logistics and inventory cost
- Improved compatibility of components and materials between projects

DNV GL has carried out a study to assess CAPEX and OPEX savings by applying "maritime execution model", e.g. using prescriptive requirements and rules as far as possible for offshore projects. A tension leg platform was used as a case study and DNV GL's estimate shows a cost saving potential of 6-16% of total facility CAPEX. This figure takes into account direct savings only, such as lower costs for equipment and materials as well as smaller site-teams and reduced teams for supplier follow-up. The potential for savings on material and equipment alone was estimated to 30% - 50% of the equipment cost. In addition, indirect savings are expected related to reduced probability for re-work, delays and over-runs as well as reduced work-hours for designers, yards and suppliers.

The main objective of the initiated Offshore Standardization JIP is to reduce cost and increase predictability without compromising quality and safety by using standardized components and equipment and qualification procedures, e.g. welding procedures, welder's qualifications and vendor pre-qualification. The JIP will aim to reduce the number and variations in requirements to the minimum necessary to reflect a common and global best practice. Geographical and operational differences related to climate and weather as well as unique country specific requirements will be addressed as separate requirements.

Initially the Offshore Standardization JIP will define requirements for simple components and equipment (e.g. handrails, grating, steel plates and valves) to establish cost effective requirements. Standardization of equipment packages and even complete modules will be considered at a later stage. The JIP will also seek to define standard requirements to fabrication procedures and processes such as welding procedures, welder's qualifications and vendor pre-qualification.

A typical challenge related to standardization is the different definition of steel material properties such as plate thickness, material composition and strength between the EN standard normally applied in Europe and JIS typically applied in the Far East. Although the JIS standard and EN standards are similar, small differences creates problems in EPC projects since vendors in the Far East normally follow the JIS standard and engineering (often carried out in Europe) uses the EN standard as basis.

Variations in grating design from different vendors are also creating bottlenecks in the projects for the Korean yards due to limited vendor capacity to deliver one specific design specification requested. There are also variations in the grating standards, e.g. ISO, NMA, ANSI and Australian regulations. As an example, different sizing of the grating holes are recommended, leading to major grating cost and weight differences from project to project.

Figure 5 shows different design requirements for handrails, illustrating several differences depending on which standards are applied. The different steps applied in the JIP are as follows:

a) Identification of different standards applied
b) Description of challenges in applying the different standards
c) Risk based evaluation (e.g. safety, delays or cost)
d) Provide recommendation for which requirements should apply.

Parameter	Standard			Regulation		
	EN ISO	ANSI	NORSOK	PSA	NMA	OSHA
Height	Min. 1100 mm	36" - 42"	Min. 1100 mm	Min. 1100 mm	Min. 1000 mm	42"
Height on Stair	900 - 1000 mm	30" - 42"	Min. 1000 mm	Min. 1000 mm		30" - 34"
Intermediate kneerail	Min. 1	Min. 1	Min. 2	Min. 2	Min. 2	Min. 1
Vertical opening	Max. 500 mm	Max. 21" (1/2)	Max. 380 mm	Max. 380 mm	Max. 380 mm (Lower Max. 230 mm)	Max. 21" (1/2)
Diameter	25 - 50 mm	Min. 1 1/2 "	25 - 50 mm	25 - 50 mm		Min. 1 1/2 "
Distance btw stanchions	Max. 1500 mm	8 ft	Max. 1500 mm	Max. 1500 mm	Max. 1500 mm	8 ft
Height of toe plate	Min. 100 mm	4"	Min. 100 mm	Min. 100 mm		4"
Toe plate vertical clearance	Max. 10 mm		Max. 10 mm	Max. 10 mm		1/4"
Toe plate horizontal clearance		Max. 1"	continuous type	continuous type		Max. 1"
sharp edges	No		No	No		

Figure 5: Different design requirements applied for handrails

4. CONCLUSIVE SUMMARY

Offshore oil & gas projects in Korea have multiple and complex challenges during design and construction. However, the challenges are very similar from one project to the other and there are ways to efficiently address and mitigate project risk avoiding delays and cost overruns as outlined in this paper. The oil & gas industry as a whole will have to face the challenge of the current business climate finding ways to reduce project cost and improve project efficiency. Standardization is proposed as one key initiative to address some of the causes for the current inefficiencies. Other quality assurance processes such as vendor prequalification, certification and verification of materials and components as well as careful preparation to manage cultural and organizational differences in multi constellation projects are other means to reduce risk in projects.

5. REFERENCES

1. INTSOK, Global Procurement Processes – Phase I, March 2014.

2. University of Stavanger, Master Thesis Managing the Efficiency of Foreign Engineering Contracts: A Study of a Norwegian and South Korean Project Interface, 2015.

3. DNV GL Issue 01 | 2015 | PERSPECTIVES

6. AUTHORS BIOGRAPHY

Hans Petter Ellingsen is a Principal Risk Management Consultant with more than 20 years of experience within safety and asset risk in the offshore industry. Mr. Ellingsen has taken on different types of management positions in DNV GL after joining DNV GL in 2002. He has gained broad experience from working in Germany, Norway, UK and the US. Ellingsen has successfully managed large projects related to technical safety, risk studies, due diligence and technology qualification. Mr. Ellingsen is currently responsible in DNV GL for offshore risk advisory in Korea.

Hyun Chang Shin has 15 years of working experience from the oil and gas and maritime industry within risk management and safety, working environment, management system and software development. Shin has gained several years of project management experience from major oil & gas projects in Korea. Mr. Shin is currently working in DNV GL for offshore risk advisory in Korea.

THE TRANSITION OF ADVANCED RISK AND SAFETY CAPABILITY TO THE OFFSHORE INDUSTRY

J.A. Waltham-Sajdak, Alion Science and Technology, USA

SUMMARY

Given the environmental, legislative, economic and media attention which surrounds the oil and gas industry, it is imperative that the exploration and production of offshore oil and gas proceeds in the safest and most reliable manner possible. To achieve this, many companies are exploring how methodologies and technologies from other industries can help ensure safe and reliable operations. One example of this type of initiative is the application of naval integrated survivability analyses in both the design and operation of offshore vessels and platforms. Based on these integrated survivability modeling and simulation technologies originally developed to assess naval combatant mission capability in a damaged state; through-time integrated failure effect and consequence classification analyses across all technical, operational, and procedural barriers can be achieved and can be used to ensure offshore operations are conducted with as low as reasonably practical risk.

1. BACKGROUND

The HMS Sheffield, a 4,100-ton Type 42 destroyer completed in 1975, was struck on 4 May 1982 by an Exocet missile on her starboard side amidships, which penetrated a fuel tank, and smashed into her machinery spaces. Though the Exocet's warhead did not detonate, the burning residual fuel from the missile ignited fuel fires, resulting in thick, black acrid smoke being spread through the ship's ventilation system. The missile and subsequent fires destroyed her communication and electronics cables as well as her single high pressure fire main. The ship's company fought the fires for 4-1/2 hours before the ship had to be abandoned. Twenty men were lost in the fire and the Sheffield subsequently sank later in heavy seas.

Five years and 13 days later, the USS Stark, a 4,100-ton Perry Class frigate completed in 1982, was struck by two Exocet missiles. The first Exocet tracking in a little over 3.0 m above the sea surface striking her port side near the bridge. Although it failed to detonate, rocket fuel ignited and caused a large fire. The second Exocet also struck and detonated on the port side leaving a 3.0 m by 4.6 m hole. The Stark lost her electronics, communications and combat system functionality. A total of 37 crew were killed in the attack. However, unlike the Sheffield, the Starks remaining crew was able to fight the fire for almost 24 hours and flood her starboard side to keep the hole on the hull's port side above water which prevented her from sinking.

What was learned through these incidents was expressed most clearly by the CO of the USS Samuel B. Roberts (FFG-58), Captain Paul X. Rinn when he stated: *"...solid damage control, good training, and sound leadership based on experience can save a ship that is on fire and sinking, to fight another day."*

Since Captain Rinn's quote, a common interpretation of naval survivability is to provide the capability of a warship to "Fight Through" damage which is often performed by increasing automation, designing more system and structure redundancy and separation, enhancing technologies or adding "smarter" systems, controls, decoys or weapon systems.

The effectiveness of any onboard "survivability enhancement system" however, depends on the ability of the crew to work with the system in its damage control efforts. Simply stated, effective survivability requires effective damage control and effective damage control (DC) requires the correct use of equipment and techniques to prevent or minimize the damage effects caused by battle, fire, collision, grounding, explosion, and so forth [1]. Though this lesson was learned on naval combatants intended to go into harm's way, its applicability is universal to any complex operation requiring and relying on human system interaction (HSI) to achieve a safe outcome. This was proven a year after the Stark incident, when the explosion aboard the Piper-Alpha resulted in the loss of 167 lives and more than two decades later, when the blow-out, explosion and fire aboard the Deepwater Horizon resulted in the loss of 11 lives. These offshore incidents further re-emphasized the importance of effective damage control through the correct use of and training on equipment and techniques to prevent or minimize damage. Because in each event, Sheffield, Stark, Piper-Alpha and Macondo, the major findings reflected inadequate instruction and untimely procedures as well as inadequate training and procedures for emergency situations.

2. LEGISLATIVE SOLUTIONS

Post the Sheffield and the Stark, the United States Congress enacted survivability legislation [2] requiring for the first time the realistic threat-based survivability testing of aircraft, surface ships and submarines [3]. This survivability legislation required full-up system-level testing under realistic threat conditions, unless suitable alternatives were approved for reasons of cost and practicality. As a result, the United States and the Royal Navy each initiated efforts to develop advanced methodologies to better assess performance based total ship survivability. In the United States, these efforts led to the development of modeling and simulation methodologies to accurately reflect the effect of the initial weapon event as well as the through time impacts to systems and overall mission capability [4].

The funded development of these modeling and simulation tools were born of the realization that traditional methodologies of assessing ship survivability via prescriptive assessment had been outpaced by the significant increase of system complexity and interdependencies present in modern warships. Simply put, static rule based damage assessments were no longer able to accurately predict mission capability after damage occurred nor was there any effective way to assess the through time consequences of that damage as systems and crew responded to, and were affected by, the initial hit and follow-on events. What developed became known as Integrated Survivability Assessments (ISAs).

Concurrent to the naval developments inspired by the Sheffield and the Stark, the Piper Alpha led to the adoption of the Offshore Installations (Safety Case) Regulations of 1992 [5] which aimed to reduce the risks from major accident hazards to the health and safety of the workforce employed on offshore installations. Safety cases aim to show that specific safety claims are substantiated and that risks are kept 'As Low As Reasonably Practicable' (ALARP) through constructs similar to ISAs called Quantitative Risk Assessments (QRAs) via application of performance based assessment tools as well as management philosophy integration.

While the UK HSE and Norwegian PSA lead the establishment of safety case analysis offshore, the US offshore regulatory system, contrary to the naval progression to ISAs, maintained a prescriptive requirement standard toward safety through Chapter I of Titles 33 and 46 of the Code of Federal Regulations (CFR) [6]. Only after the Macondo incident has the prescriptive standard approach received significant attention and rebuke.

Though prescriptive standards tend to be a distillation of past experience, seek to include the best engineering practice at the time they were written, and give clear guidance on what is required, they are not very flexible and can rapidly become deficient where best practice is changing with evolving technologies and innovations [7].

As a result, the USCG-2013-0316 Interim Voluntary Guidelines "... urged the Coast Guard to evaluate the need for fire and explosion risk analyses to ensure an adequate level of protection is provided..." [8]. Further [8] continued to state "Recommendations from the 2009 IMO MODU Code are insufficiently specific... [The Code] needs to ... better define what the 'engineering evaluation' should include and what performance criteria should be met to ensure 'adequate protection' ... and ... the survivability of the facility and the personnel onboard". To this end:

- *The 'engineering evaluation' should identify hazards and the potential damage of all major accident events.*
- *Minimum allowable explosion design loads should be determined and used as performance criteria for blast resistant structures in each particular protection arrangement.*
- *Allowable radiant heat flux and exposures produced by fire hazards should be prescribed and used as performance criteria for protection of all safety critical spaces and elements (including escape routes, assembly and muster stations and survival craft launching stations).*
- *Where explosion and heat loads exceed the performance criteria, mitigation measures should be analyzed for effectiveness and, if sufficient, incorporated.*

With these recommendations, the USCG Interim Voluntary Guidelines laid the foundation for ISA style performance based assessments in the US commercial offshore market, bringing full circle the necessity to develop advanced methodologies to better assess performance based safety and survivability.

3. TECHNICAL PROBLEMS

Though the UK HSE and Norwegian PSA QRA based safety case is readily adoptable, the interaction of complex systems within the accepted QRA practices are not adequately addressed to achieve the level of integrated HSI assessments needed to measure the performance of effective damage control through the correct use of and training on equipment and techniques to prevent or minimize damage and determine ALARP risk conditions. The current safety case QRA methodologies utilize a wide variety of standalone analysis ranging from the classic "BOGSAT" (bunch of guys sitting around a table) to complex FEA and CFD analyses that do not take system affects into account and may minimize or negate structural effects. Consequence classification / equipment criticality assessments using fault / event trees or BOW TIE tools do not link to physical effects while crew routing / tasking / egress assessments may use a process management tool that does not integrate changing physical environment or man-in-the-loop operations. Together, the current safety case QRAs use multiple tools, methods, and processes that are sometimes hastily assembled by separate organizations in an ad-hoc, non-integrated, and non-repeatable manner to generate a report that is obsolete when written and immediately consigned to a shelf.

Basically stated, the risk levels determined within QRAs are developed based on the assumption that the personnel on board are properly trained, possess the required competence for their type of work, and are able to respond according to existing procedures for any emergency response scenario.

As noted by the real world events of Piper Alpha and Macondo however, these assumptions are not necessarily unproblematic. In fact, several studies, PSA accident investigations and audit reports show that between 50% and 80% of all accidents and near misses in the petroleum industry can be traced back to human error [9, 10 and 11]. A significant proportion of these errors are related to mistakes due to shortage in crew competence (i.e. theoretical knowledge, practical experience, training and personal qualities) and/or communication failures [12]. The possession of relevant crew competence is therefore crucial in carrying out safety critical tasks according to best practice.

Unfortunately, the effect of crew competence is not quantified specifically within QRAs even though the potential increase in risk if competence requirements are not met is regarded as significant. Within QRAs this significance is demonstrated with qualitative examples purporting quantification such as "*the probability for successful move-off is 80% rather than 90% for blowout events due to shortage of competence such that the probability of unsuccessful move-of increases from 10% to 20% and consequently the risk from such events is doubled*".

4. TECHNICAL SOLUTIONS

The ISA methods and analysis technologies that have been proven effective on complex naval combatants offer the offshore industry the opportunity to create Integrated Barrier Assessments (IBAs) as the next evolution of QRAs integrating the man-in-the-loop operations (MITLO) within the assessments to better quantify the effect of crew competence on the risk levels determined as well as establishing realistic emergency performance timelines and time requirements to achieve ALARP risk conditions.

ISA software's for naval combatant survivability are designed to simulate the ships susceptibility (realistic battle engagements including the tracking and prediction of missile and torpedo seekers against the ships signatures and the targeting of the ships combat systems to counter the threat), vulnerability (physics based prediction of battle damage inclusive of ballistic impact, blast, shaped-jetting, structural failure, equipment failure, system damage and crew injury), and recoverability (assessing cascading system failures, fire and smoke spread, and flooding). With more advanced programs like MOTISS additionally assessing man-in-the-loop operations and human systems interactions such as fire-fighting, system isolation, jumpering and damage control repair bringing together the total integrated survivability program which allows navies to understand the combat capability and performance characteristics completely [13, 14, 15, 16, and 17].

Like the ISAs for naval combatants, IBA technologies consolidate multiple disparate analyses (e.g. root cause, FMEA, CCA, bow-tie, accidental load analysis, physics based damage routines with operational activity, etc...) within a single tool allowing evaluation of "Performance Based" metrics through time and with the ability to assess, based on platform specific system and structural limitations, the quantitative capability of a platforms technical and operational barriers working in concert (i.e. integrated) to mitigate or minimize 1) the impairment of any one or combination of the Main Safety Functions, 2) the loss of life, 3) the loss of the platform (both physical and / or operational loss), and 4) the environmental impact.

4.1 ISAs TO IBAs

The U.S. MOTISS ISA software is a physics-based, integrated survivability modeling and simulation package that permits integrated assessment of hazard and accident events together with response actions. MOTISS was originally developed as a commercially available modeling and simulation technology to evaluate complex naval ship survivability weapons effects incidents. In its original application, MOTISS provided a single survivability measure enabling direct vessel to vessel comparisons of survivability requirement achievement and allowed for the inclusion of optimization and "best buy" evaluation of design changes and survivability features, within a balanced and mathematically accurate integration of susceptibility, vulnerability, and recoverability in order to guide resource allocations. Like risk, survivability is a statistical measure of performance, requiring many (~1000's) analyses in order to provide high statistical confidence. In order to maintain manageable computation times, MOTISS takes advantage of simplified "Empirical" and "First Order Physics" routines in order to reduce the computational time while maintaining statistical confidence thus mitigating the potential for higher error through the use of experimental method error corrective routines, producing sufficient statistical data to magnify the signal (or accuracy of the result) while mitigating the error through random error noise reduction. Thus, like all ISAs, MOTISS is a supra-simulation tool designed to precede (supra) CFD and FEA assessments in order to identify those specific events needing consideration of High Accuracy and High Precision – i.e. the "Critical Consequence" survivability events (Figure 1).

Following an intense verification and validation effort to over 800 tests, MOTISS ISA software was accredited by the United States Navy and has been used to support both U.S. and international naval combatant and noncombatant ship designs, ISAs and survivability assessment reports (SARs).

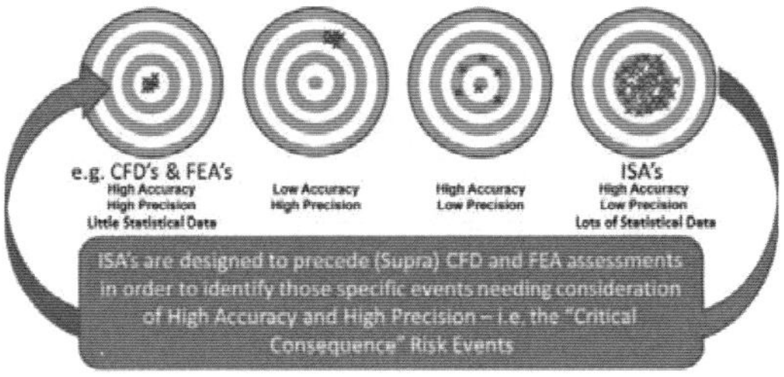

Figure 1: ISA Supra-Simulation Statistical Explanation

Following the events at Macondo Prospect, a cooperative engineering arrangement with an international drilling contractor initiated a "Pilot Study" to determine the applicability of ISAs technology to provide continuous improvement of safety and risk reduction in the offshore oil industry [18]. In particular the MOTISS ISA software was deemed appropriate as it offered an existing, commercially available technology that would directly demonstrate technology transfer between the defense and offshore industries. As part of this effort, enhancements were incorporated into the ISA framework in order to support more effective modeling of systems and to incorporate additional and enhanced man-in-the-loop evaluations.

The developed MOTISS IBA software is therefore a probability of occurrence assessment tool that works similar to the "bow-tie" analysis methodology. As a probability of occurrence assessment tool, IBA provides not only the critical link path (standard bow-tie result) but also quantifies the criticality of each path element (component, system or operation criticality), isolates the system and component failure mode effects and identifies the "root cause" for each loss of control event and consequence. Within the analysis methodology, the user can define either initiating events or loss of control scenarios individually, or permit a Monte-Carlo "jitter" to better assess random probability. Prevention and control barriers (typically the left hand side of a classis bow-tie) and mitigation barriers (the right hand side) are modeled in the system network logic and crew action models and can be conditionally monitored, deterministically through time for a single event, probabilistically at a set time, or through time for multiple events. Initiating events and barrier failures can be probabilistically derived for each consequence including which induced failures correlate to critical consequences.

As with ISAs, IBAs are built as a federation of software subroutines or modules linked by a common physical model and controlled by a single master routine (Figure 2).

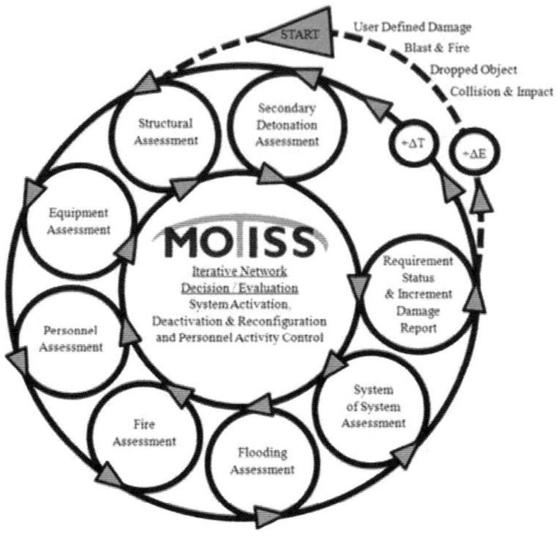

Figure 2: IBA Federated Architecture

The MOTISS IBA utilizes a 3D meta-model representation of the system coupled with detailed logic models of selected subsystems (Figure 3). This "one model and one tool" approach enables through-time integrated failure effect and consequence classification analyses across all barriers (technical, operational, and procedural). The benefits of this approach include reduced costs to perform the analysis through the use of less software and the development of fewer separate system models, reduced time to conduct analysis by requiring fewer models to be built and less total run-time to generate results, and enhanced capability and understanding of the total system of systems through significantly enhanced integration of efforts and a greater understanding of system complexities.

The IBA methodology enables a better understanding of the system's response to causality events while incorporating HSI and HMI. This understanding in turn offers the opportunity to modify the design in the most optimal fashion to achieve the specified requirements or to incorporate operational or procedural changes (with or without associated design changes) that will enhance barrier effectiveness.

IBAs like ISAs can be used significantly through design to assess the evolving safety and survivability of a vessel, though ISAs, like MOTISS, were further funded to support commanding officers and emergency response leadership (ERL) in developing, conducting and assessing emergency response team (ERT) emergency preparedness trials and training – i.e. enabling the effect of crew competence to be quantified specifically within ISAs and QRAs.

Figure 3: MOTISS IBA 3D Meta-model Representation

In this context, during a casualty condition, the offshore installation manager (OIM) or ships commanding officer (CO) is interested in four pieces of information so that he may make command decisions in order to minimize the damage effects, reduce the probability of lives lost and possibly retain operations, these are:

- What equipment, systems and capabilities has the ship lost?
- What equipment, systems and capabilities does the ship still have?
- What equipment, systems and capabilities will the ship lose and when will they be lost?
- Can the ship regain the capabilities that were lost and if so, when will they be regained?

To answer these questions, the OIM/CO relies upon his knowledge of the vessel as well as the capability of his damage control officer (DCO) or ERL. The DCO/ERL must therefore fully understand the ship system dependencies and be able to convey the state of the vessel in order to answer the OIM/CO's questions. More importantly however, the DCO/ERL must know:

- How to isolate the damage in order to minimize the cascading damage,
- What equipment, systems and capabilities are impacted by the isolation, and
- How to restore or reconfigure systems in order to regain lost capabilities as they are needed.

To assist OIM/CO's and ERL/DCO's in answering these questions, they often rely on a casualty control management system (CCMS) such as L3 MAPPS Ship Safety Assessment System or L3 MAPPS Battle Damage Control System (BDCS). These systems assist the DCO/ERL by:

- Tracking equipment and system damage based on sensors or human-machine interface (HMI) inputs,
- Tracking fire, flooding, casualty damage based on sensors and HMI inputs, and
- Calculating and / or predicting stability.

In order to effectively use CCMS, Damage Control Isolation Instructions (DCIIs) or Emergency Response Plans (ERPs) are established so as to provide methodical instruction for the mitigation of cascading damage and maintenance of operational capabilities.

The primary objective of an ERP/DCII is to provide information and procedures which are of critical importance to quickly contain and suppress fires (isolate flammable fluids piping), as well as contain and stop flooding (isolate ruptured fluid piping) within a compartment or space within the primary damage zone (PDZ). ERP/DCIIs are developed for each accessible compartment on the ship, and on notification of damage, fire or flooding, the DCII for the space is accessed through the CCMS by the DCO/ERL. If damage conditions extend beyond a single space, the DCII/ERP for all the affected spaces are accessed. The DCII/ERP provide information on primary and secondary fire, flood and smoke boundaries for a given compartment. Additionally, the DCII/ERP cards lists distributive piping systems passing through the compartment, detailed information about what components require action in order to facilitate electrical isolation, mechanical isolation, piping systems isolation, and ventilation control to minimize to effects of smoke and reduce combustion air [19].

Throughout design and planning of operations, ISAs and IBAs are employed to develop and optimize DCII and

ERP through analyses by subjecting the 3D Meta-model including all technical, operational, and procedural barriers to situations of hazard. Within ISAs these hazard situations include weapons effects such as missile impact, detonation, fragmentation, fire, flooding etc... while within IBAs these hazard situations include loss of well control, gas explosion, dropped objects, collision, fire, etc...

Within the MOTISS ISA and IBA constructs, each crewmember is provided one or more capability skill sets (e.g. fire-fighter, electrician, etc...), a speed of movement, an initial start location and an assigned tasking priority based on the damage events ERP/DCII. With the MITLO capability embedded within the ISA and IBA packages, the effectiveness of shipboard operations in reconfiguration, personnel task loading, data and communication connectivity, as well as passive and active damage control can be measured to ensure graceful degradation and enhanced operational risk reduction. Furthermore, the MOTISS ISA and IBA computational resource enables full evaluation of the crew damage control procedures and training; allowing full predictive simulations of total ship safety survivability trials.

In both ISAs and IBAs the simulations are used to determine optimal time to correct, most critical paths, and cascading damage expectations (what systems, equipment and capabilities are lost when other equipment or system components are lost or isolated). After analysis, the pertinent information (i.e. the critical paths, 'downstream' effects and ALARP operational responses) as identified are condensed within the DCII/ERPs and integrated within the CCMS (Figure 4).

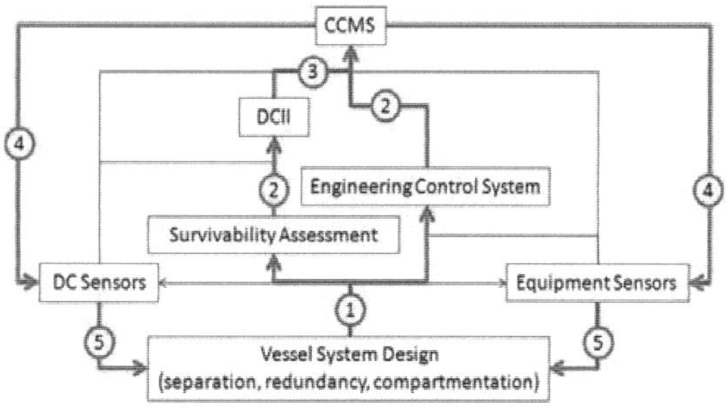

Figure 4: Integration of Design, Survivability Analyses, DCII and CCMS

Because the IBA and ISA methodologies utilize a detailed model of the system that can be modified to reflect system changes as well as different procedures or updated emergency response plans, they offer the holistic assessment capability to assess barriers and technical management processes in order to determine the probabilistic risk of major or minor accidents, identify the exposure to individuals or groups, and identify potential impacts to the environment.

4.2 IBA APPLICATION

To support the risk quantification of an operational drillship, an IBA was conducted to assess the potential of impairment to any one or combination of the Main Safety Functions (MSFs) when subjected to a defined situation of hazard and accident (DSHA). For the drillship, the MSFs to be assessed included: 1. Escape Routes, 2. Temporary Refuges, 3. Supporting Structure, 4. Evacuation Systems, and 5. Vital Systems (Figure 5). The IBA for the drillship was applied to update the QRA Fire and Explosion Risk Assessment (FERA) and Escape, Temporary Refuge, and Evacuation Risk Assessment (ETRERA).

MSF 1 (Escape Route):

- At least one escape route must be available from all compartments within the proximity of the Primary Damage Area (PDA) to the Temporary Refuge (Crew Mess)
- At least one route from the Temporary Refuge to the muster station, lifeboat station or helicopter landing area must be habitable and permit transit for 30 minutes following an event

MSF 2 (Temporary Refuges):

- The Temporary Refuge must remain habitable (maintain a temperature of less than 40°C) for 30 minutes following the damage event

MSF 3 (Supporting Structure):

- Rupture or structural integrity loss shall not occur on any of the primary load bearing structure

MSF 4 (Evacuation Systems):

- At least one escape route from the Temporary Refuge leading to the forward lifeboat station must be habitable and permit transit for 30 minutes following an event and the total capacity of functional appliances located at the forward lifeboat station must be greater than the total number of personnel onboard; or
- At least one escape route from the Temporary Refuge leading to the aft lifeboat station and to the aft lifeboat station must be habitable and permit transit for 30 minutes following an event and the total capacity of functional appliances located at the aft lifeboat station must be greater than the total number of personnel onboard

MSF 5 (Evacuation Systems):

- Assess the probability of vital system impairment or loss for:
    - Power System
    - Alarm System
    - Fire System
    - Watertight System
    - Ventilation System
    - Well Control System
    - Drilling System
    - Anchoring System
    - Dynamic Positioning System
    - Communication System
    - Ballast System
    - Emergency Shutdown System

Figure 5: Drillship MSFs

In order to implement the MOTISS IBA, a 3D model of the drillship was constructed and included structure, insulation, access ways, personnel and all MSF 1 through MSF 5 equipment, components and systems. For the FERA and ETRERA a total of 9 variable scenarios comprising a total of 900 DSHA events were assessed to determine the impact of the DSHA's (Accidental Loads) of Explosion Loads (in accordance with DNV-OS-A101, Section 2, Article D 600) and Heat Loads (in accordance with DNV-OS-A101, Section 2, Article D 700) on the drillship.

Within the drillship IBA, the specific measured outputs (reported items) related with the impact of the DSHA's included the capability of the as designed primary structural barriers to absorb, without rupture, the maximum applied probabilistic explosive loading (blast overpressure and duration) and to maintain fire integrity (prohibit flame passage / ignition in the space on the non-flame loaded side of the boundary), thermal integrity (prohibit greater than 12.5 kW/m^2 heating in the space on the non-flame loaded side of the boundary) and Structural integrity (prohibit the average surface temperature {i.e. core plate temperature} from exceeding 400°C for Steel, or 200°C for aluminum) for a period of 60 minutes. Additional measured outputs (reported items) related with the impact of the DSHA's included the MSF 1 through MSF 5's probability of impairment or loss, provided an assessment of Damage Occurring on / to the Oil and Chemical Spill Source / Response Equipment / Systems, and predicted personnel fatalities.

Within one scenario, 30 detonation points were evaluated for the resultant damage from fire initiation across each of three gas volume release rates, totaling 90 analysis, generated by fuel ignition at flow rates of 35, 150 and 200 kg/s. All 30 detonation points were located in the Mud Module (Figure 3, bottom) with ignition varied between 1 and 3 minutes post release. The following provides a summary of the IBA ETRERA conducted wherein the total probability of fire breach by compartment was determined (Figure 6), the probability of main living space / refuge aft bulkhead failure was determined (Figure 7), crew route checks were dynamically analyzed every 1 minute over a 60 minute span from onset of fire ignition to determine the availability of egress routes from each habitable compartment to the temporary refuge (TR) and from the TR to the secondary muster station and to each of the 6 lifeboats (Figure 8), and the probability of evacuation appliance loss was determined (Figure 8).

Compartment	Probability of Fire Breach
CCB Air Compressor Room	18
Mechanical WS	16
Mud Pump Room	62
Mud Module Electrical Room	52
Energy Storage Room	48
Ware House	54
HVAC Plant Room	49
Big Bag Storage Area	40
HVAC Plant Room	30
Toilet 2nd	24
Battery Room	87
Cement Room	87
Active Mud Tank Room	80
Mud Control Room	38
Mud Lab	39
Passage MM Mud Lab	71

Figure 6: Drillship IBA Scenario 3 - Mud Module Gas Release, Explosion and Fire DSHA – Probability of Fire Breach by Compartment

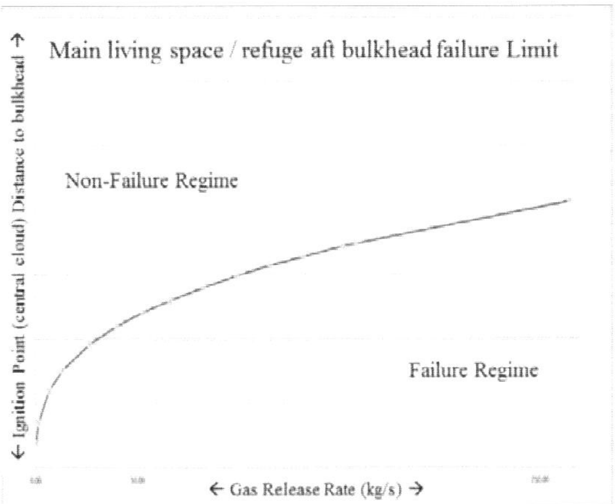

Boundary	Percent Structural Failure from Mud Module Iginition (Residual Strengths <5-50%)			
	Overal Average	Gas Release Rate		
		35	150	200
Main Living Quart Aft Bulkhead	11	0	12	22

Figure 7: Drillship IBA Scenario 3 - Mud Module Gas Release, Explosion and Fire DSHA – Probability of main living quarter aft bulkhead rupture

Route		Probability of Route Loss		
Initial Area	*Final Area*	*At 1 Minute*	*At 30 Minutes*	*Over 30 Minutes*
Any Habitable Compartment	Temporary Refuge	7	10	9
Temporary Refuge	Secondary Muster Station	10	10	10
Temporary Refuge	Helo Deck	0	0	0
Temporary Refuge	Lifeboat Station 1	3	9	7
Temporary Refuge	Lifeboat Station 2	3	9	7
Temporary Refuge	Lifeboat Station 3	3	9	7
Temporary Refuge	Lifeboat Station 4	3	9	7
Temporary Refuge	Lifeboat Station 5	3	6	5
Temporary Refuge	Lifeboat Station 6	3	6	5

Evacuation Appliance	Average Percent Loss from Mud Module Detonation			
	Overall Average	*Release Rate (kg/s)*		
		35	*150*	*200*
Lifeboat Port Aft	0	0	0	0
Lifeboat Port Fwd	4	0	4	7
Lifeboat Stbd Aft	0	0	0	0
Lifeboat Stbd Fwd	4	0	4	7
Rescue Boat	4	0	5	8

Figure 8: Drillship IBA Scenario 3 - Mud Module Gas Release, Explosion and Fire DSHA – Probability of route loss and evacuation appliance loss

Based on the drillship IBA ETRERA results evaluating the drillship technical and operational barriers and risk mitigation plans including the activity specific emergency response requirements (ASERRs) emergency response plans and escape route plans (ERPs), the net probability of loss of any of MSF 1 through 4 resulting in an increase of lives lost as a result of the Mud Module Gas Release, Explosion and Fire DSHA scenario was determined to be 13%.

Applied recommendations for drillship technical and operational barriers including modification of the forward lifeboat thermal shielding, modification of the main living quarter / refuge aft bulkhead blast shielding, modification of escape route plans and increasing operational importance of securing ventilation within the drillship ASERR reduced the IBA ETRERA net probability of loss of any of MSF 1 through 4 resulting in an increase of lives lost as a result of the Mud Module Gas Release, Explosion and Fire DSHA scenario was determined to be 4% corresponding with the 200kg/s release rate ignited 3 minutes post detonation in the forward center portions of the mud module inducing impulsive loads greater than the design load requirement (Figure 9).

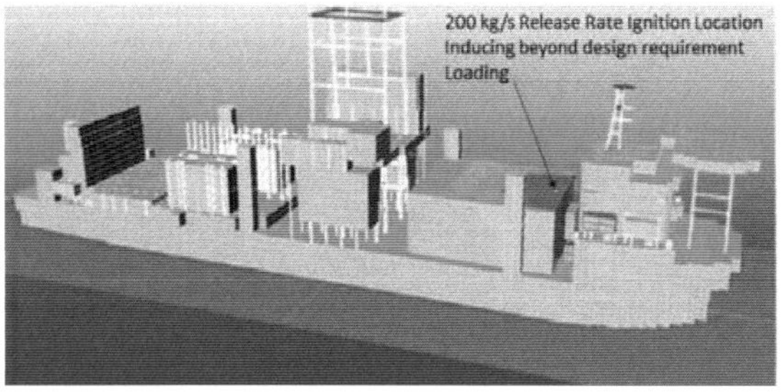

Figure 9: Drillship IBA Scenario 3 - Mud Module Gas Release, Explosion and Fire DSHA – ETRERA post correction critical risk to TR zoning (red)

5. CONCLUSIONS

The risk levels determined within QRAs for the offshore industry are developed based on the assumption that the personnel on board are properly trained, possess the required competence for their type of work, and are able to respond according to existing procedures for any emergency response scenario. As noted by the real world events of Piper Alpha and Macondo however, these assumptions are not necessarily unproblematic - nor are they restricted to only the offshore industry as noted by the events of the Sheffield and the Stark within the naval industry.

The development of technologies to enhance the survivability of naval combatants through integrated dynamic assessments of physical and operational barriers combined to achieve reduced operational risk are applicable to the offshore industry and offer the offshore industry the opportunity within the evolving performance based legislation to create use IBAs as the next evolution of QRAs integrating MITLOs within the assessments and optimized ERPs within a CCMS like L3 MAPPS Ship Safety Assessment System in order to better quantify the effect of crew competence on the risk levels determined as well as establishing realistic emergency performance timelines and time requirements to achieve ALARP risk conditions as evidenced through the exampled application of an operational drillship.

6. ACKNOWLEDGEMENTS

The author would like to acknowledge Mr. Goodfriend of Alion Science and Technology for his modelling and simulation efforts of the drillship IBA as discussed within this paper as well as Mr. Yao of Alion Science and Technology for his editorial assistance.

7. REFERENCES

1. WALTHAM-SAJDAK, J.A., 'Survivability System Effectiveness Assessments for Naval Combatants', *ROKN Naval Ship Technology Seminar*, 2011.

2. Title 10, US Code of Federal Regulations (CFR).

3. BLOOM, REESE AND HOPKINS, 'Live Fire Test and Evaluation for Ships', *Naval Engineers Journal*, May 1994.

4. MAIR, REESE AND HARTSOUGH, 'Simulated Ship Shock Tests/Trials?' *http://www.ida.org/LFTEsimulation.*

5. 1992 Safety Case Regulations.

6. Titles 33 and 46, US Code of Federal Regulations (CFR).

7. LANTZ, LITTLE, NADEAU, AND REYNOLDS, '"Lessons Learned" following Macondo - Safety Enhancement on the U.S. Outer Continental Shelf', *Offshore Technology Conference*, 30 April–3 May 2012.

8. USCG-2013-0316 Interim Voluntary Guidelines.

9. HAUGEN, SELJELID, SKLET, VINNEM, AND AVEN, 'Operational risk analysis: total analysis of physical and non-physical barriers', *H3.1 Generalisation Report, Rev. 1*, 2007.

10. PETROLEUMSTILSYNET (NORWEGIAN PETROLEUM SAFETY AUTHORITY), 'Tilsynsaktivitet med Statoils planlegging av brønn 34/10-C-06A', Revisjonsrapport, 2010.

11. PETROLEUMSTILSYNET (NORWEGIAN PETROLEUM SAFETY AUTHORITY), 'Granskningsrapport etter ulykken på Oseberg B', 7.5.2009.

12. SKLET, S. ET AL., 'Monitoring of Human and Organizational Factors Influencing the Risk of Major Accidents', *SPE International Conference on Health, Safety and Environment in Oil and Gas Exploration and Production*, 2010.

13. WALTHAM-SAJDAK, J.A. ET AL., 'MOTISS Beta Version 3.0 User Manual', Alion Science and Technology, 2011.

14. WALTHAM-SAJDAK AND KARNI, 'Determination of a Measure of Total Integrated System Survivability', *RINA Warship 2006: Future Surface Ships Papers*, 2006.

15. JUCKETT AND WALTHAM-SAJDAK, 'Modeling Techniques for Measure of Total Integrated System Survivability (MOTISS)', *Maritime Systems and Technology (MAST)*, 2007.

16. WALTHAM-SAJDAK, J.A., 'MOTISS Integrated Ship Survivability Simulation - Enhancing the Next Generation of Naval War-Fighters', *ROKN Naval Ship Technology Seminar*, 2009.

17. SHIN, KWON, CHUNG AND WALTHAM-SAJDAK, 'Development of Vulnerability Analysis Program for Review of Warship Survivability at Concept Design Stage', *Design and Analysis of Protective Structures (DAPS)*, 2012.

18. WALTHAM-SAJDAK AND COWARDIN, 'The Use of Integrated Barrier Assessments to Improve Safety and Reduce Risk in Offshore Oil and Gas Exploration and Production', *SNAME 18th Offshore Symposium*, 2013.

19. WALTHAM-SAJDAK, HADLEY AND STARK, 'Shipboard Damage Control and Recoverability Simulations', *ROKN Naval Ship Technology Seminar*, 2013.

8. AUTHORS BIOGRAPHY

Dr. Waltham-Sajdak holds the current position of Technical Director of Ship Survivability at Alion Science and Technology. He is responsible for analysis, assessment and enhancement of naval survivability and commercial safety applications for domestic and international clientele. His previous experience includes the assessment and enhancement of both design and operational survivability of more than 15 naval combatants and auxiliaries as well as the assessment and enhancement of the design and operational safety of 3 offshore drilling units.

TEST FACILITIES FOR SAFETY STUDIES OF SHIPS AND OFFSHORE STRUCTURES ASSOCIATED WITH EXTREME AND ACCIDENTAL CONDITIONS

Jeom Kee Paik* and **Jung Kwan Seo** The Korea Ship and Offshore Research Institute (The Lloyd's Register Foundation Research Centre of Excellence), Pusan National University, Busan, Korea (*Also, Department of Mechanical Engineering, University College London, London, UK)

SUMMARY

Ships and offshore structures can face extreme and accidental events that can result in catastrophic consequences in association with casualties, property damages and pollution. Because the mechanism and its responses of structures in extreme and accidental events are highly nonlinear, it is essential to take advantage of experimental approaches as well as computational approaches in terms of identifying such nonlinear responses. Relevant test facilities are then required to meet the needs where full scale or at least large scale models should be dealt with. The objective of this article is to introduce test facilities recently built in the Korea Ship and Offshore Research Institute (KOSORI) at Pusan National University in Korea in terms of specifications and capabilities of various test facilities. Those facilities include test infrastructures in association with ultra-high subsea pressure, fires, explosions, structural failure and dropped object as well as high speed material test.

1. INTRODUCTION

Ships and offshore platforms usually operate under normal conditions while in service, but can face extreme and even accidental conditions, as shown in Figure 1. They are subject to various action effects resulting from these extreme and accidental conditions[1-4].

The sources of such actions and action effects include (without consideration of any specific order): welding induced high temperature causing initial imperfections (e.g., initial distortions, residual stress or softening in the heat-affected zones of welded aluminium structures); abnormal waves/winds/currents; dynamic pressure loads arising from sloshing, slamming or green water; low temperature in Arctic operations; cryogenic conditions resulting from liquefied natural gas cargo; ultra-high pressure in ultra-deep waters; elevated temperature due to fire; blast loads due to explosion; impact loads associated with collision, grounding or dropped objects; and age-related degradation such as corrosion, fatigue cracking and local denting damage. Such events sometimes result in catastrophic consequences that lead to casualties, property damage, and pollution.

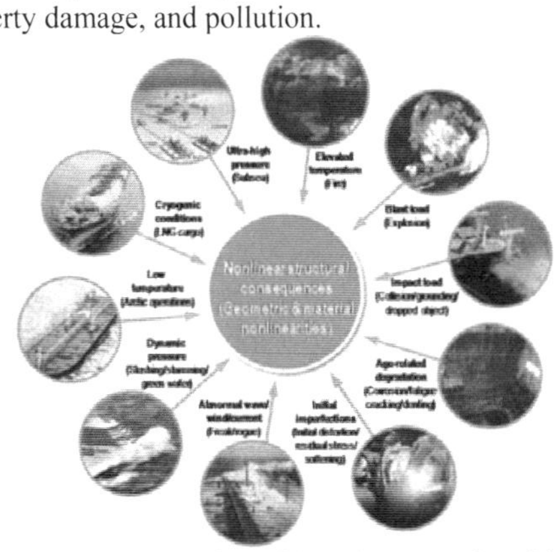

Figure 1. Extreme and accidental events for ships and offshore installations[5]

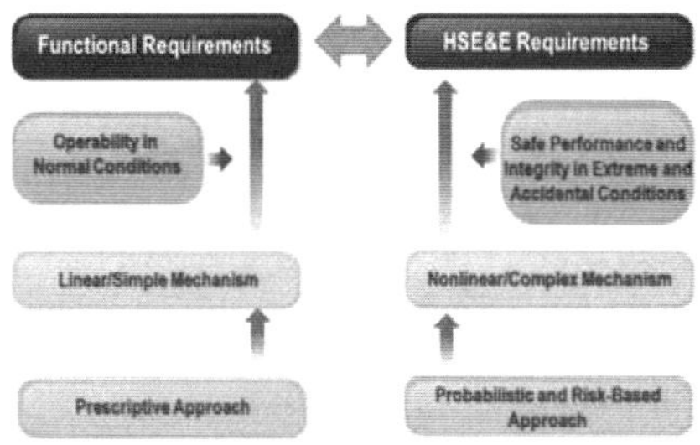

Figure 2. Paradigm change in engineering and design of ships and offshore structures[5]

Figure 3. Suggestions of modifying the IMO's original concept for goal-based standards with six tiers [5]

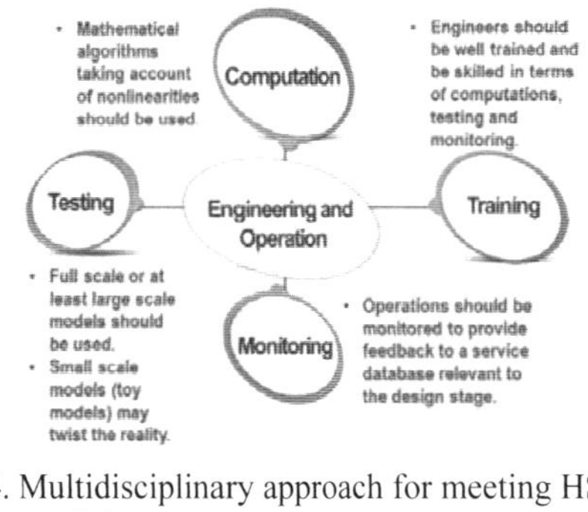

Figure 4. Multidisciplinary approach for meeting HSE&E requirements [5]

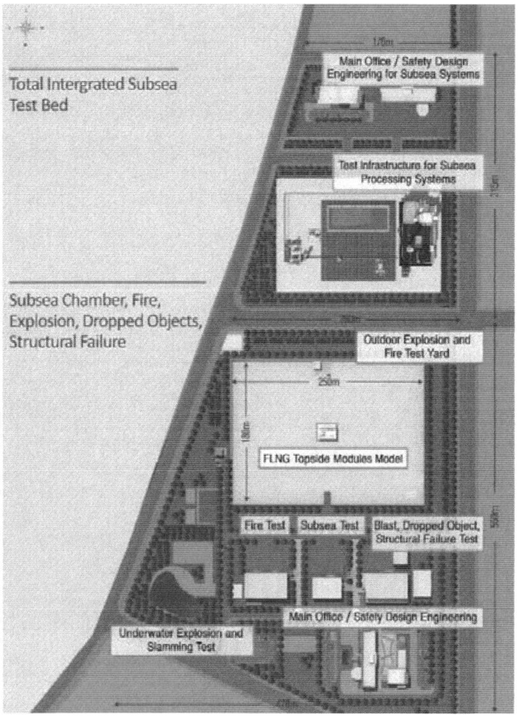

Figure 5. Layout of KOSORI test facilities with the total area of 23.1ha in Hadong, Korea

Successful design and engineering should meet not only functional requirements but also health, safety, environment and ergonomics (HSE&E) requirements. Functional requirements represent operability under normal conditions, and HSE&E requirements address safe performance and integrity in extreme and accidental conditions.

The paradigm change is currently pertinent in design and engineering, as shown in Figure 2. Paik [5] suggested to modify the IMO's original concept for goal-based standards by including a new tier dealing with HSE&E requirements as shown in Figure 3. Integrated and multidisciplinary approaches should then be applied as illustrated in Figure 4. Computation by itself is not enough, and testing is essential with large- and full-scale test models. Also, operations must be monitored to provide feedback to a service database relevant to the design stage. Relevant test facilities are then required to meet the needs where full scale or at least large scale models should be dealt with.

This article introduces test facilities recently built in the Korea Ship and Offshore Research Institute (KOSORI) at Pusan National University in Hadong, Korea, with the area of 23.1 ha. Figure 4 shows the layout the KOSORI test facilities in Hadong, Korea. Those include test infrastructures in association with ultra-high subsea pressure, fires, explosions, structural failure and dropped object as well as high speed material test.

The test facilities at KOSORI have just been built and no specific test results are available yet, while some projects of testing have already been planned. Also, this article does not attempt to review existing similar test facilities of other organizations because of limited access to them in detail. Test setups and test procedures of individual facilities shall be presented in separate articles in near future.

This article is an expanded version of the first author's article presented in the April 2014 issue of (mt) magazine of the Society of Naval Architects and Marine Engineers, USA [6].

2. ONGOING PARADIGM CHANGE IN ENGINEERING AND DESIGN

In the process of engineering and design, safety studies are required to assure the integrity of offshore installations in case of the extreme or accidental events noted above. Today, it is well recognized that a risk-based approach is the best practice for safety studies. Qualitative risk approaches such as hazard and operability studies (HAZOP), safety and operability studies (SAFOP), safe job analysis (SJA), preliminary hazard analysis (PHA) or failure mode and effect analysis (FMEA) need to be extensively supplemented for quantitatively characterizing hazards, properly managing them, or lowering risks to the levels acceptable for best practice. It is obvious that quantitative risk approaches should be applied for a more refined assessment and management of hazards.

Extreme and accidental events result in nonlinear structural consequences of both the geometric and material kinds. The effects of such events are highly nonlinear, and involve non-Gaussian, multi-physics, multi-scale, and multi-criteria dimensions. As risk is defined as a product of the frequency of hazards and their consequences, it is highly desirable to quantitatively characterize both the frequencies and consequences involved.

The application of numerical simulations derived from mathematical algorithms is often inadequate for identifying consequences that are highly nonlinear. It is challenging to obtain accurate simulations using mathematical or engineering models, such as nonlinear computational fluid dynamics (CFD) modeling or nonlinear structural modeling. Therefore, it is not surprising that some benchmark studies on CFD simulations associated with fires or explosions can show results that differ by more than a factor of 5.

In such cases, experimental investigations on prototype models should be applied to validate the numerical simulations and assure the safe performance of structures and equipment. The prototype models should be full scale, or at least large scale in relation to the real structures, because the scaling effects are generally significant and quite uncertain; and even more so in nonlinear problems.

3. TYPES OF HAZARDS IN SHIPS AND OFFSHORE INSTALLATIONS

Ships and offshore installations are usually subject to normal conditions, but may face extreme or accidental events in association with site-specific metocean and operational conditions. Such events sometimes result in catastrophic consequences that lead to casualties, property damage and pollution.

The main types of serious hazards for offshore installations include the following [7].

- Process leaks of gas and/or oil, leading to fires or explosions, which can arise from various kinds of equipment including separators and other process devices, compressors and other gas treatment devices, process pipes/flanges/valves/pumps, topsides and flowlines, wellhead equipment, flare/vent systems, storage tanks, loading/unloading systems or turret swivel systems.
- Riser/pipeline leaks of gas and/or oil that lead to fires or explosions arising from import flowlines, export risers, subsea pipelines or subsea wellhead manifolds.
- Collisions causing structural damage that involve offshore loading tankers (shuttle tankers), drifting offshore vessels (e.g., semi-subs, barges, storage vessels), supply vessels, icebergs, drilling rigs, drilling support vessels (e.g., jack-ups or barges), stand-by vessels, other support vessels (e.g., diving vessels or barges), passing merchant, fishing, or naval vessels (including submarines) or floatels.
- Dropped objects causing structural damage, which can occur during construction, crane operations, cargo transfer, drilling, or rigging-up derricks.
- Structural failures causing local or total collapse (e.g., buckling, crash) and/or fracture (e.g., mooring line failure, pipeline burst), which can arise from extreme weather and/or operation, corrosion, fatigue, design error or earthquake.
- Blowouts leading to spills or explosions, which occur during drilling, completion, production, workover or abandonment.

Essential safety measures or risk control options should be applied to mitigate the degree of consequence or damage from such hazards. Such risk control options include the following:

- Blast walls to protect against explosion.
- Fire walls to control fire.
- Passive fire protection (PFP).
- Escape route.
- Deluge or water-spray systems to control fire or explosions.
- Structural strengthening to improve structural crashworthiness and protect against collisions, dropped objects or structural collapses.
- Subsea equipment safety systems to prevent blowouts, fatigue, erosion or corrosion, and enable flow assurance or interaction between individual subsea components.

4. TEST FACILITIES FOR MATERIAL PROPERTIES IN CASE OF IMPACT AND/OR LOW/ELEVATED TEMPERATURE

Dynamic strain rate related structural phenomena are often classified into five groups, namely, creep , where the strain rate is lower than 10^{-5}/sec; static or quasi-static (e.g., dead or live loading of a ship hull), where the strain rate is 10^{-5}/s to 10^{-1}/sec; dynamic (e.g., dynamic pressure loads arising from sloshing or slamming), where the strain rate is 10^{-1}/s to $10^{1.5}$/sec; impact (e.g., explosions, collisions, dropped objects), where the strain rate is $10^{1.5}$/sec to 10^{4}/sec; and hyper-velocity impact (e.g., bombing), where the strain rate is greater than 10^{4}/sec. Among these groups, most hazards in offshore installations are of the impact type.

The effects of low or elevated temperatures are also significant for material properties and, ultimately, for structural response. The hazards of extremely low temperature conditions apply to operations in Arctic regions and to liquefied petroleum gas (LPG) and liquefied natural gas (LNG) facilities. The hazards from elevated temperature arise mainly from fires or explosions.

Structural responses under impact loads or temperature extremes are in nature quite different from responses under static or quasi-static loads at room temperature. A major factor in such differences is that the mechanical property of structural steel can be significantly affected by loading speed or strain rate and temperature. Therefore, it is extremely important in safety studies to take the correct as-changed material properties into account, to correctly include the effects of strain rates and temperatures on nonlinear structural responses. As the actual material properties can be defined only by experimental investigation, it is essential that test facilities be designed to accommodate this reality.

The Split-Hopkinson pressure bar (SHPB) is often used in studying the strain rate effect on material properties (http://en.wikipedia.org/wiki/SplitHopkinson_pressure_bar), in which the striker is usually propelled by a gas gun. The test specimen is placed between the ends of two straight bars, called the incident bar and the transmitted bar. At the end of the incident bar (some distance away from the specimen, typically at the far end), a stress wave is created that propagates through the bar toward the specimen. This wave is referred to as the incident wave, and upon reaching the specimen it splits into two smaller waves. The first wave, called the transmitted wave, travels through the specimen and into the transmitted bar, causing plastic deformation in the specimen. The second wave, called the reflected wave, is reflected away from the specimen and travels back down the incident bar. Most modern test setups with SHPB use strain gages on the bars to measure the strain caused by the waves. Assuming that deformation in the specimen is uniform, the stress and strain can then be calculated from the amplitudes of the incident and the transmitted and reflected waves.

The disadvantage of the SHPB device is that it is unable to treat test specimens with thick walls, or walls in the actual shapes or sizes that are usual for the basic components of offshore installations.

To overcome this disadvantage, an innovative device is now available at KOSORI in which the impact of a striker generated by hydraulic power is transferred to the test specimen, as shown in Figure 6. The weight of the striker can be changed depending on the desired impact kinetic energy. The device successfully generates a maximum load of 1,000 kN at the impact speed of 20 m/sec for a tensile coupon test specimen with a thickness of 20 mm, which is equivalent to a strain rate of 800/sec. In such test chambers, the temperature will often need to be precisely controlled to be anywhere between -170 and 700 degrees Celsius.

Figure 6. High speed material test facility at KOSORI

5. TEST FACILITIES FOR FIRES

Outdoor or indoor test facilities are used for safety studies associated with fires. Outdoor fire test facilities as shown in Figure 7 are used where it is needed to apply high fire loads to larger process modules or equipment under jet or pool fires. Jet fire test facilities will typically include high-pressure gas storage and networks of above- and below-ground pipework with valves and pressure or flow control systems. These systems enable a wide range of pressurized gas releases under controlled conditions. Liquid hydrocarbons such as LNG, LPG or oils, either alone or mixed with gases, are released to form jet or pool fires. Supplier is available at maximum speed of 30kg/s with gas and 20(ℓ/min)/m^2 with water.

Typical outdoor fire test projects aim to study the effects that topology or layout, geometry, congestion, confinement, gas concentration, ventilation, ignition position, leak position/rate/direction/duration, or wind speed/direction have on the expansion of heat fluxes; and on the temperatures associated with radiation and convection. Such tests allow the detection of nonlinear structural responses in fires.

Figure 7. Outdoor test module for fires and explosions at KOSORI (20m×15m×10m)

Indoor jet or pool fire test facilities as shown in Figure 8 include a large floor area with a high ceiling, and systems for wall cooling, smoke evacuation, air supply, air inlet, and wind simulation with a flexible door. The test hall should be able to withstand test fires with or without extra wall protection. Furnaces heated by gas are also used to test equipment or structures in fires. Because structural responses in fire may differ according to posture and orientation (among other factors), international standards and good practices require furnace testing in both the vertical and horizontal positions. Temperature changes in the test models over time are measured using thermo-couplers. The test results are used to quantify the characteristics of gas cloud temperatures in fires around the models. The temperature of steel in the models can also be measured in these tests.

Structures in fires are also usually subjected to mechanical loads due to the weight of equipment and utilities. Heated structures grow weaker, and are vulnerable to collapse. Test facilities are therefore used to investigate structural performance under combined mechanical and fire loads. Hydraulic jacks are used to apply mechanical loads in fixed or variable static conditions onto test structures positioned inside the furnace. Such test facilities can be used to study structural collapse behavior under various combinations of fire and mechanical loads.

Examples of outdoor fire test projects include studies on the performance of structures (e.g., topside modules, buildings or pressure vessels), structural components (e.g., walls or plate panels, windows and doors) and equipment or fire-risk control options (e.g., water-deluge systems, fire walls, passive fire protection, escape routes). Such tests can be conducted on near full-scale models. In the jet fire tests that are widely used by the oil and gas industry and regulatory authorities, a typical primary concern is whether the passive fire protection system can endure a jet fire for a period of 2 hours at up to 3 kg/sec and 60 bars.

Figure 8(a). Horizontal type fire wall test facility (4.5m×6.5m×1.5m) with furnace of 1,250 deg. C and loading actuator of 3MN at KOSORI

Figure 8(b). Vertical type fire wall test facility (4.5m×1.5m×4.5m) with furnace of 1,250 deg. C and loading actuator of 2MN at KOSORI

Figure 8(c). Fire collapse test facility (ϕ 4.5m×5.5m) with furnace of 1,250 deg. C and loading actuator of 10MN at KOSORI

Figure 8(d). Anti-fire test facility with gas supplier of 1kg/s and deluge system of 20(ℓ /min)/m^2 at KOSORI (30m×25m×15m)

Figure 9. Blast wall test facility at KOSORI (12×6×4m)

Other examples of outdoor projects include testing the performance of fire walls to protect against jet fire attack and the response of gas storage vessels under fires that result in boiling liquid–expanding vapor explosions (BLEVE) such as of fuel tanks. The related explosion tests can help indicate how to prevent catastrophic failures through appropriate ventilation.

Wind tunnel tests are usually used to study smoke ingress and ventilation problems on offshore platforms. Examples of wind tunnel test projects associated with fires include the study of wind effects on gas releases and fire scenarios. Such tests can facilitate the identification of regions with poor ventilation and subsequent remedies.

6. TEST FACILITIES FOR EXPLOSIONS

Outdoor or indoor explosion test facilities are used to study the performance of structures (e.g., topside modules, buildings), structural components (walls or plate panels, windows, doors), or equipment; or the characteristics and effectiveness of explosion risk control options (e.g., blast walls, deluge systems, escape routes). As in fire test facilities, the effects of various parameters can be studied, but in this case with a focus on gas dispersion or accumulation, overpressure or impulse profiles, drag forces, and structural response to explosions.

Offshore module test rigs and prototype models of real structures as shown in Figure 7 are often used for safety studies regarding explosions. Offshore module test rigs are made with walls and roofs that enable wide flexibility in confinement configurations. Deluge systems are installed to supply water, which is a typical option for explosion risk control.

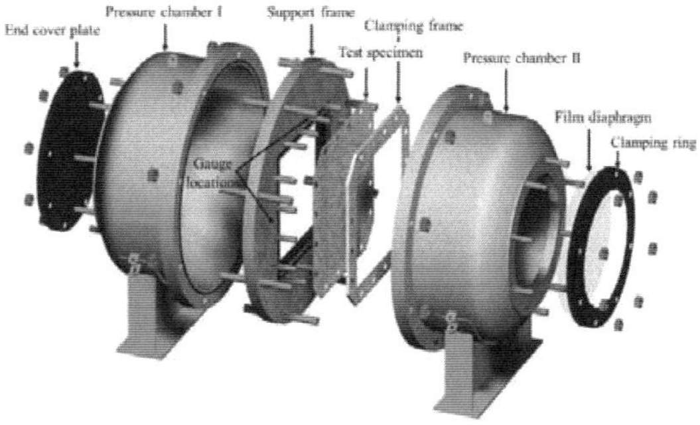

Figure 10. Pulse pressure loading rig at KOSORI

A highly pressurized gas supply system enables dynamic release in terms of speed and pressure that is suited to designated explosion scenarios. Internal pipework and vessels are added or removed to provide different levels of congestion as necessary. Outdoor explosion test facilities as shown in Figure 9 are used to study the performance of blast walls in explosions that occur due to the ignition of released gas in a chamber. The maximum size of test blast walls is 8m×4m.

Outdoor test facilities for similar purposes are usually instrumented to measure overpressure by using transducers that record on high-speed data acquisition systems. Such systems can measure flame arrival times. Other instruments, including displacement transducers or accelerometers, high-speed video recorders, and strain gauges are used to detect structural responses.

Indoor explosion test facilities are useful for studying the structural response or performance of small-scale or even full-scale components and equipment under blast pressure. The pulse pressure loading rig (PPLR) as shown in Figure 10 is a unique test facility for studying the performance of small-scale walls or plate panels under blast pressure loads. The PPLR was originally developed by the Impact Research Centre at the University of Liverpool, and is now implemented at the Korea Ship and Offshore Research Institute (KOSORI) in Pusan National University, South Korea.

This rig is composed of two pressure chambers with a film diaphragm, clamping ring and end-cover plates. The PPLR is able to produce a repeatable and uniform transient pressure loading over an exposed area. The test plate is attached to the support plate with a clamping frame. The chamber is then sealed with a rigid cover plate over a nozzle at one end, and a thin film diaphragm covering a nozzle at the other end. The chamber is then pressurized to the desired pressure on each side of the support plate so that no net load is applied to the test plate. At firing, the diaphragm is melted along its circumference by energizing a fuse wire. This initiates the rupture of the diaphragm and a rapid blow-down of the chamber.

The pressure difference arises because the pressurized air evacuates much more quickly through the nozzle than through the holes in the plate. The rig can also be operated in static mode by having the end cover attached at both ends. More details can be found in G.K. Schleyer et al. [8].

7. TEST FACILITIES FOR DROPPED OBJECTS AND COLLISIONS

Test facilities are now available to study the performance of full- and small-scale structures, structural components and equipment under the impact of dropped objects. These facilities are composed of a jacket-type frame and a striker.

Two sets of dropped object test facilities with small scale and full or large scale models are available at KOSORI as shown in Figures 11(a) and 11(b). For the full or large scale test facility, the maximum weight of the striker is 3 tonnes, the maximum height of its fall is 10 m and the maximum speed of impact is 14 m/sec. Such testing can generate an impact energy of upto 300 kJ.

The test will also need to be conducted at temperatures ranging between -170 and 700 degree Celsius representing either LNG induced cryogenic condition or fire-induced high temperature. This facility can accommodate a test model with a maximum size of 3m×3m×2m.

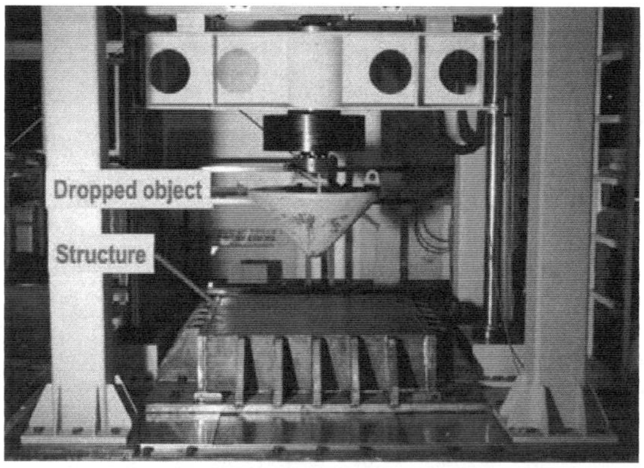

Figure 11(a). Dropped object test facility with small scale models at KOSORI

Figure 11(b). Dropped object test facility on full or large scale models at KOSORI with 30kN weight× 10m height and the test bed (floor) size of 3m × 3m together with temperature control chamber in the range of -170 to 600 deg. C

Figure 12. Structural failure test facilities at KOSORI with three dynamic loading actuators for loading capacity of 10,000kN each and total loading capacity of 30,000kN

8. TEST FACILITIES FOR STRUCTURAL FAILURE UNDER EXTREME LOADS

It is highly desirable to study structural failure under extreme loads using full-scale prototype test models. Mooring line failure in extreme tension and pipeline burst in extreme internal-pressure as well as buckling and plastic collapse of hull structures and structural components are good examples of such tests.

Test facilities for this purpose are composed of a loading frame and hydraulic jacks to manage combined loads and/or high speed loading in room and low or elevated temperatures. The facilities available at KOSORI has a maximum test floor size of 42m× 13m× 5m as shown in Figure 12. The maximum loading capacity of the hydraulic jacks is 30,000 kN in tension or compression, 5,000 kN in bending under all quasi-static conditions, and 5,000 kN under dynamic conditions.

9. TOTAL INTEGRATED SUBSEA TEST BED

Various subsea equipment and components, including as pipelines, risers or flowlines, are subjected to high pressure both inside and outside that may cause blowouts or burst. The maximum pressure outside the equipment or component depends on the water depth. Today, it is not surprising that the external pressure can reach several hundred bars, considering that the operating water depths have grown increasingly deeper, nearing almost 6,000 m. The maximum pressure inside pipelines is often over 700 bars. The temperatures inside pipelines are often over 200 degrees Celsius.

Figure 13. Subsea test chamber at KOSORI with Φ1.8m diameter × 4m deep with maximum pressure loading capacity of 825bar

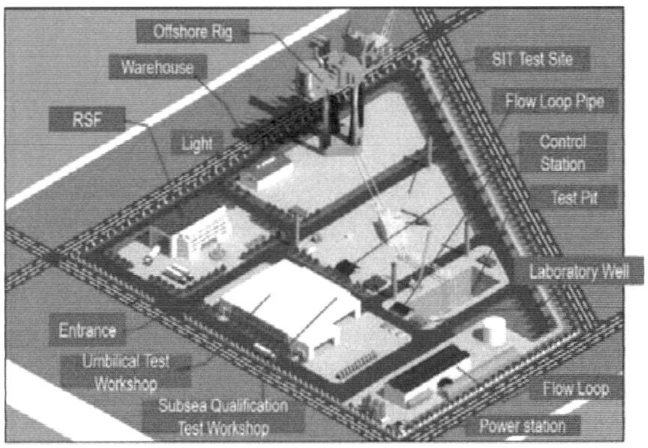

Figure 14. Integrated (assembled) subsea production equipment test bed under construction (2014-2017) at KOSORI

An additional important consideration is that subsea equipment and components are remote from the topsides of offshore installations where its functions are controlled. Therefore, before starting the operation of offshore installations, it is necessary to assure the performance of various individual subsea components, not only as standalone units, but also as assembled systems, to determine their total functional and HSE requirements in terms of risk management, flow assurance, remote-control security, and fatigue performance.

A cylindrical chamber available at KOSORI as shown in Figure 13 is used to study the performance of subsea equipment under high external pressure. For example, one of the available facilities has a chamber 1.8 m in diameter and 4 m deep. The maximum pressure that this facility can apply is 825 bars equivalent to water depth of 8,250 m.

Subsea test beds on land as shown in Figure 14 are very useful for studying the performance of subsea components in association with integrated safety systems. In such test beds, the environmental conditions inside the components (in terms of pressure and temperature) can be manipulated to match those in subsea, with pressures external to the test at atmospheric conditions. The principal advantage of a comprehensive total subsea test bed is that the integrated performance of full-scale subsea equipment can be tested and assured with all of the components assembled before the equipment is installed at the operating site. Such testing under real conditions enables risk levels in design and operation to be greatly reduced.

10. CONCLUDING REMARKS

Application of integrated and multidisciplinary approaches is essential to meet HSE&E requirements in association with extreme and accidental events which involve highly nonlinear problems. This is because computation by itself is not enough to identify such highly nonlinear problems and testing is essential with large- and full-scale test models.

KOSORI has constructed test facilities in Hadong, Korea that can be used for testing full- or large scale models of ships and offshore structures in extreme and accidental conditions.

Joint industry projects are on-going and also planned by taking advantage of the test facilities as well as nonlinear computational techniques.

11. ACKNOWLEDGMENTS

The construction of the test facilities at KOSORI (The Lloyd's Register Foundation Research Centre of Excellence) has been funded by central and local governments of Korea and industries, including Ministry of Trade, Industry and Energy, Ministry of Science, ICT and Planning, GyeongNam Province, Busan Metropolitan City, Ulsan Metropolitan City, and Hadong-gun as well as Lloyd's Register Foundation (LRF), a UK registered charity and sole shareholder of Lloyd's Register Group Ltd, that invests in science, engineering and technology for public benefit, worldwide. The original version of this article was presented at OMAE 2015 Conference in St. John's, Canada.

12. REFERENCES

1. Paik, J.K. and Thayamballi, A.K., Ultimate limit state design of steel plated structures, Wiley, Chichester, UK, 2003.
2. Paik, J.K. and Thayamballi, A.K., Ship-shaped offshore installations: Design, building, and operation, Cambridge University Press, Cambridge, UK, 2007.
3. Paik, J.K. and Melchers, R.E., Condition assessment of aged structures, CRC Press, New York, USA, 2009.
4. Hughes, O.F. and Paik, J.K., Ship structural analysis and design, Alexandria, VA, USA, 2013.Paik.
5. Paik, J.K., Making the case for adding variety to goal-based standards, The Naval Architect, Royal Institution of Naval Architects, London, UK, January 2015, pp.22-24.
6. Paik, J.K., Putting deep water safety to the test, Marine Technology, The Society of Naval Architects and Marine Engineers, Washington DC, USA, April 2014, pp.12-17.
7. Spouge, J., A guide to quantitative risk assessment for offshore installations, The Centre for Marine and Petroleum Technology (CMPT), Oslo, Norway, 1999.
8. Schleyer, G.K., Underwood, N.J., Do, H.M., Paik, J.K. and Kim, B.J., On pulse pressure loading of plates with holes, Central European Journal of Engineering, Vol.2, No.4, 2012, pp.496-508.

Professor Jeom Kee Paik is a professor of the Department of Naval Architecture and Ocean Engineering and serves as president of the Korea Ship and Offshore Research Institute (KOSORI) and as director of the Lloyd's Register Foundation Research Centre of Excellence at Pusan National University (PNU) in Korea. In 2015, he also joined University College London's Department of Mechanical Engineering as a professor and faculty member. He is a Fellow and Vice President of SNAME, USA and a Fellow and Council Member of RINA, UK. He was awarded William Froude Medal from RINA in 2015, the David W. Taylor Medal from SNAME in 2013, and a Doctor Honoris Causa from the University of Liege in Belgium in 2012. His research interests are safety engineering and design of ships and offshore structures in extreme and accidental conditions associated with risk assessment and management.

Dr Jung Kwan Seo is an associate professor of the Korea Ship and Offshore Research Institute (KOSORI) and as director of the Lloyd's Register Foundation Research Centre of Excellence at Pusan National University (PNU) in Korea. He serves as director of Subsea Research Centre of the KOSORI. His research interests are safety engineering of ships and offshore and subsea structures associated with risk assessment and management. He has been a member of the specialist committee V.8 Risers and Pipelines of International Ship and Offshore Structures Congress (ISSC).

ICSOT Korea: Safety of Offshore and Subsea Structures in Extreme and Accidental Conditions, 15-16 September 2015, Busan, Korea

THE MARS AND CHIRP REPORTING SCHEMES

AD Loynd, MARS and CHIRP Ambassador, Hong Kong SAR

SUMMARY

In this paper, the author considers the changing conditions aboard modern ships, and suggests that the MARS and CHIRP reporting schemes are vital in ensuring that young seafarers can learn from the mistakes of others, before they run the risk of repeating those mistakes themselves.

The paper discusses the reporting schemes in detail and suggests that everyone involved in the shipping industry can contribute to their success.

1. INTRODUCTION

Seafaring has changed dramatically in recent decades – crews are smaller, trips are shorter and officers serve for shorter periods before seeking employment ashore. As a result, crews tend to be less experienced than they were in the past.

Thirty years ago, young people embarking on a career at sea generally received their practical training on board from people with vast experience. They learned how to avoid accidents and unsafe practices from older crew members who could pass on the accumulated wisdom of many years.

Today the chances are that young seafarers will not be surrounded by so much experience, so the industry is striving to put in place new resources which will enable them to operate safely in an often harsh and dangerous environment.

These new resources include company safety circulars, the excellent guidance documents issued by the P&I Clubs, and industry publications such as *The Navigator* from the Nautical Institute, but there are two other resources which are equally valuable although they are often overlooked – the MARS and CHIRP reporting schemes.

Essentially, both schemes consist of actual case studies which are submitted by one of the persons or parties involved. These submissions are then verified and edited to make them clear and concise, and all names and identifying characteristics are removed to preserve confidentiality. The reports are then published to reach as wide an audience as possible, so that lessons can be learned and similar mistakes can be avoided in future. They are an invaluable resource for all seafarers, especially newcomers to the industry.

Despite their similarities the two schemes differ in some important aspects, so they will be discussed separately in the following sections.

2. THE MARINERS' ALERTING AND REPORTING SCHEME (MARS)

MARS is a confidential reporting system run by the Nautical Institute to permit full reporting of accidents and near misses without fear of identification and litigation. It is administered by Captain Paul Drouin, a serving seafarer with vast experience of accident investigation and safety management.

The primary purpose of MARS is to identify lessons learned from near-miss and recent incident reports and relay these to mariners afloat and ashore through the pages of *Seaways* magazine and the searchable database on the MARS website.

In addition to reports submitted by members of the industry, MARS also condenses reports from official industry sources such as accident investigation bureaux. All reports are held on a publicly-accessible database which can be searched using keywords, and this provides a valuable tool for risk assessment, work planning, loss prevention and training of both crew and management.

MARS is confidential, and operates a no-blame culture aimed simply at making the industry safer. It is a free service which may be accessed via *Seaways* magazine or www.nautinst.org .

Reports can be submitted by anyone who witnesses a dangerous situation, a close call or an accident. Reporters can use the dedicated form on the website to submit a brief description – preferably with photographs – or can simply send an email to mars@nautinst.org .

When a report is received, Captain Drouin asks any follow-up questions which may be appropriate, then prepares the report for publication in the standard format. All names are removed so that people, vessels or organisations cannot be identified. If it is impossible to maintain complete anonymity and confidentiality, then the report will NOT be published.

Reports will be rejected if they contain no safety lessons, or if they involve personality conflicts, industrial relations matters or employment conditions.

Great care is taken to preserve the integrity of the MARS reporting system so it remains a trusted and reliable tool for the maritime industry. A typical recent report is appended below:

201533 Five deaths in a tank

A 'largo' pontoon was beached at an isolated location for repairs. At one point the person in charge of the pontoon entered a compartment approximately 5 metres deep; there were no checks on the air quality of that compartment before entry. Within a very short time after entering the tank he became unconscious and fell, face down, onto the plates below. One of the two co-workers that had remained outside attempted a rescue and was also rendered unconscious after entering the tank. A third worker then entered the tank and also succumbed. Seeing the commotion and wanting to help, a man on the beach also entered the space to carry out a rescue and he too became unconscious. In short succession two others entered the tank but also succumbed. After more than an hour a successful rescue attempt was made and all bodies were removed. Resuscitation was attempted at length but, of the six who entered the compartment, there was only one survivor.

Lessons Learned

- *A false sense of security may have been acquired by the person in charge as he had entered other tanks earlier without negative consequences.*
- *Never enter a confined space without first testing the atmosphere from top to bottom.*
- *Always wear a portable four-gas detector while in a confined space, even if it has passed the initial testing prior to entry.*
- *Confined space training and practice is essential in order to prevent tragedies such as the one above.*

Editor's Note

While the barge in question was not a SOLAS convention vessel, the workers would have benefited from proper training and awareness of enclosed spaces. Since January 2015, enclosed space entry drills and practices are required every two on SOLAS vessels.

3. THE CONFIDENTIAL HAZARDOUS INCIDENT REPORTING PROGRAMME (CHIRP)

International CHIRP Maritime was established in 2003 with the aim of contributing to maritime safety worldwide by providing a totally independent confidential reporting system for the maritime industries. It is administered by the CHIRP Director (Maritime), Captain John Rose, who had a distinguished career afloat and ashore and is one of the Younger Brethren of Trinity House. He is assisted by a Maritime Advisory Board composed of leading figures from respected maritime bodies such as Lloyd's Register, the P&I Clubs and the Nautical Institute.

Reports may be submitted by any interested persons by mail, by confidential telephone call or via the reporting form on www.chirp.co.uk

When a report is received, Captain Rose validates it. Anonymous reports are not accepted, but strenuous efforts are made to maintain the confidentiality of the reporter. Only de-personalised reports are used in future discussions.

Like the MARS scheme, CHIRP does not seek to apportion blame to companies or individuals, and will not handle reports which deal with personality conflicts, industrial relations or employment matters.

The Director (Maritime) may then conduct an investigation if the circumstances warrant it, to determine what lessons can be learned or whether any remedial actions are required.

Once a report is verified, de-personalised and investigated, it is passed to the Maritime Advisory Board for their consideration. The Board will decide on subsequent action and whether the report is suitable for publication. It will only be published or made available to those who can remedy the problem with the permission of the reporter.

Final results are fed back to the reporter and all personal information is deleted from all files. Only key information is retained which will help establish systemic trends or root causes linked to safety and human element behaviours. This information can be made available to professional bodies for safety research.

CHIRP reports are available at www.chirp.co.uk and are published in printed form in *Maritime FEEDBACK* magazine, which is circulated to over 100,000 shipping companies, organisations and individuals. A typical recent report is appended below:

TRIPPING OVER TOWING WIRE DURING MOORING OPERATIONS

Report Text

During mooring operations the Second Officer in charge at the aft mooring station tripped on the towing wire that was permanently deployed from a storage reel to the aft chock. No warning markings existed to draw attention to draw attention to this obstruction on the deck. The underlying cause was the lack of proper marking.

Lessons Learned

Permanent fittings which may cause obstruction and which may be dangerous should be marked with black and yellow alternate stripes to draw attention to the obstacle. Where necessary, warning notices should be posted, and some obstructions at head height may need to be padded.The company instructed the Master to discuss the above at the ship's next safety committee meeting and order the crew to inspect the ship more carefully and report on any areas for improvement.

CHIRP Comment
Company superintendents should ensure that warning marks are in place and maintained. The risks of falls onto steel decks can be seen in a recent IMCA Flash 14/14 in which a crew member tripped, lost balance and fell forwards, hitting the deck.He instinctively raised his arms in an effort to lessen the impact of the fall, which resulted in him sustaining two broken elbows.

4. HOW DO THE SCHEMES DIFFER?

On the surface, there may not appear to be much difference between the CHIRP and MARS reporting schemes. They are both confidential, they both focus purely on safety matters, and they both aim to publish reports showing the lessons learned. In short, they are both dedicated to improving safety at sea and assisting mariners by educating them to avoid unsafe practices.

However, there are some notable differences between them.

The primary purpose of MARS is to identify the lessons learned from recent incident and near-miss reports and relay these to mariners ashore and at sea.

CHIRP Maritime, on the other hand, allows proactive follow-up and investigation of individual safety issues. The organisation also attempts to identify trends and bring these to the attention of the relevant authorities.

The schemes may be considered complimentary, and they are both attempting to establish a robust reporting culture as part of a systemic approach to reducing risks to levels which are as low as reasonably possible.

5. LONG-TERM SURVIVAL OF THE SCHEMES

Both CHIRP and MARS have secured sponsorship from a variety of respected international bodies which will cover their costs for the immediate future, but they will need additional funding to guarantee their long-term survival. If any delegates control charitable, welfare or CSR budgets then the author assures them contributions to both schemes would be gratefully received.

It is not the purpose of this paper to solicit funding, however, but to address the other requirement for the continuation of the schemes – the reports submitted by the industry.

It goes without saying that CHIRP and MARS can only fulfil their aims if they receive and disseminate a sufficient number of useful reports, and it is unfortunate that very few reports are received from the Asian region.

Both schemes are administered from London, so the exposure they have received in Asia has been minimal. Until recently, there were no funds for travel or promotion of the schemes, so many in Asia were unaware of them. It is only very recently that a number of volunteers (including the author of this paper) were recruited to act as unpaid ambassadors for CHIRP and MARS in places outside Europe, and even more recently that a limited travel budget became available to the CHIRP scheme.

It is also true that some cultures in Asia do not encourage discussion of mistakes by individuals or organisations.

The author would argue that the advantages of the MARS and CHIRP reporting schemes should outweigh any cultural reservations. They are confidential and operate a no-blame culture, so there is no danger of reporters suffering or being embarassed if they contribute to the schemes. In addition, the positive benefits to the industry, particularly to new entrants to the industry, are too valuable to ignore.

6. THE RELEVANCE TO *ICSOT* DELEGATES

Asia is rapidly becoming the most important area for maritime enterprises. We have many of the world's major ports, probably the largest concentration of ship builders, ship owners and ship managers, the busiest waters, and we supply the majority of the officers and crew who man the world fleet. We should therefore be providing the majority of MARS and CHIRP reports, but we are not.

The administrators of the schemes and their ambassadors are starting to make strenuous efforts to encourage greater participation from Asian owners, managers and seafarers, but *ICSOT* delegates might be forgiven for asking how this is relevant to them?

The majority of speakers and delegates are scientists, engineers and naval architects who are not directly involved in the running of ships, but the author believes that almost all of the participants at this conference are in a position to contribute to the schemes.

Many of you will find yourselves on board vessels or in shipyards as part of your research, whilst others will be involved in trials of new equipment or data collection when new equipment is in use. Some delegates may be involved in the investigation of structural failures or machinery malfunctions.

On behalf of the CHIRP and MARS schemes, I ask you all to report accidents, incidents and near-misses and let us know the causes of such incidents. Even incidents which might appear relatively obscure can have important consequences.

For example, there was recently a case where the reporter discovered that an equipment malfunction was partly caused because instruction manuals for the equipment were not written in a language which was used by the crew so they were unable to read the manuals. A report was submitted to CHIRP, and Captain Rose noted that similar reports had been received from other reporters. Upon investigation, it was discovered that many ships suffered from the same problem – manuals written for the original owners of a ship were not appropriate for the crews of another owner who subsequently purchased the vessel. Once the problem was noticed, it was possible to enact new legislation to ensure manuals were always available in a language which the crew could understand.

On a more topical note, we are now beginning to see ship groundings caused by improper use of electronic charting systems. If an observer becomes aware of such a case and reports it to MARS and/or CHIRP, then the information can be promulgated and young officers will be made aware of the dangers of not using the equipment properly.

When any delegate becomes aware of such events, they should report them. Lessons will be learned and, quite possibly, lives will be saved if we can make the industry aware of the dangers.

The author is not suggesting that all of us will be in a position to submit reports on a regular basis, but occasionally each of us might be in a position to do so. Such opportunities should not be ignored. We should all do whatever we can to promote safety at sea, and to ensure that seafarers are aware of the dangers inherent in their profession.

There is an added urgency at the moment due to the international trend of criminalisation of seafarers. Young officers who, perhaps, have never sailed with grizzled old hands who could give them guidance, now face the prospect of being imprisoned for their mistakes. This is a tragedy, not least because the majority of young officers are smart, keen, and brimming with enthusiasm. If they are told about potential dangers they are quite capable of absorbing the information and avoiding the mistakes of others. However, if they are not aware of a potential problem, they are not in a position to ensure they avoid it. There is a real danger that they will find themselves in prison simply because there was nobody to warn them of the potential dangers.

7. CONCLUSIONS

The shipping industry has changed dramatically in recent years, and seafaring is no longer seen as a lifelong vocation. As soon as officers reach senior positions they tend to seek employment ashore.

As a result, young seafarers are no longer surrounded by vastly experienced older hands who can pass on safety lessons and keep the younger people out of harm's way. It is therefore vital that the industry finds new ways of teaching young seafarers the lessons which will keep them out of danger and permit them to operate safely and effectively.

The MARS and CHIRP schemes are invaluable tools in helping to achieve this aim, and everyone in the industry should contribute to ensure the schemes survive and continue to contribute to maritime safety.

8. ACKNOWLEDGEMENTS

The author acknowledges with thanks the assistance of MARS and CHIRP personnel in preparing this paper, although all opinions are entirely his own.

9. AUTHORS BIOGRAPHY

Alan Loynd is a master mariner who is Managing Director of his own company, Branscombe Marine Consultants, and an honorary ambassador for the CHIRP and MARS reporting schemes. He spent many years at sea with the Royal Fleet Auxiliary, China Navigation Co., Swire Pacific Offshore and Hongkong Salvage & Towage (HKST). He was Senior Salvage Master with HKST before he took early retirement to form his own company.

Capt. Loynd is a Fellow of the Nautical Institute and a member of Council. He is also a founding member and Fellow of the International Tugmasters Association, where he served as Chairman between 2012 and 2015.

ICSOT Korea: Safety of Offshore and Subsea Structures in Extreme and Accidental Conditions,
15-16 September 2015, Busan, Korea

THE NEW METHODOLOGY OF FMEA FOR PROCESS INDUSTRY

Jinhyung Park, Yokogawa Electric Korea, South Korea

SUMMARY

FMEA is one of the most representative hazard analysis methodologies especially. Failure Modes and Effect Analysis (FMEA) is a systematic procedure for the analysis of a system to identify the potential failure modes, their causes and effects on system performance (performance of the immediate assembly and the entire system or process). FMEA for turret safety system was done for each part like UPS unit & Power Distribution, HIS, Engineering Workstation, Network, Controller, Marshalling Cabinet, Extension Alarm Panel. This FMEA methodology was slightly quantitative by adopting the likelihood and the severity. The criticality is decided by the criticality matrix configured with likelihood and severity. The detection was also included in this FMEA methodology to evaluate if the failure in each component can result in a system-level failure. The methodology of FMEA can vary in different industry types, but the methodology of FMEA in this research project was developed and optimized for process industry as far as possible

NOMENCLATURE

FMEA Failure Mode Effect Analysis

1. INTRODUCTION

There are many methodologies of FMEA. In this paper, the appropriate methodology for process industry was developed and this methodology is applied to Turret Safety System.

2. THE DEFINITION OF FMEA

FMEA is one of the most representative hazard analysis methodologies especially. The Failure Modes and Effect Analysis is a systematic procedure for the analysis of a system to identify the potential failure modes, their causes and effects on system performance (performance of the immediate assembly and the entire system or process). [1]

FMEA is an inductive reasoning approach that is best suited to reviews of mechanical and electrical hardware systems. The FMEA technique considers how the failure modes of each system component can result in system performance problems and ensures that appropriate safeguards against such problems are in place. A quantitative version of FMEA is known as failure modes, effects and criticality analysis (FMECA). [2]

3. THE COMMON USE OF FMEA

FMEA is not appropriate to broader marine issues such as harbour transit or overall vessel safety. This technique is applicable to any well-defined system, but the primary use is for reviews of mechanical and electrical systems (e.g., fire suppression systems, vessel steering/propulsion systems). It also is used as the basis for defining and optimizing planned maintenance for equipment because the method systematically focuses directly and individually on equipment failure modes. FMEA generates qualitative descriptions of potential performance problems (failure modes, root causes, effects, and safeguards) and can be expanded to include quantitative failure frequency and/or consequence estimates. [2]

Some techniques get bogged down when used to analyze extremely complicated problems. The complexity and size of a problem are functions of the number of activities or systems, the number of pieces of equipment, the number of operating steps and the number and types of events being analyzed. For most analysis techniques, considering a larger number of equipment items or operating steps will linearly increase the time and effort needed to perform a study. For example, using the FMEA technique will generally take five times more effort for a system containing 100 equipment items than for a system containing 20 items. Thus, the types and number of events and effects being evaluated are proportional to the effort required to perform a risk analysis .[2]

The FMEA approach has a well-deserved reputation for efficiently analyzing electronic and computer systems, whereas the HAZOP analysis approach is typically applied to fluid transport or processing systems.[2]

4. THE HISTORY AND TYPES OF FMEA

The aerospace industry during the Apollo missions employed FMEA in the 1960s. The US Navy in 1974 developed a tool which discussed the proper use of the technique. Today, FMEA is universally used by many different industries. There are three main types of FMEA in use today:

- System FMEA: concept stage design system and sub-system analysis.
- Design FMEA: product design analysis before release to manufacturers.
- Process FMEA: manufacturing assembly process
- Analysis [3]

5. THE STRENGTHS AND WEAKNESSES OF FMEA

The strengths of FMEA are:

- It is widely-used and well-understood

- It can be performed by a single analyst
- It is systematic and comprehensive, and should identify hazards with an electrical or mechanical basis
- It identifies safety-critical equipment where a single failure would be critical for the system [4]

Its weaknesses are:

- Its benefit depends on the experience of the analyst.
- It requires a hierarchical system drawing as the basis for the analysis, which the analyst usually has to develop before the analysis can start.
- It is optimised for mechanical and electrical equipment, and does not apply to procedures or process equipment.
- It is difficult for it to cover multiple failures and human errors.
- It does not produce a simple list of failure cases.[4]

6. THE PROCEDURE OF FMEA

The FMEA procedure consists of the following four main stages:
a) establishment of the basic ground rules for the FMEA and planning and scheduling to ensure that the time and expertise is available to do the analysis;
b) executing the FMEA using the appropriate worksheet or other means such as a logic diagrams or fault trees;
c) summarizing and reporting of the analysis to include any conclusions and recommendations made;
d) updating the FMEA as the development activity progresses. [1]

FMEA detailed process is shown in the below.

1. Initiate FMEA of an item
2. Select a component of the item to analyse
3. Identify failure modes of the selected component
4. Select the failure mode to analyse
5. Identify immediate effect and the final effect of the failure mode
6. Determine severity of the final effect
7. Identify potential causes of that failure mode
8. Estimate frequency or probability of occurrence for the failure mode during the predetermined time period
9. Propose mitigation method, corrective actions or compensating provisions. Identify actions and responsible personnel
10. Document notes, recommendations, actions and remarks [1]

7. DEFINING SYSTEM BOUNDARY FOR THE ANALYSIS

The definition of the system boundary is more likely to be influenced by design, intended use, source of supply, or commercial criteria rather than the optimum requirements of the FMEA. However, where it is possible to define the boundaries to facilitate the system FMEA and its integration with other related studies in the program, such action is preferable. This is especially so if the system is functionally complex with multiple interconnections between items within the boundary and multiple outputs crossing the boundary. In such cases it could be advantageous to define a study boundary from functional rather than hardware and software point of view to limit the number of input and output links to other systems. This would tend to reduce the number of system failure effects. [1]

8. LEVELS OF ANALYSIS

It is important to determine the indenture level in the system that will be used for the analysis. For example, systems can be broken down by function or into subsystems, replaceable units, or individual components (see Figure 1). Ground rules for selecting the system indenture levels for analysis depend on the results desired and the availability of design information. [1]

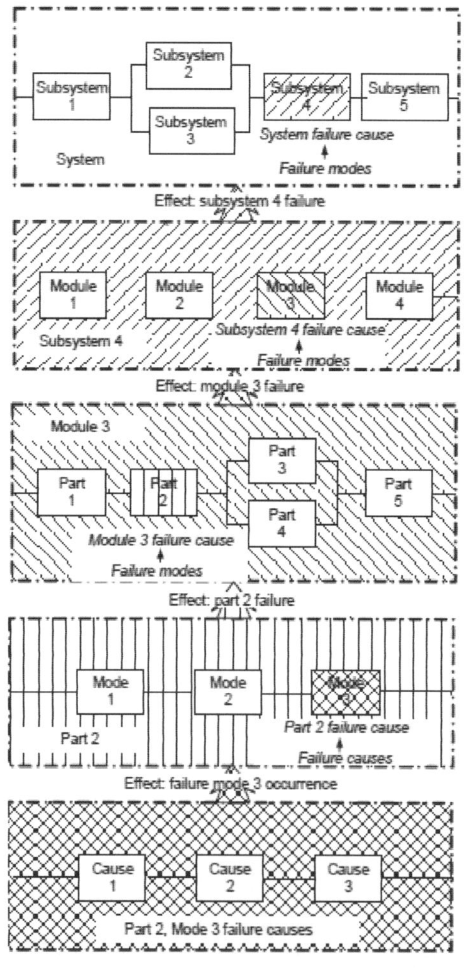

Figure 1: Relationship between failure modes and failure effects in a system hierarchy

In the FMEA, the definitions of failure modes, failure causes and failure effects depend on the level of analysis and system failure criteria. As the analysis progresses, the failure effects identified at the lower level may become failure modes at the higher level. The failure modes at the lower level may become the failure causes at the higher level, and so on. [1]

When a system is broken down into its elements, effects of one or more of the failure mode causes make a failure mode, which in turn is a cause of the higher level effect, a part failure. Part failure is then the cause of a module failure (effect), which in itself is a cause of a subsystem failure. The effect of a cause of one system level thus becomes a cause of another effect at a higher level. The above rationale is shown in Figure 1. [1]

9. FAILURE EFFECTS

A failure effect is the consequence of a failure mode in terms of the operation, function or status of a system. A failure effect may be caused by one or more failure modes of one or more items. [1]

The consequences of each failure mode on system element operation, function, or status need to be identified, evaluated and recorded. Maintenance activities and system objectives should also be considered whenever pertinent. A failure effect may also influence the next level up and ultimately the highest level under analysis. Therefore, at each level, the effect of failures on the level above should be evaluated. [1]

10. LOCAL FAILURE EFFECTS

The expression "local effects" refers to the effects of the failure mode on the system element under consideration. The consequences of each possible failure on the output of the item should be described. The purpose of identifying the local effects is to provide a basis for judgement when evaluating existing alternative provisions or devising recommended corrective actions. In certain instances, there may not be a local effect beyond the failure mode itself. [1]

11. END FAILURE EFFECTS AT THE SYSTEM LEVEL

When identifying end effects, the impact of a possible failure on the highest system level is defined and evaluated by the analysis of all intermediate levels. The end effect described may be the result of multiple failures. (For example, failure of a safety device results in a catastrophic end effect only in the event that both the safety device fails and the prime function for which the safety device is designed goes beyond allowed limits.) These end effects resulting from a multiple failure should be indicated on the worksheets. [1]

12. LIKELIHOOD OF OCCURRENCE FOR PROCESS INDUSTRY

The criteria of likelihood of occurrence can be determined by the safety policy of end user and local regulation, but the appropriate likelihood of occurrence for process industry can be proposed as the below.

1: Not expected to occur during system life
2: Could occur once during system life
3: Could occur several times during system life
4: Could occur on an annual basis
5: Could occur on more than an annual basis

13. SEVERITY OF CONSEQUENCES FOR PROCESS INDUSTRY

The criteria of severity of consequences can be determined by the safety policy of end user and local regulation, but the appropriate severity of consequences for process industry can be proposed as the below.

1: Safety – Minor injury (first aid)
Environment – Temporary release and cleanup within days
Reputation – Immediate community not affected
2:Safety – Minor injury or minor health impacts (Lost time recordable, Medical treatment case)
Environment – Temporary release and cleanup within weeks
Reputation – Immediate community affected
3:Safety – Injury or moderate health impacts (Permanent injury)
Environment – Temporary damage to the facility and cleanup within months
Reputation – Affects more than one communities/ state
4: Safety – Single fatality
Environment – Permanent damage to facility
Reputation – Affects national communities

5: Safety – Multiple fatalities
Environment – Permanent damage to facility and offsite environment
Reputation – Affects regional/ international community

14. CRITICALITY RANKING MATRIX

The criteria of criticality ranking matrix can be determined by the safety policy of end user and local regulation, but the appropriate criticality ranking matrix for process industry can be proposed in the figure 2.

Criticality ranking matrix		SEVERITY OF CONSEQUENCES				
		1	2	3	4	5
LIKELIHOOD OF OCCURENCE	1	1	1	1	2	2
	2	1	2	2	2	3
	3	1	2	2	3	3
	4	2	2	3	3	3
	5	2	3	3	3	3

Figure 2: Criticality ranking matrix

FAILURE MODE EFFECT ANALYSIS												
System:				Sub-system:								
Failure Analysis					Failure Effect							Page:
No.	Component	Function	Failure Mode	Failure Cause	Local	End	S	L	C	Detection	Recommendation	Comment

Figure 3: FMEA format for process industry

In failure effect, the evaluation by the supplier called "Supplier critical level" corresponds to:
Level 1 (Minor): Failure with no effect
Level 2 (Serious): Loss of production will be leaded
Level 3 (Critical): Turret shutdown will be leaded

15. FMEA FORMAT FOR PROCESS INDUSTRY

The FMEA format can be determined by the safety policy of end user and local regulation, but the appropriate FMEA format for process industry can be proposed as figure 3.

16. CONCLUSIONS

FMEA is a systematic and comprehensive methodology and best suited to reviews of mechanical and electrical hardware systems. FMEA generates qualitative descriptions of potential performance problems (failure modes, root causes, effects, and safeguards) and can be expanded to include quantitative failure frequency and/or consequence estimates. The methodologies of FMEA vary according to the type of system to be analysed. In this paper, the appropriate likelihood of occurrence and appropriate severity of consequences and appropriate criticality ranking matrix of FMEA for process industry was developed and applied to Turret Safety System. The block diagram is one of most important procedure to define scope of FMEA and clarify the target of FMEA.

17. ACKNOWLEDGEMENTS

This research work is done as activity of national research project titled "Development of Turret System for FPSO Operating at Hs 15m by the support of Korea Evaluation Institute of Industrial Technology (KEIT). I appreciate the support of KEIT for this research work.

18. REFERENCES

1. IEC, IEC 60812 – Analysis techniques for system reliability – Procedure for failure mode and effects analysis (FMEA), IEC, 2006

2. ABS, 'RISK ASSESSMENT APPLICATIONS FOR THE MARINE AND OFFSHORE OIL AND GAS INDUSTRIES', ABS, JUNE 2000

3. A.H. Saharuddin, O.O. Sulaiman and A.S.A. Kader, 'Risk and Hazard Operability Process of Deep Water Marine System', IACSIT International Journal of Engineering and Technology, August 2011

4. DNV, 'Marine risk assessment, OFFSHORE TECHNOLOGY REPORT', DNV, 2001

19. AUTHORS BIOGRAPHY

Jin Hyung Park holds the safety expert at Yokogawa Electric Korea and holds Certified Functional Safety Expert, TUV Rheinland Functional Safety Expert. He is responsible for the functional safety consulting including HAZOP, SIL classification and SIL calculation and responsible for the trainer of TUV Rheinland Functional Safety Engineer Course and the lecturer of Korea OSHA training center and responsible for editorial member of 'Ships And Offshore Structure'. His previous experience includes the SIS engineering and the DCS engineering including design, implementation, manufacture review, site modification and the project engineering including schedule management and document management.

PHYSIOLOGICAL COMPUTING FOR MARITIME ERGONOMICS APPLICATIONS

YB WU, T Miwa and **M Uchida**, Kobe University, Japan

SUMMARY

In safety-critical systems such as marine engine plant operation, we always hope that the operator could persistently monitor system parameters (e.g. engine revolution) and adapt to unforeseen changes (e.g. lubrication oil pressure low) and avoid any misses or incidents (e.g. engine shut down), in other words, to perform their jobs perfectly. With the improvement of hardware reliability, it is the abnormal psychophysiological state of the operator that more likely lead to performance degradation of the system. Thus it is necessary to address how to maintain the optimal operator functional state (OFS) in the context of human-machine systems, where the risk of accident is closely related to human component. Physiological computing is to correlate an operator's physiological changes and their functional states based on simple selected features or complex algorithms such as support vector machine and fuzzy modelling. The physiological evaluation indices mainly include heart rate variability (HRV) recorded from electrocardiogram (ECG), blood pressure, respiration, eye blinks and pupil diameter, skin potential, hemodynamic indices and cerebral cortex indices such as electroencephalogram (EEG) and event-related potentials (ERPs). This paper aims to review the psychophysiological indices that are widely used to predict human cognitive states in both cognitive psychology research and applied industries. Possible applications of physiological computing in maritime human factors are suggested. A marine engine power-plant simulator operation study is done to evaluate operators' mental workload based on chosen psychophysiological indices and physiological computing methods. Significance tests show that alpha band wave ratio is effective for referring subject's mental workload.

1. INTRODUCTION

As design companies see the benefits of adopting human-oriented design methods instead of technology-oriented ones, Human Computer interaction (HCI) researchers focus on designing interaction methods and systems that are 1) more usable (cause lower stress and distraction) 2) satisfactory (provide pleasant user experience) and 3)adaptable (adaptive according to user's needs and state) [1]. At the meantime, with the on-board updating of technology in maritime domain, HCI issues continually arise, yet remain unsolved. For instance, in order to avoid the drowsiness of officers on watch, the international convention of Safety of Life at Sea (SOLAS) currently requires ships above 150 tonnage to install a Bridge Navigation Watch Alarm System (BNWAS), to which the officer on watch has to response by either directly pressing specific buttons or letting their movements be detected within a pre-set time interval, say, 12 minutes. However, the high intrusiveness caused by BNWAS and the usability of BNWAS are problematic from the view of ergonomics.

Physiological information, such as brain wave extracted from EEG, pupil diameter and blink frequency measured by view trackers, and heart rate, provide an unobtrusive and objective method to infer the operator's physical health state as well as affective and cognitive state, although these relationships are mostly implicit. Physiological computing refers to correlate these explicit physiological information with implicit operator functional states. There are numerous types of physiological indices that can be measured in real-time by wearable devices or cameras. Most of the existed research use these evaluation indices on mental workload (MWL) or cognitive workload measurements. Mental workload is considered as one of the core elements of OFS, with
exorbitant transitional MWL result in stress, and an accumulation of lower workload or task disengagement contribute to drowsiness. MWL is a causal factor of several different OFSs. Fig.1 shows the flowchart of physiological computing.

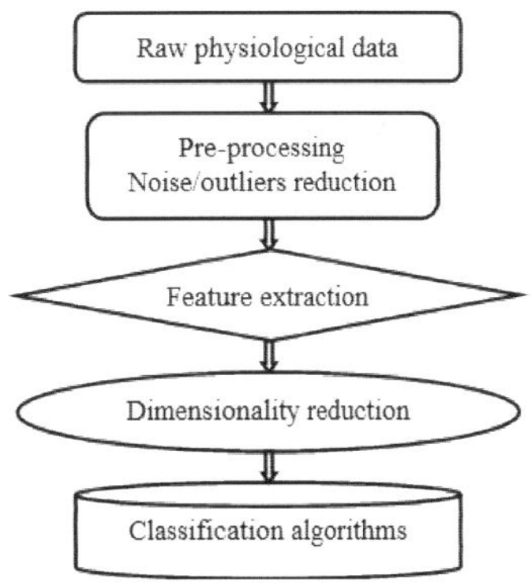

Figure 1 Flow chart of physiological computing

Concerning the working environment of offshore vehicles or platforms, there are several factors that may affect the feasibility of applying physiological computing in these environment. The factors include: 1) unlike civil aviation and driving, operator has to move around inside their routing workplaces, either the bridge or the engine control room; 2) vibration and noise caused by working machineries; 3) the ship motion may be in six degree of freedom caused by wave and wind; 4) cooperative working context for ship crew, the physiological computing should be in team scale rather than individual level (related to function allocation between team members, e.g. when on watch navigation officer are in abnormal state, allocate the duty of officer to the on watch sailor). Therefore, ambulatory and portable devices that are immune to noise are necessary for applying physiological computing in maritime operation. Chest type heart beat sensor, portable view tracker, and portable EEG device are considered in this paper.

2. PHYSIOLOGICAL INDICES

Lean and Fu [2] briefly reviewed the physiological and biochemical evaluations of human mental workload and proposed a taxonomy of physiological metrics (Table 1).

Table 1 Physiological and biochemical indices

Physiological metrics		Biochemical metrics
Peripheral	Central	
Heart rate and heart rate variability Blood pressure, respiration, eye blinks, skin potential, hemodynamic indices.	Cerebral cortex; EEG and ERP	Catecholamine, cortisol, immunoglobulin A collected from blood, salivary or urinary samples.

2.1 Heart rate related indices

Heart rate and heart rate variability analysis are generally used for evaluating autonomic nervous system and are peripheral physiological indices. As heart rate sensors are generally low-cost, simple and user friendly causing less obtrusiveness, the application of heart rate related indices are becoming more and more popular in many industries such as athlete's monitoring and pilot's state monitoring. Except the absolute value of heart rate, many time domain, frequency domain, and nonlinear indices are used as physiological computing inputs. For instance, mean value of heart beat interval, standard deviation of heart beat interval, LF/HF ration based on Discrete Fast Fourier Transformation (DFFT). Low Frequency (LF) was defined as 0.04-0.15Hz and High Frequency was defined as 0.15-0.4Hz.

2.2 EEG

The electroencephalogram (EEG) is the recording of electrical activity on the scalp surface through a set of electrodes. The studies of applying EEG signal in innovative HCI fall into two categories: time domain analysis and frequency domain analysis. ERPs are measured brain response fluctuation that is related to a specific sensory, cognitive or motor event in a certain time delay. Prinzel et al. [3] used the P300 component of ERP to assess participants' task engagement and performance in an adaptive automation situation. The EEG waveforms are usually analysed by wavelet transformation or DFFT. The power spectral of EEG are dived into several bands: delta (1-4 Hz), theta (4-7 Hz), alpha (8-15 Hz), beta (16-31 Hz) and gamma (32+). The taxonomy sometimes can be slightly different. The power of these bands or their relative percentage of total power bands are used as physiological indices.

2.3 VIEW TRACKERS

View trackers are used to record the pupil diameter, blink interval and gaze fixation of operators. This technology is generally based on video and image analysis of eyeball variation and movement. One popular metrics is saccade, the fast unconscious movement of pupil fixation. Siegenthaler et al. [4] found that task difficulty in mental arithmetic affects micro saccadic rates and magnitudes. Pupil diameter and blink interval are found effective to infer mental workload in some research.

3. POTENTIAL APPLICATIONS

3.1 OFS-ADAPTIVE AUTOMATION

The bio cybernetic loop proposed by Pope et al. is considered to be seminal research in OFS-AA. The bio cybernetic loop [5] is formed by adjusting the mode of operation of a task set (e.g., level of automation) based on the brain activity criterion (EEG-based index of engagement) reflecting an operator's engagement in the task set.

The conceptual model of OFS-AA is composed of two main parts, physiological computing and system adaption. Physiological computing is the determining foundation of adaption strategies and performance-oriented adaption strategies provide feedback to improve the robustness of the former.

3.2 USABILITY TEST

Figure 2 Modern ship's bridge, 18 screens in an operation room (adapted from Pan et al. 2015[6])

Human-oriented systems have emerged in modern ship design. Fig. 2 shows multiple, distributed interaction interface with several interacting screens on a ship bridge. For a ship, favourable usability means that operators can accomplish required tasks using the limited on-board resources with efficiency, effectiveness and self-satisfaction. In the development of new innovative maritime systems, usability experts need to avoid creating distributed interfaces that can become technology "barriers". The object is to support the operator and make technology as user friendly as possible.

Usability testing of maritime products should be undertaken not only by designers but also by the operators (ship crew) who would use it. This is especially true for scenario-based products for situations of high uncertainty, such as collision avoidance systems. Robert et al. using a ship control simulator found that higher levels of collision threat and impaired performance on the secondary oil pressure monitoring were associated with markedly increased ratings of MWL. Their work has relevance for the design of collision avoidance systems [7]. Gould et al. [8] examined the effects of two different navigation methods, the conventional system using paper charts, and an electronic chart display and information system (ECDIS), on workload and performance in simulated high-speed ship navigation. They reported that ECDIS navigation significantly improved course-keeping performance. They also used heart rate variability (HRV) and skin conductance as MWL measurements and the results indicated higher workload in conventional navigation, although the difference between the groups was not significant. In the engine room of a ship, the on watch engineer is required to simultaneously monitor and control a few complex sub-systems within one or several interaction interfaces, i.e. central cooling system and lubrication oil system. With the increase of human-centred control systems, it should be a necessary requirement to conduct usability testing concerning MWL before adoption.

3.3 MARITIME TRAINING

In professional education such as in the training of pilots, advanced seafarers and surgery operators, high fidelity simulators are widely used to help the trainees practice routine and emergency operation procedures with lower cost and shorter time. There are also cases when simulators can provide learning experience beyond that which can be learned in real systems, for example, in teaching students systematic trouble shooting skills, a simulator that can malfunction any single component in a system may provide more effective training than a real system. Therefore, although many companies focus on developing the fidelity of simulators, high fidelity does not necessarily bring high effectiveness of training. Training procedures and instructions should be carried out to pace and train the trainees to maintain engagement with the operational process, and keep their MWL at the optimal point, where the learner is neither overloaded nor under loaded. To improve the training regime, training institutions need to adjust their syllabus and pedagogics based on individual mental capacity and students' engagement throughout the learning process. A record of psychophysiological data such as brain activity and pupil movement can be used for a mental workload model which can provide more reliable and objective data in addition to the usual subjective or experts' ratings. It may also improve the following training related issues:

- evaluate the effectiveness of training regimes
- compare differences and similarities of simulator operation and real operation
- elaborate on responses of experienced operators and novices
- choose optimized pedagogics for individualized teaching

4. EXPERIMENT STUDY

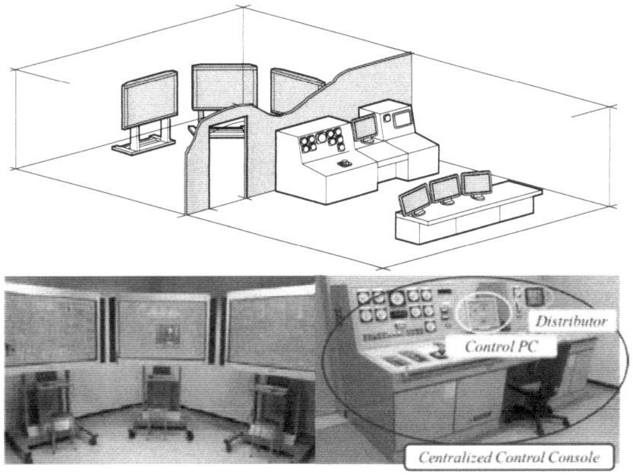

Figure 3 MEPS (a) outline of MEPS; (b) real picture of simulated engine room; (c) control console

To evaluate the feasibility and limitations of physiological computing models for maritime operation we have to examine the models in real or quasi-real operation scenario. Marine engine system operation can be clarified as complex human-machine interaction system. Marine Engine Plant Simulator (MEPS) is a full scale simulator of the engine system of a real ocean-going container ship. Four routine tasks of ship engine department with different level of difficulty were designed: 1. Transfer diesel oil from settling tank to service tank; 2. Prepare and start the central cooling system; 3. Start diesel engine of NO.2 generator; 4. Start lubrication oil purifier. These four tasks were denoted as MEPS-1, MEPS-2, MEPS-3 and MEPS-4 respectively. MEPS-1 required 5 procedures within fuel oil pipes, MEPS-2 required 14 procedures within sea water pipes and fresh water pipes, MEPS-3 required 14 procedures within lubrication oil pipes, fuel oil pipes, compressed air pipes and cooling water pipes, and MEPS-4 required 12 procedures within fresh water pipes, sludge pipes, steam pipes and lubrication oil pipes. Marine operation tasks were complex and multi-attributed, operators therefore had different subjective feelings regarding the complexity and requirement of these tasks. Despite the inaccuracy to hastily conclude that one task was absolutely more demanding than another, we generally expected that the task with less operation procedures and pipes were less demanding than the one with more of those. As all the subjects had the knowledge of marine operation, we did not arrange practice session beyond an instruction in MEPS tasks.

Ten male and one female subjects voluntarily participated in this experiment, their average age was 23.2 years (SD=2.4). They all have the knowledge of maritime engine system operation.

5. RESULT AND DICUSSION

Alpha wave rate was defined as $r_\alpha = \frac{P_\alpha}{P_\theta + P_\alpha + P_\beta}$, in which P_α was the power of alpha band in one second. The average alpha wave rate, $\frac{\sum r_\alpha}{l}$ (l was time length of each task) was chosen as EEG feature of each task.

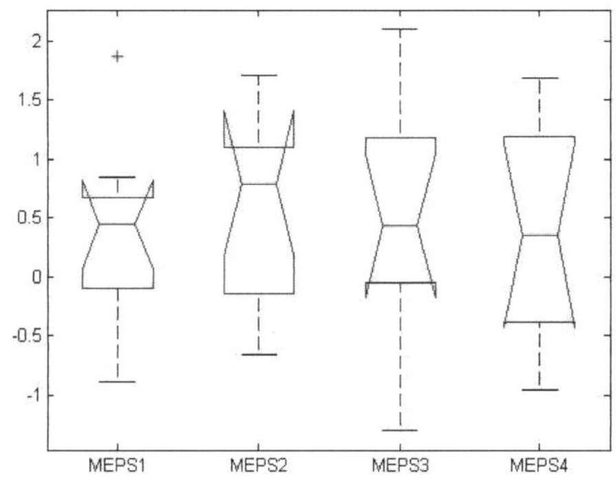

Figure 4 ANOVA results of LF/HF, insignificant ($p>0.05$)

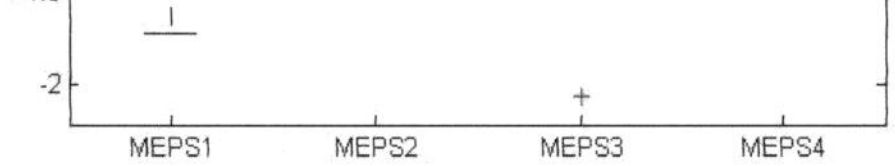

Figure 5 ANOVA results of alpha wave rate, significant

For MEPS tasks, a main effect of four tasks on alpha wave rate was found, with $F(3,36)=11.83$, $p<<0.001$. Post-hoc tests for each pair of tasks showed that except MEPS1-MEPS3 ($p=0.55$), there were significant difference between the other 5 pairs all with $p<0.05$. Main effect of task difficulty on all three HRV related features, MHBI, SDHBI, and LF/HF were not significant. Post-hoc tests also found no significant difference regarding MHBI and SDHBI. Fig. 6 shows the average value of alpha wave rate and NASA-TLX [9]. NASA-TLX is subjective MWL rating scale used as reference, which reflects the subject's feeling towards the difficulty of each MEPS task. In Fig.6 all subjects had lower alpha wave rate in MEPS-1, which was reversely considered to be less demanding. One of the reason may be that there was no practice session in MEPS tasks, the subjects unconsciously feel nervous and stressful in first MEPS operation, of which they failed to report this effect in NASA-TLX.

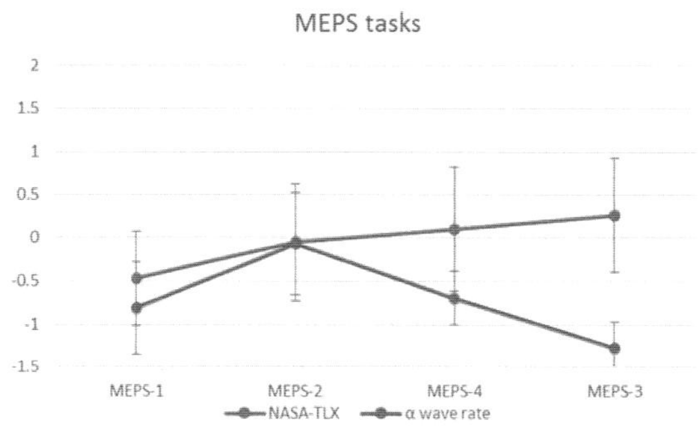

Figure 6 Average value of NASA-TLX and α wave. (Error bar is standard deviation)

The conclusions of this paper include:

a. Physiological computing models have potential applications in shipping industry, but these applications require reliable, sensitive, and online physiological data processing
b. Alpha band wave suppression are sensitive to task demand while heart beat related features are not. LF/HF ration is sensitive only to large difficulty level disparity.
c. Ceiling effects of using physiological metrics to infer human cognitive state are found and this effect raises new requirements in OFS-AA system design.

Limitations of the experiment study include that the number of valid participates in this study is relatively small, the statistical results may be weakened. Besides, although we realize real-time monitoring subject's EEG and Heart beat data, these features relies on post-experiment analysis. We expect to solve the above limitations in the future research.

6. REFERENCES

1. DIRICAN, A. C., & GÖKTÜRK, M. (2011). Psychophysiological measures of human cognitive states applied in human computer interaction. Procedia Computer Science, 3, 1361-1367.

2. LEAN, Y., & SHAN, F. (2012). Brief review on physiological and biochemical evaluations of human mental workload. Human Factors and Ergonomics in Manufacturing & Service Industries, 22(3), 177-187.

3. PRINZEL, L. J., FREEMAN, F. G., SCERBO, M. W., MIKULKA, P. J., & POPE, A. T. (2003). Effects of a psychophysiological system for adaptive automation on performance, workload, and the event-related potential P300 component. Human Factors: The Journal of the Human Factors and Ergonomics Society, 45(4), 601-614.

4. SIEGENTHALER, E., COSTELA, F. M., MCCAMY, M. B., DI STASI, L. L., OTERO - MILLAN, J., SONDEREGGER, A., ... & MARTINEZ - CONDE, S. (2014). Task difficulty in mental arithmetic affects microsaccadic rates and magnitudes. European Journal of Neuroscience, 39(2), 287-294.

5. POPE, A. T., BOGART, E. H., & BARTOLOME, D. S. (1995). Biocybernetic system evaluates indices of operator engagement in automated task. Biological psychology, 40(1), 187-195.

6. PAN, Y., KOMANDUR, S., & FINKEN, S. (2015). Complex Systems, Cooperative Work, and Usability. Journal of Usability Studies, 10(3), 100-112.

7. ROBERT, G., HOCKEY, J., HEALEY, A., CRAWSHAW, M., WASTELL, D. G., & SAUER, J. (2003). Cognitive demands of collision avoidance in simulated ship control. Human Factors: The Journal of the Human Factors and Ergonomics Society, 45(2), 252-265.

8. GOULD, K. S., RØED, B. K., SAUS, E. R., KOEFOED, V. F., BRIDGER, R. S., & MOEN, B. E. (2009). Effects of navigation method on workload and performance in simulated high-speed ship navigation. Applied ergonomics, 40(1), 103-114.

9. HART, SANDRA G., and LOWELL E. Staveland (1988). Development of NASA-TLX (Task Load Index): Results of empirical and theoretical research. Advances in psychology 52: 139-183.

7. AUTHORS BIOGRAPHY

Yanbin Wu is a PhD candidate majoring in Maritime Management in Kobe University. He is focusing on the research of reliable and effective engine-room management method.

Takashi Miwa, lecturer of Kobe University, Speciality is Marine Engineering, Ph.D. in Engineering (Kobe Univ., 2012).

Makoto Uchida is a professor working for Graduate School of Maritime Sciences, Kobe University. He obtained a PhD in Osaka University, 1998. His main research area is ship propulsion engineering and marine engineering management.

ICSOT Korea: Safety of Offshore and Subsea Structures in Extreme and Accidental Conditions, 15-16 September 2015, Busan, Korea

INTEGRATING HUMAN CENTRED DESIGN (HCD) WITH THE DEVELOPMENT AND QUALITY ASSURANCE PROCESS OF INTEGRATED SOFTWARE DEPENDENT SYSTEMS (ISDS)

S Keränen, DNV GL, South Korea

SUMMARY

DNVGL has initiated a process to integrate Human Centred design (HCD) with the DNVGL class notation OS D-203. This paper reports the initial results of this process where the ISDS process areas together with the six related subjects has been used as a basis for integration of HCD principles defined in ISO 9241-201. It can be concluded that probably no entirely new ISDS requirements will be required by the integration, but the application of HCD will be somewhat limited by the scope that addresses semi-finalized systems.

1. INTRODUCTION

Human Factor is not integrated to its full potential in engineering design [4, 13], this also seems to be the case for the design of control systems in the Oil and Gas industry. ISO 11064 and activities defined herein are applied, at least to some extent, but there is often a lack of integration of these principles with overall engineering. At the same time the control of offshore units is becoming increasingly dependent on complex software-based controls and automation. The Human Machine Interface is an important information source used by the operators to build their situation awareness (SA) [4]. Lacking or inadequate SA is one of the primary factors in accidents attributed to human error [5].

The DNVGL class notation OS D-203 sets requirements to the development and quality assurance process of Integrated Software Dependent Systems (ISDS) [15]. ISDS are integrated systems where the overall behaviour depends on the behaviour of the systems' software components. The notation was first introduced in 2008 with the objective to:

- provide an internationally acceptable standard for integrated software dependent systems by defining requirements for the work processes during design, construction, commissioning and operation,
- serve as a contractual reference document between suppliers and purchasers,
- serve as a guideline for designers, suppliers, purchasers and regulators,
- Specify processes and requirements for units or installations subject to DNV certification and classification services.

Usability is regarded as an essential part of software quality [9, 12] but the practical implementation often turns out to be a challenge. One of the key success factors for integrating usability engineering (UE) and software engineering (SA), is to integrate the methods and activities of UE, into the existing structures of an organization and its established process model for product development and implementation [12]. Standards for HCD also require that the related activities are included in the overall project plan and that HCD principles are integrated in activities, as applicable [7].

The use of an established standard and its required processes as a basis for the integration means that less new processes needs to be introduced. In addition, it would be possible to meet the ISDS and additional HCD requirements by applying one defined DNVGL process only.

DNVGL has initiated a process to integrate Human Centred design (HCD) principles with the ISDS standard. The efforts are also intended to support the IMO E-navigation initiative where guidelines for software quality assurance and human centred design are developed.

2. THE HUMAN-CENTRED DESIGN PROCESS

Usability is not a quality that originates accidentally; it cannot be based on common sense only and it's not a matter of opinion. As much as ensuring the proper structural integrity of a ship requires certain expertise and analyses, quality of use cannot be achieved without the right expertise and the application of good practices. User centred design is a methodology to design, review and build the interaction between a human and a machine. It describes the means to arrive at an effective, efficient and satisfying interaction between these two [6, 7]. ISO 9241-201 defines requirements and recommendations for human-centred design principles and activities throughout the life cycle of computer-based interactive systems. The standard addresses the following six principles as the basis of HCD.

1) The design is based upon an explicit understanding of users, tasks and environments.
2) Users are involved throughout design and development.
3) The design is driven and refined by user-centred evaluation
4) The process is iterative
5) The design addresses the whole user experience
6) The design team includes multidisciplinary skills and perspectives

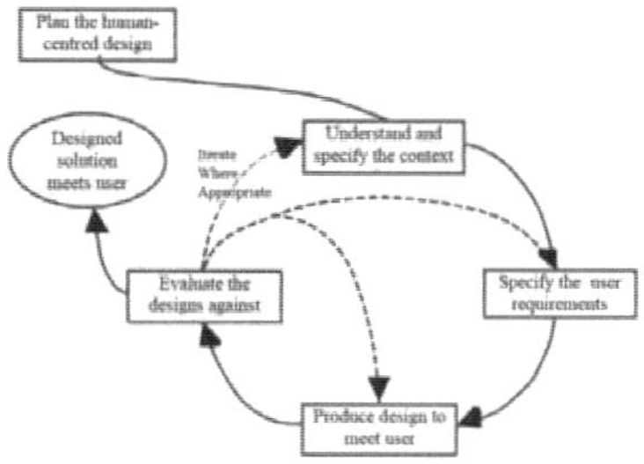

Figure 1: HCD activities

The activities included in the HCD process presented in figure 1 are fulfilled iteratively until the particular usability objectives have been reached.

2.1 SPECIFYING THE CONTEXT OF USE

The Human Centred Design process starts with activities to understand and specify the context of use. The outcome of these activities defines the users of the system, their particular needs and tasks as well as the context they operate in. Context of use studies analyses the factors related to the physical, technical and organisational environment which all can constrain or affect the use of a system to be designed. Examples of context of use factors are; the users system experience, user task detailed in frequency and steps, the technical environment consisting of the hardware, software as well as other equipment, physical environmental factors related to noise and vibration, user posture and health hazards etc., and the organisational factors such as the work organization, work characteristics, and work practices etc. [10].

The context of use specification is detailed further to user requirements. The context of use information should be detailed to such level, that it enables it to be used as measurable requirements in engineering. These requirements will cover all relevant aspect of the context of use factors, including safety and health requirements, user tasks, and work design.

2.2 DESIGNING THE SYSTEM

The process of producing design solutions and prototypes in order to meet the requirements defined starts as soon as these are available. The HCD process is iterative and if in any phase the need for more details related to an activity is revealed; the context of use analysis may be detailed further. The idea of making and evaluating prototypes first, instead of starting the real implementation supports the principle of communicating the concepts to the users and getting their feedback in the early phase.

Concepts and design of the system are evaluated against requirements and iterated until the requirements are met. Prototyping provides an relatively inexpensive method for finding errors and flaws compared to if these same defects are found after implementation. Changes done late in the process might be very expensive and time consuming to fix. The users are an integral part of this process and evaluation is included in all stages in the system life cycle. User-based testing is however not always practical or cost effective and other ways of evaluating might fit the purpose better. Input from evaluations and user feedback on previous versions of the system provides important input when starting a new version of a system.

To ensure that HCD is followed through and implemented effectively, the plan for HCD has to be a part of the overall system development plan. HCD should also be subject to the same project disciplines (e.g. responsibilities, change controls) as other key activities. Human-centred design activities are required to start at the earliest stage of the project and include time for iteration, the incorporation of user feedback, and for evaluating whether design solutions satisfies the user requirements. The relative importance of human factors/ergonomics in the project should be evaluated as a part of the planning activities. This evaluation should be based on aspect of the levels of the various types of risk that might result from poor usability, as well as how usability relates to the purpose of use of the system.

3. THE ISDS ACTIVITIES AND PROCESSES

The DNVGL class notation OS D-203 sets requirements for the development and quality assurance process of Integrated Software Dependent Systems. The ISDS requirements are formulated as activities assigned for a specific process area and role, distributed over a lifecycle of five phases (figure 2): basic engineering (A), engineering (B), construction (C), acceptance (D), operations (E). The responsibilities are assigned to four different roles: owner, system integrator, (Shipyard), supplier and the independent verifier (Class). The process areas include activities related to requirements engineering, design, implementation, integration and verification and validation. The requirements are categorized in six subjects related to risk management, project management, process and quality assurance, acquisition, configuration and management and Reliability, availability, maintainability and safety (RAMS).

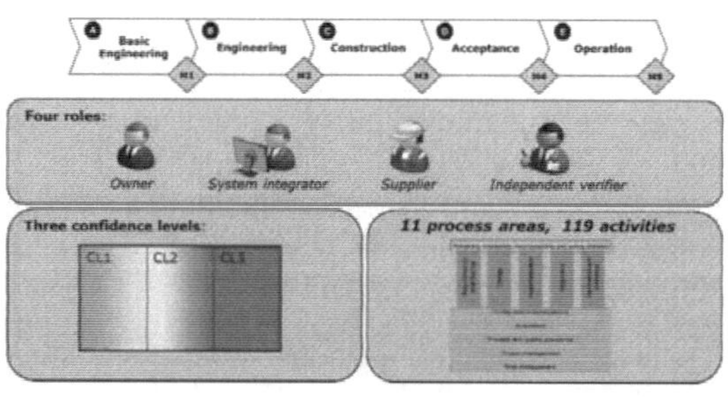

Figure 2: ISDS activities and processes

In the process of integrating usability and software engineering, similarities in objectives and proceedings may be identified on a high level, when examining the descriptions of each activity, by relating tasks and outputs with each other [12]. The ISDS process areas together with the six related subjects provided an applicable level of detail for initiation of the integration of HCD and ISDS.

4. ISDS PROCESS AREAS AND SUBJECTS AND HOW THEY RELATE TO HCD

The ISDS project management process area covers activities to ensure that the project plans for the different organizations involved are created, synchronized and followed up. A project plan is established and maintained, in coordination with other stakeholders, for engineering and all succeeding phases. The plan includes at least schedules and resource allocation of software related activities and takes into account the activities for the different phases of the project. ISO 9241-201 requires that the planned HCD activities are included in the overall system development plan and that they are organised and managed the same way as other key activities. The choice and planning of HCD activities should be based on the relative importance of human factors/ergonomics in the project. The relative importance should be based on aspect of the levels of the various types of risk that might result from poor usability, as well as how usability relates to the purpose of use of the system. The ISDS reliability, availability and safety (RAMS) will have to be identified together with the risk mitigation actions. The risk aspects of HCD would need to be integrated with the RAMS requirements.

4.1 REQUIREMENTS ENGINEERING

The requirements engineering process area, covers activities needed to define, document and manage the requirements on the unit, the systems and the related software. This includes defining the unit design intention and philosophy: The vision of the unit, descriptions of the unit's overall behaviour and the expected business/safety/environmental performance are formalized and distributed to all stakeholders. The HCD requirements related to the operational context needs to be integrated in this. In order to fulfil the HCD requirements the unit design intention and philosophy would need to be complemented with a high level operational philosophy, describing the users of the systems, their roles and responsibilities. Requirements related to HCD would need to be identified as a part of the identification of applicable regulatory requirements for the envisioned operational context.

The requirement specification activities defined in ISO 9241-201 are divided into the specification of user requirements on the one hand and gathering requirements arising from relevant ergonomics and user interface knowledge, standards and guidelines, on the other. ISDS also includes generic system requirements collected in activities supported by prototyping, gathered in FEED studies etc. This area needs to be extended to also cover all relevant analyses related to HCD conducted prior to the start of the ISDS process. Examples of such activities could be FEED studies conducted in order to comply with the NCS regulatory regulations, such as organization and manning studies, function and task analysis etc. [3, 6, 11, 14].

ISDS requires a documented traceability from the requirements to design to validation. This should also apply for the HCD related activities as they should be subject to the same change control as other key activities. ISO 9241-201 also sets requirements related to the quality of the user requirements. One important aspect related to traceability is that the user requirements need to be defined in such a way that they are quantifiable and testable. It is also required that the rationales, factors and the weighting of human-system issues for use in any trade-off are documented for future understanding.

4.2 OPERATIONAL SCENARIOS

The owner is responsible of formulating operational modes and scenarios in the scope of ISDS. These are supposed to capture expected behaviour of the unit and systems for key scenarios and put the system requirements into an operational context. The contexts of use descriptions of HCD, are supposed to provide descriptions of the systems in a level of detail that support the requirements, design and evaluation activities. The intended context of use should be a part of the user requirements specification to clearly identify the conditions under which the requirements apply. The information should provide an explicit statement of user requirements in relation to the intended context of use and the business objectives of the system. The context of use information is overlapping with the information required in the operational scenarios. The selection of key scenarios of ISDS can be done by evaluating the scenarios towards criteria like the novelty of the scenario, novelty of the systems involved, associated risks etc. An evaluation of the levels of the various types of risk that might results from poor usability should be included in the selection criteria.

4.3 DESIGNING THE SYSTEM

The design activities in HCD include designing the system, user-system interaction and the user interface in order to meet the user requirements. In ISDS, the major subsystems and components and how they interact is documented. The design describes the software components of the system together with an indication of which parts are developed for the specific project and which parts are reused/ configured.

In order to integrate the HCD in the ISDS activities requirements related to the inclusion and follow up of user requirements in the configuration activities as well as developed of project specific software would be required. ISO 9241-210 refers to seven principles of the quality of use that should be followed ins system design; suitability for the task, self-descriptiveness, conformity with use expectations, suitability for learning, controllability, error tolerance and suitability for individualization. The integration of the HCD requirements in this phase will however face some limitations as the systems to be integrated usually are supplied as semi-finalized packages. Further analysis and application will be required in order to define the extent to which the HCD requirements may be integrated.

4.4 VERIFICTAION AND VALIDATION

The verification and validation activities in ISDS addresses the quality assurance activities needed to ensure that each software component, each system and the integrated systems perform as expected and fulfils the needs of the intended use and defined targets. A verification strategy should be defined, and the basic verification activities like Factory Acceptance Tests (FAT), peer reviews, and qualification of reused software are prepared and performed according to this strategy. The ISDS requirements focus on functionality and performance of the integrated systems working together, and on early verification of the correctness and the completeness of requirements, interface specifications, user interfaces, and design documentation.

In HCD and its user centred evaluation, the concepts and design is evaluated against requirements and iterated until the requirements are met. The users are an integral part of this process and the ISDS requirements would need to be extended to include the involvement of users. User-based testing is however not practical or cost effective at all stages of a project and other ways of evaluations might be needed. The context of use descriptions as well as the operational scenarios is supposed to support the evaluation of the systems. This information can be used as further input to more detailed use-cases and tests-cases and further used in evaluations of the interaction between different systems. Scenario based verification and validation is used as a part of current HCD activities [8, 14] and is also a regulatory requirement on NCS. Using the detailed operational scenarios incorporating relevant HCD requirements, would thus support the compliance with multiple requirements.

User-based testing would fit well together with the factory acceptance test (FAT) where the system is evaluated for compliance with selected operational requirements, including normal modes, error situations, and degraded modes. It is, however, important not to postpone all the user centred testing to this final stage in the project. Quality cannot be tested in; it has to be designed in. Means of involving users in and validating user requirements in earlier phases would need to be defined in further detail.

5. CONCLUSION

From the high level mapping of ISO 9421-201 requirements to the ISDS standard, it is possible to conclude that no major changes in the ISDS standards would be required in order to incorporate the HCD principles. The integration would probably not require any entirely new ISDS requirements to be introduced. The required changes would mostly affect the suppliers of systems that don't address HCD at present.

The unit design intention and philosophy defined by the owner would need to be complemented with some high level context of use information. All applicable HCD standards and requirements would also need to be identified in the start-phase. It would be essential to use these requirements in the vendor specification in order to ensure the consistency of HMI between the systems to be integrated.

The requirements for consistency in the HMI will however face limitations and one core task for the future is to find out to which level the different systems can be adjusted to meet common HCD requirement. This however would only be possible to do in a pilot.

A greater emphasis on operational scenarios and use cases would be required in the ISDS process as these would be used both in order to set the user requirements into an operational context as well as in some of the required user centred evaluation and validation activities. The aspect of iterations would be somewhat limited as the systems to a major extent already exist and the amount of new design would be limited.

The ISDS process starts when a contract is established between the owner and the system integrator/shipyard for construction of a vessel. This means that the FEED phase is not covered by the current ISDS requirements. HCD principles should be applied early in design instead of being an afterthought. Integration with ISDS would not help to ensure that this is the case. HCD also sets requirements addressing the quality of requirements. Currently the ISDS scope doesn't set any requirements of this type. ISDS assumes that all requirements are ready for use; the ISDS process specifies the need to gather requirements, not to formulate them.

6. REFERENCES

1. BEVAN, N. International Standards for HCI and Usability. In International Journal of Human Computer Studies, 2001. *http://www.usabilitynet.org/papers/HCI-Usability_standards.pdf,* 2001.

2. DUL, J., BRUDER, R., BUCKLE, P., CARAYON, P., FALZON, P., MARRAS, W. S., ... & VAD DER DOELEN, B. A strategy for human factors/ergonomics: developing the

discipline and profession. *Ergonomics, 55(4), 377-395,* 2012.

3. Facilities regulations § 20-21, *ptil.no*, 2015

4. GRECH, M., & HORBERRY, T. Human error in maritime operations: situation awareness and accident reports. *In 5th International Workshop on Human Error, Safety and Systems Development, Newcastle, Australia,* 2002, June.

5. GRUHN, P. Human machine interface (HMI) design: the good, the bad, and the ugly (and what makes them so). *In 66th Annual Instrumentation Symposium for the Process Industries (pp. 27-29)*, 2011, January.

6. ISO 11064- Ergonomic design of control centres

7. ISO 9241-210 Ergonomics of human-system interaction -Part 210: Human-centred design for interactive systems

8. JOHNSEN, S. O., BJØRKLI, C., STEIRO, T., FARTUM, H., HAUKENES, H., RAMBERG, J., ... & ASA, N. H. CRIOP®: A scenario method for Crisis Intervention and Operability analysis. *Report No. STF38 A03424*, 2004.

9. JURISTO, N., WINDL, H., & CONSTANTINE, L. Guest Editors' Introduction: Introducing Usability. IEEE Software, (1), 20-21, 2001.

10. MAGUIRE, M., Context of use within usability activities. *International Journal of Human Computer Studies, 55, pp. 453-483*, 2001.

11. Management regulations § 18, *ptil.no*, 2015

12. NEBE, K., ZIMMERMAN, D., & PAELKE, V. Integrating Software Engineering and Usability Engineering. *INTECH Open Access Publisher*, 2008.

13. NORROS, L., 'Developing human factors/ergonomics as a design discipline', *Applied Ergonomics 45 (1), 61-71,* 2014.

14. NORSOK S-002 Working Environment, *standardnorge.no*, 2015

15. Offshore Standard DNV-OS-D203, *dnvgl.com*, 2015

7. AUTHORS BIOGRAPHY

Saara Keränen holds the current position of Senior Engineer at DNV GL. She is the DNV GL practice lead for Human Factors and Working Environment in Asia working as a consultant applying HFE in projects within the oil and gas and maritime industry.

ICSOT Korea: Safety of Offshore and Subsea Structures in Extreme and Accidental Conditions, 15-16 September 2015, Busan, Korea

IMPLEMENTATION OF RISK AND WORKING ENVIRONMENT ANALYSES FOR OFFSHORE INSTALLATIONS ON THE NORWEGIAN CONTINENTAL SHELF

K. W. Yang, DNV GL Oil & Gas, Korea

SUMMARY

According to the Petroleum Safety Authority Norway (PSA) regulations, risk and working environment analyses shall be carried out to manage major accidents, environmental and other risk, to ensure a sound working environment and to provide support for decision making related to design, construction and operation phases of offshore facilities operated on the Norwegian Continental Shelf (NCS). Due to strict requirements for the risk and working environment analyses in NORSOK standards referred to in the PSA regulations and lack of experience of users, there have been challenges on implementing the analyses and utilizing the results of the analyses. This paper will provide an introduction to implementation and application of the risk and working environment analyses for offshore facilities on the NCS based on the PSA regulations and NORSOK standards.

NOMENCLATURE

u	Averaged wind speed (m s^{-1})
T	Temperature (K)
WCI	Wind chill index (kg m^2 s^{-3})

1. INTRODUCTION

1.1 BACKGROUND

According to the Petroleum Safety Authority (PSA) regulations, risk and working environment analyses shall be carried out to manage major accidents and to ensure a sound working environment and to provide support for decision making in design, construction and operation phases of offshore facilities operated on the Norwegian Continental Shelf (NCS).

Due to strict requirements for the risk and working environment analyses stipulated in the NORSOK standards and various internationally recognized standards which are referred to in the PSA regulations and lack of experience in the analyses and applicable requirements, there have been challenges on implementing analyses and interpreting and utilizing results of the analyses.

This paper will provide an introduction to implementation and application of the risk and working environment analyses for offshore facilities on the NCS based on the PSA regulations and NORSOK standards.

1.2 TERMS & DEFINITIONS

This subsection provides main definitions which are addressed in this paper:

- Design accidental load: Chosen accidental load to be used as the basis for design;
- Dimensioning accidental load (DAL): Most severe accidental load that the function or system shall be able to withstand during a required period of time;
- Hazard: Potential source of harm;
- Main safety function: Safety functions which should be intact in order to ensure the safety for personnel and/or to limit pollution;
- Major accidents: An accidental event such as a major emission, fire, or explosion, which leads to serious consequences to fatalities, environment and/or assets;
- Risk: Combination of the frequency of event and the consequence of that event;
- Risk acceptance criteria: Criteria which are used to express a risk level that is considered as the upper limit for the activity in question to be tolerable;
- Risk analyses: Structured use of available information to identify hazards and to describe risk; and
- Working environment: totality of all physical, chemical, biological and psychological factors at work place that may affect workers' health and well-being through acute trauma and lasting exposure.

2. REGULATIONS & STANDARDS

2.1 PSA REGULATIONS

PSA is an independent government regulator, responsible for safety, emergency preparedness and the working environment for petroleum activities on the NCS and delegated by the Ministry of Labour to issue detail regulations for safety and the working environment in the petroleum industry.

The PSA issues key regulations for Health, Safety and the Environment (HSE) and the working environment respectively. Five sets of the HSE regulations, i.e. The Framework Regulations [1], The Management Regulations [2], The Activity Regulations [3], The Facility Regulations [4], and The Technical and Operational Regulations [5] have been applied for petroleum activities on the NCS and at selected petroleum plants on land. The PSA has also issued five guidelines [6], [7], [8], [9] and [10] to the regulations, demonstrating how provisions in the regulations can be met and which standards could be used.

The working environment regulations pursuant to the Norwegian Working Environment Act [11] are issued by the Ministry of Labour and enforced by the Norwegian Labour Inspection Authority and PSA within their respective areas of authority. Provisions on the working environment regulations shall be complied with as an integrated part of the requirements in the HSE regulations.

2.2 NORSOK STANDARDS

The first editions of the NORSOK standards were established in 1993 based on a joint effort from the Norwegian petroleum industry to ensure adequate safety and cost effectiveness for petroleum industry developments and operations by replacing company specifications where possible, and serving as references in the authorities' regulations. The NORSOK standards are based on various recognised international standards, adding the provisions deemed necessary to cover the needs of the Norwegian petroleum industry.

3. RISK ANALYSES

3.1 GENERAL

In the guidelines to Section 17 of the Management Regulations [7], it is recommended that the NORSOK Z-013 standard [12] should be used amongst other standards to comply the requirements for the risk analyses for offshore facilities operated on the NCS. This standard covers the risk assessment process, including methodologies to establish the total risk picture and for assessing risk reducing measures as shown in Figure 1.

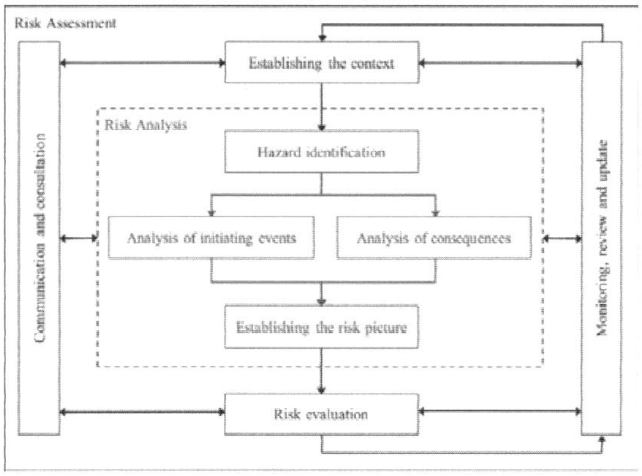

Figure 1: Risk Assessment Process [12]

The NORSOK Z-013 standard is focused on detailed risk assessment of major accidents with serious consequences to personnel, environment and/or assets. While occupational accidents are included in the estimation of personnel risk, detailed analysis of occupational fatalities and injuries is not covered in the quantitative risk analyses (QRA) or in NORSOK Z-013 [12].

In following subsections, general approaches and methodologies for the risk assessment are addressed based on the NORSOK Z-13 standard.

3.2 RISK ASSESSMENT PROCESS

3.2 (a) Establish the context

The objectives of the establishment of the context are to define the purpose of the risk assessment, scope, system description, battery limits including limit conditions and criteria for the risk assessment process [12]. Normally, a set of assumptions including the parameters, scope, methodologies and criteria is prepared and agreed by responsible parties, i.e. owner, operator, engineering companies, etc. for the risk assessment process. It needs to be continuously updated with the best available information, as a working document during the risk assessment process.

The risk acceptance criteria which are considered as the maximum limits of risk levels and used for As Low As Reasonably Practicable (ALARP) principle shall be established prior to the risk assessment process, based on requirements of the applicable regulations and standards. The quantitative risk acceptance criteria should include risks to people, assets, the environment and impairment frequencies of main safety functions [12].

The risk to people normally covers risk of fatality. In order to generate a total picture of the risk, multiple risk criteria may need to be applied, depending on risk exposed groups. Various risk expression, e.g. fatal accidental rate (FAR), individual risk per annum (IRPA), potential loss of life (PLL), etc. can be used for the personnel risk [12], [13].

The following main safety functions assisting in ensuring the safety of personnel shall be maintained in the event of major accidents [4]:

- Preventing escalation of accident situations so that personnel outside the immediate accident area are not injured;
- Maintaining the capacity of main load-bearing structures until the facility has been evacuated;
- Protecting rooms of significance to combatting accidents so that they remain operative until the facility has been evacuated;
- Protecting the facility's safe areas so that they remain intact until the facility has been evacuated; and
- Maintaining at least one escape route from every area where personnel are present until evacuation to the facility's safe areas and rescue of personnel have been completed.

With regard to impairment frequencies of the main safety functions, systems and equipment included in the main safety functions shall be designed such that dimensioning accidental loads (DALs) or dimensioning environmental loads with an annual frequency greater than or equal to

1×10^{-4} per year shall not result in loss of a main safety function [4]. The following accidental and environment load categories are to be assessed against project risk acceptance criteria for loss of the main safety functions [12]:

- Heat/radiation loads due to hydrocarbon leaks;
- Smoke and toxic loads due to hydrocarbon leaks;
- Explosion loads due to explosion;
- Impact loads due to ship collision, dropped objects, helicopter collision, etc.; and
- Extreme environmental loads.

In other words, the frequency of impairment of any of main safety functions shall not exceed 1×10^{-4} per year for each accidental load listed above. It should be noted that this may not be applicable for facilities to be operated on other continental shelves as this is an example based on the risk level aimed at operation on the NCS.

Environment risk acceptance criteria should include frequencies and consequence of discharges of pollutants to the environment. The environmental consequence can be defined as recovery time, effect on populations and habitats or volume of pollutants discharged [12].

3.2 (b) Hazard Identification (HAZID)

The objectives of the HAZID are identification of possible hazards, sources of the hazards and their potential consequences, screening of critical hazards for subsequent risk analyses, identification of the safety measures in place (or planned to) including assessment of the adequacy of the existing measures, and identification of possible risk reducing measures.

HAZIDs are normally conducted by workshop activities based on pre-defined and populated checklists for facilities or checklists in internationally recognized standards, e.g. ISO 17776 [14], accident statistics and experience of teams from previous similar analyses or assessments. The HAZIDs should involve key disciplines in order to provide correct inputs and assure that all relevant hazards are identified.

The identification of hazards through the HAZID is critical as hazards not identified at this stage will be excluded from the risk assessment and as a result potentially leads to an inadequate risk assessment and results. It is also important to document the criteria used in the screening of the hazards and the reasons why the hazards are classified as non-critical to demonstrate that the hazards could be safely disregarded [13].

The results from the HAZID should be followed up by responsible parties, and utilized as input to the design and engineering processes and the rest of the risk assessment process.

3.2 (c) Analysis of Initiating Events

The objectives of the analysis of initiating events are to analyse and identify potential causes of the initiating events determined by the HAZID, and to assess the frequency of initiating events occurring [12].

The analyses and identification of causes of the initiating events is normally conducted by analysis techniques are as follows [13]:

- Hazard and Operability Analysis (HAZOP);
- Fault Tree Analysis (FTA);
- Preliminary Hazard Analysis (PHA); and
- Failure Mode and Effect Analysis (FMEA).

Calculation of frequencies of the initiating events based on relevant historical statistics or databases is the most commonly used approach. The data used for analyses shall be suitable in relation to location of field, type of installations or units, the context of the method, models and tools used for the risk assessment. Recommended data sources for the analyses are given in the Annex D of the NORSOK Z-013 standard [12].

The analyses of the initiating events shall cover, but not limited to, the following events if the hazard is relevant [12]:

- Process accidents;
- Risers/landfall and pipeline accidents;
- Storage accidents;
- Loading/offloading accidents;
- Blowouts and well releases;
- Accidents in utility systems;
- Accidents caused by external impact and environment loads, e.g. collision, falling load, etc.;
- Structural failure; and
- Loss of stability and/or buoyancy.

Engineering inputs to be provided for the analyses of the initiating events depends on the type of the event analysed. For example, in order to analyse the event related to process accidents, process flow diagram, piping and instrumentation diagrams and operation philosophies should as a minimum be provided with relevant process data in order to identify leakage sources and calculate leak frequencies. In addition, site specific data such as ship traffic, helicopter traffic and environmental data is required for the analyses of the initiating events caused by external impact.

3.2 (d) Analysis of Potential Consequences

The objectives of the analysis of potential consequence are to assess possible consequences of the identified initiating events, to analyse of potential sequences, to determine the effect on the existing barriers, the magnitude of the physical effects and the extent of damage to personnel, environment and assets [12].

The sequences of the events shall be modelled using suitable tools or methods such as event tree analysis (ETA) or similar. If the ETA is used, node probabilities shall be based on barrier systems, physical effects and physical outcomes [12].

The magnitude of the physical effects that determine sequence and consequences are to be modelled using recognised models e.g. empirically based, analytical or simulation data [12]. The type of consequence model depends on the actual hazard considered and the project phase and the data/information available.

3.2 (e) Establishing the Risk Picture

The objective of the establishing the risk picture is to establish a synthesis of the risk assessment in order to provide information to decision making. This step includes reporting of the risk assessment process performed. The risk picture shall include description of methodologies, models and tools used for the risk assessments, evaluations of the risk level and the main risk contributors, and comparison of the risk level with the risk acceptance criteria [12]

3.2 (f) Risk Evaluation

The objective of the risk evaluation is to establish a basis for decision making with the results of the analyses.

Identification of possible risk reducing measure is to be carried out with the following priority [12]:

- Risk reducing measures providing inherently safer design;
- Risk reducing measures reducing the frequencies of accidental events and;
- Risk reducing measures reducing the consequences.

Evaluation of risk reducing measures should be based on quantitative or qualitative analyses. The evaluation processes are to be documented with description of the risk reduction process.

4. WORKING ENVIRONMENT ACTIVITIES

4.1 GENERAL

Working environment is defined as totality of all physical, chemical, biological and psychological factors at work place that may affect workers' health and well-being through acute trauma and lasting exposure [15].

Special attention should be given to the working environment issues as they are considered as important input for engineering, design, construction, operation, commissioning and decommissioning activities of offshore facilities and units on the NCS.

NORSOK S-002 Standard [15] which specifies design requirements, as well as requirement regarding systematic management of working environment issues should be used in order to meet requirements related to working environment [6], [7].

In the following subsections, general approaches to working environment design basis, analyses, verification and validation are discussed based on the NORSOK S-002 Standard [15].

4.2 WORKING ENVIRONMENT DESIGN BASIS

4.2 (a) Objectives

During early phase of engineering, specific working environment requirements specified in authorities' regulations, company specifications and applicable standards are identified and documented in working environment design basis, i.e. working environment area limits, organisation and manning study, means of access, etc., as inputs for engineering and design.

4.2 (b) Working Environment Area Limits

Detailed specifications of working environment area limits for each work area and room of an installation are established as input to design and engineering based on manning level, purpose of each work area and requirements stipulated by authorities and company. Applicable area limits for illumination, temperature, vibration, total noise levels for typical areas are specified in the NORSOK S-002 standard [15].

Implementation of the area limits are to be verified by appropriate measurement methods, documented and updated in the working environment area chart for each work area and room, with description of any deviations or findings identified from the measurements.

4.2 (c) Organisation and Manning Study

An organisation and manning study is performed to provide key inputs in establishing working environment area limits and working environment analyses. Personnel categories, operations and maintenance tasks, responsibilities and the distribution of working hours are described in the organisation and manning study [15].

4.2 (d) Means of Access

Requirements to access for operation and maintenance of equipment (e.g. control devices, valves, etc.) are defined and documented as inputs to design and engineering of work areas, with the access frequency and the needs for access in the event of an emergency [15].

4.3 WORKING ENVIRONMENT ANALYSES

4.3 (a) Objectives

Working environment analyses are to be performed to identify and assess occupational risks and potential problems associated with various working environment factors as inputs for design and engineering [15].

The analyses are to be conducted by personnel or teams who have adequate expertise for working environment (e.g. machinery safety, ergonomics, human factors, noise, vibration, etc.) and various disciplines of a project to ensure that relevant input data is properly used for the analyses. Operations' representatives for the project and worker representatives (e.g. safety delegates) from the company's organisation need to participate in the analyses.

4.3 (b) Job Hazard Analysis

A coarse job hazard analysis needs to be carried out for each workplace to identify and evaluate risks of injuries and fatalities due to task associated with operation, maintenance, repair, material handling and other types of tasks. A detailed job hazard analysis is to be performed for critical workplaces where tasks with a high risk of accidents are required. Criteria for the workplaces for the detailed job hazard analysis are repeated manual tasks, manual tasks in the risk zone of equipment and involving hazards with consequences [15].

The analysis normally includes various activities such as review of documents and drawings, workshop activities and documentation of findings and deviations. Typical inputs for the analyses are arrangement drawings for work areas/rooms and vendor packages, philosophies for operation, repair and maintenance. Especially, it is important to have documentation specifying tasks for each work area as much as detail.

The results from the analyses including the finding and deviations shall be utilized as input in design and engineering processes and followed up by responsible parties according to requirements for systematic management of working environment issues [15].

4.3 (c) Ergonomic Task Analysis

The ergonomic task analysis is performed to identify potential problems in areas with high risks of musculoskeletal injuries due to tasks and to ensure that the maximum workload requirements can be met. The analyses cover evaluations of layout, access and clearances for tasks required, location of work functions, material handling and repetitive movements. Information and documentation including manning, work sequences, frequency of operation/maintenance, type of tasks, equipment for performance of the tasks, and experience in similar tasks should be provided as input to the analysis [15].
The analysis normally includes various activities such as review of documents and drawings, workshop activities and documentation of findings and deviations.

The results from the analyses including the finding and deviations shall be utilized as input in design and engineering processes and followed up by responsible parties according to requirements for systematic management of working environment issues [15].

4.3 (d) Human Factor Analysis

The human factor analyses are performed to ensure that the designs are in line with philosophies and requirements, to minimise potential human error in work system which control safety critical activities. Human factor principles specified for the facilities shall be applied to control centre designs such as central control room and the driller's control room [15].

Furthermore, a scenario analysis shall be performed to validate the design, manning and procedures and ability to control disturbances and emergency situations of the control rooms [15]. One of the most commonly used method for the scenario analysis is a Crisis Intervention and Operability (CRIOP®) analysis used for the validation and evaluation of the control centre according to the SINTEF report [16]. A CRIOP analysis consists of two main parts, review of the static characteristics of the control centre and scenario analysis including task analysis. The review of the static characteristics covers layout, working environment, control and safety system, job organisation, procedure and work descriptions, training and competence based on checklists specified. In the scenario analysis, step diagrams of potential accident scenarios are established [16].

4.3 (e) Chemical Health Risk Assessment

The chemicals health risk assessment is performed to identify evaluate and control chemical health risks to acceptable levels specified in applicable requirements for all activities where hazardous chemicals are used during operations or maintenance, and for all processes containing hazardous chemicals that can be released to the working environment [15].

The analysis normally includes various activities such as review of documents including list of chemicals and material safety data sheets (MSDSs), workshop activities and documentation of findings and deviations. Typical inputs for the analysis are arrangement drawings for work areas and rooms and vendor packages, philosophies for operation, repair and maintenance and MSDSs. Especially, it is important to have documentation specifying list of chemicals used for or stored in each work area as much as detail.

The results from the analyses including the finding and deviations shall be utilized as input in design and engineering processes and followed up by responsible parties according to requirements for systematic management of working environment issues [15].

4.3 (f) Outdoor Operations and Cold Stress

Outdoor operations analyses is carried out for open work areas exposed to ambient condition without obstacles and semi-open work areas partially exposed to ambient conditions with weather protection, in order to identify and control potential problems due to exposure to low ambient temperature, wind, icing and precipitation [15].

The analysis covers investigating the need for weather protection to comply with wind chill index (WCI) and other applicable requirements. The WCI means effective heat loss per time felt by human body and is defined as a function of temperature of human skin, ambient temperature and wind speed. The WCI is calculated based on the formula stipulated in ISO Technical Report 11079 [17] and meteorological data. With assumptions such as average human body temperature, metabolic power and emissivity of clothing, empirical relation for the WCI is determined according to the following formula [17]:

$$WCI = 1.16 \times (10.45 + 10u^{0.5} - u)(306.15 - T)$$

The percentage of time that operation or maintenance persons is exposed to a WCI higher than 1000 kg $m^2 s^{-3}$ shall be reduced for work places where frequent work with a duration 10 minutes or more is required. Available working time with WCI is represented in Figure 2. The unavailability shall be less than 2 % on a yearly basis [15].

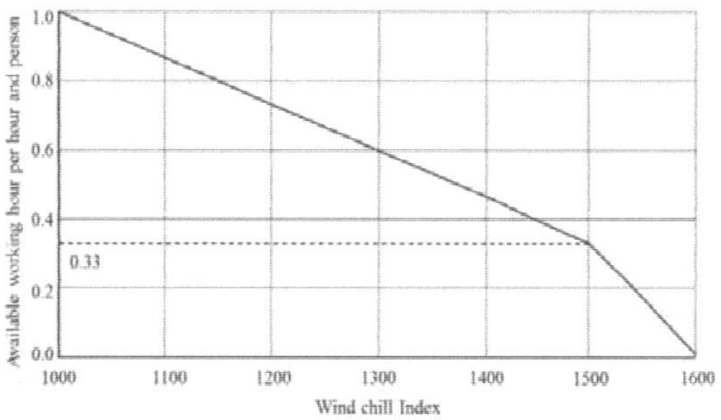

Figure 2: Availability of outdoor operations

The calculation of WCI for facilities is normally conducted by 3-dimensional models of the facilities and computational fluid dynamics tools.

If unavailability is higher than the criteria, the need of measures to avoid exposed workplaces or reduce the exposure to low temperature, wind, icing and precipitation are evaluated [15].

4.4 VERIFICATION AND VALIDATION

4.4 (a) Working Environment Design Review

The working environment design review is performed to verify that design is in compliance with working environment requirements applicable to installations or units in design and engineering phases and prior to construction. The design review covers verification of layout, arrangement, material handling, accessibility to equipment, machinery and vendor packages considered critical in aspect of working environment [15].

The working environment design review is normally conducted by workshop activities and document review. Typical inputs for the design review are arrangement drawings for work areas and rooms and vendor packages, philosophies for operation, repair and maintenance. 3-dimensional models of facilities are also used.

The results from the analyses including findings and deviations shall be utilized as input in the design and engineering processes and followed up by responsible parties according to requirements for systematic management of working environment issues [15].

4.4 (b) Working Environment Inspection

During the construction phase, working environment inspections are carried out to verify that the fabricated facilities are in compliance with working environment requirements, and that deviations identified from the working environment analyses are implemented. It is recommended that inspections during construction should be carried out while rectification work is still possible [15].

The results from the inspections including deviations shall be followed up by responsible parties according to requirements for systematic management of working environment issues [15].

5. CHALLENGES

5.1 General

In following subsections, typical challenges of the risk and working environment analyses for facilities to be operated on the NCS are addressed based on previous project experience.

5.2 Project Contract and Scope

The project contract and scope for each party who holds responsibility for which scope has a high significance on how and when work is performed and competence involved. The scope of the risk and working environment analyses mainly depends on characteristics of the facility, location of field, environmental conditions and applicable requirements.

Especially, managing interfaces and ensuring consistency is of special importance in large and complex projects with multiple parties, throughout engineering and construction

5.3 EXPERIENCE AND COMPETENCE

As the requirements specified in the NORSOK Z-013 and S-002 standards have only been applied to offshore facilities operated on the NCS and some few facilities to be operated on other continental shelves, users are not familiar with the requirements in the standards. This lack of experience and competence of the users may cause various problems in the execution of analyses and design solutions based on interpretation of the results of the analyses.

Experience transfer from personnel or 3rd party with proper competence and experience on the process of the execution and interpretation of the risk and working environment analyses is of importance to establish adequate inputs to design and engineering.

5.4 MANAGEMENT OF ISSUES IDENTIFIED FROM RISK AND WORKING ENVIRONMENT

Some of challenges are related to management of issues addressed from risk and working environment analyses in the project phases as the management and follow-up normally require a joint effort from various engineering disciplines as well as operations.

In order to avoid that lack of attention on the risk and working environment issue causes critical problem at later phases, a follow up system is to be set with clear indication of responsibilities. The follow up system is to ensure proper documentation and handling of findings and that the responsibilities for following up on deviations identified from the risk and working environment analyses as well as verification and validation activities are clearly defined.

5.5 ALARP PRINCIPLE

Risk which is lower than the maximum limit of the risk acceptance criteria can be easily misinterpreted by users, as tolerable risk. However, risk shall be further reduced even though calculated risk level is lower than the maximum limit of the risk acceptance criteria [1].

As represented in Figure 3, there is no lower level for acceptable risk in the ALARP principle according to the Norwegian and UK regulations. In case the calculated risk falls in 'ALARP or Tolerable' region, the risk shall be demonstrated to be ALARP regardless of the risk level [12].

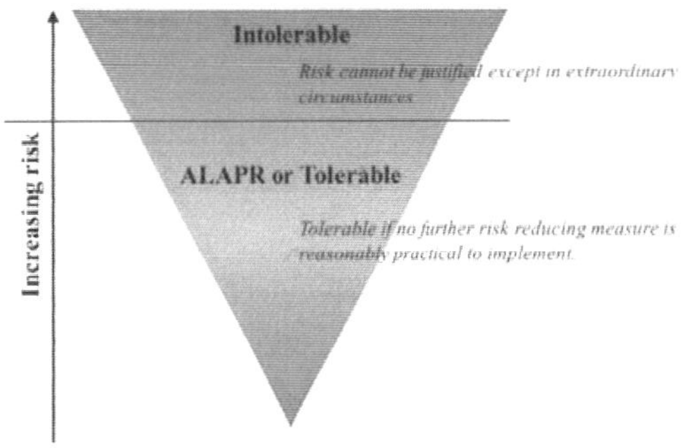

Figure 3: ALARP principle [12]

If no further risk reducing measures is reasonably practical to implement, risk can be considered as tolerable risk. This also implies that one should continuously strive to minimise the risk during the all project phases, from concept to decommissioning.

5.6 INPUTS FOR ANALYSES

In principle, the latest available engineering data should be used for the risk and working environment analyses. However, in many cases, engineering data used for the analyses, e.g. process design, layout drawings, shutdown logics, detector layout, etc. are continuously developed and updated during the design and engineering phases and it can make significant deviations between the actual data and the data used for the analyses which may lead to incorrect results in the risk and working environment analyses. Depending on the level of the deviation, the risk and working environment analyses may have to be revised or be redone with the up-to-date engineering data.

In order to avoid or minimize the negative impact of the deviation between the engineering inputs, special attention should be given by the responsible parties to the list of assumptions and inputs which forms the basis for the risk and working environment analyses.

5.7 TIMING AND SEQUENCE OF ANALYSES

In many cases, the project is often delayed in starting up the analyses work, creating a milestone impact. Timing and execution sequence of the risk and working environment analyses is important to properly provide input in project phases as wrong timing or sequence of the activities may cause inadequate or incorrect input to design and engineering.

The timing, duration and sequences of each analysis should be clearly defined in early phase of project, with consideration of design and engineering schedule and be monitored by responsible parties. For example, typical timing of the working environment activities is shown Figure 4. However, it should be noted that the timing of the activities may vary due to project characteristics.

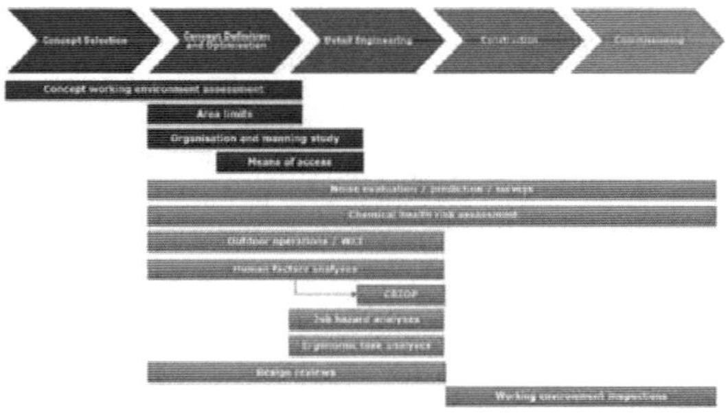

Figure 4: Typical timing and execution sequence of working environment activities [15]

6. CONCLUSIONS

In this paper, the approaches and methodologies for the risk and working environment analyses for offshore facilities on the NCS based on the PSA regulations and NORSOK standards are addressed with typical challenges observed. It should be noted that the scope, methodologies and approaches of the risk and working environment analyses of each project can vary as they depend on types of facilities, applicable requirements and company specifications.

It should be noted that any updates to the NORSOK Standards discussed in this paper may change the basic framework, expectations in projects, and challenges for application.

7. REFERENCES

1. Petroleum Safety Authority Norway, 'The Framework Regulations", 2013.

2. Petroleum Safety Authority Norway, 'The Management Regulations', 2014.

3. Petroleum Safety Authority Norway, 'The Activities Regulations', 2014.

4. Petroleum Safety Authority Norway, 'The Facilities Regulations', 2014.

5. Petroleum Safety Authority Norway, 'The Technical and Operational Regulations', 2014.

6. Petroleum Safety Authority Norway, 'Guidelines regarding The Framework Regulations', 2014.

7. Petroleum Safety Authority Norway, 'Guidelines regarding The Management Regulations', 2014.

8. Petroleum Safety Authority Norway, 'Guidelines regarding The Activities Regulations', 2014.

9. Petroleum Safety Authority Norway, 'Guidelines regarding The Facilities Regulations', 2014.

10. Petroleum Safety Authority Norway, 'Guidelines regarding The Technical and Operational Regulations', 2014.

11. Directorate of Labour Inspection, 'Working Environment Act', 2013.

12. Standards Norway, 'NORSOK Standard Z-013 Risk and Emergency Preparedness Assessment', 2010.

13. JAN ERIK VINNEM, 'Offshore Risk Assessment – Principles, Modelling and Applications of QRA Studies', 2007.

14. The International Organization for Standardization, 'ISO 17776 – Petroleum and Natural Gas Industries – Offshore Production Installations – Guidelines on Tools and Techniques for Hazard Identification and Risk Assessment', 2007.

15. Standards Norway, 'NORSOK Standard S-002 Working Environment', 2004.

16. SINTEF, 'CRIOP®: A Scenario Method for Crisis Intervention and Operability Analysis', 2008.

17. The International Organization for Standardization, 'ISO/TR 11079 – Evaluation of Cold Environments – Determination of Required Clothing Insulations (IREQ), 1993.

8. AUTHORS BIOGRAPHY

K. W. Yang holds the current position of senior engineer at offshore risk advisory group, DNV GL Oil & Gas. He is responsible for project management and technical activities for oil & gas related projects. His previous experience includes verification and safety risk management, e.g. HAZID, HAZOP, QRA, working environment studies for offshore oil & gas facilities and advisory services, e.g. feasibility study, conception evaluation, technology qualification for onshore & offshore LNG facilities.

ICSOT Korea: Safety of Offshore and Subsea Structures in Extreme and Accidental Conditions, 15-16 September 2015, Busan, Korea

STUDY ON WORKLOAD MEASUREMENTS AT THE MARINE ENGINEERING EDUCATION UNDER THE MARINE ENGINE SIMULATOR ENVIRONMENT

T Miwa, **K Shimamoto**, **YB Wu**, Kobe University, Japan, **M Nakamura,** Yuge National College of Maritime Technology, Japan and **M Uchida**, Kobe University, Japan

SUMMARY

For the safety navigation of the ship, also engineers have to keep the safety operation and management of the engine with good teamwork appropriately. Therefore, every engineer must exert full their own ability, and they are required to realize the minimum basic skills for the engine operation before they commence to be onboard. Then, at the maritime educational institutions, the requirement knowledge and skills as a ship engineer are given to students through lectures and exercises. In this study, in order to carry out the properly education training, authors pay attention to understand the mental state of the student. Then authors tried to measure the workload of student who operates the engine system under marine engine simulator environment. The EEG and the heart rate of trainees during the experiment were measured and analyzed. And the trainee's mental states were evaluated by comparing with n-back task experiment.

1. INTRODUCTION

J.Reason's the GEMS model [1] is integrated the SRK model of human behaviour classification levels proposed by Rasmussen and unsafe behaviours. J.Reason classified unsafe acts as "slip" and "lapse" which are performed inadvertently, and "mistake" and "violation" as acts conducted with intention. It is not uncommon for people to act unconsciously in daily life.
Under unusual conditions, people sometimes perform behaviour incorrectly. They may act appropriately for everyday conditions, but act erroneously when special circumstances arise. In behavioural studies of ship navigators [2] [3], in a simulation training environment of trainee's behaviour, the researchers evaluated omission errors, necessary tasks not carried out for the safety of a voyage.

Then, at the maritime educational institutions, the the required knowledge and skills as a ship engineer are given to students through lectures and exercises. At the engine department, it is not possible to carry out training using real machines to reproduce or create an accident. However, in the engine room simulator, without damaging any equipment, it is easy to repeat a reproduced situation using the same condition Settings.

When trainee tries to train by using the engine simulator, they will be flustered under the unaccustomed environment, or will be nervous. In training under such circumstances, the trainee cannot judge properly, and this judgment leads an error. In order to carry out the effective training, instructor is to understand the mental state of trainees during the training, and it should be better consider to keep the appropriate the mental load.

In this study, authors will record the behaviour of trainees carrying out watch-keeping in the marine engine simulator environment. By measuring the EEG and heart rate, we will be able to estimate the mental state and possible reasons for behaviour.

2. MATERIALS AND METHOD

Authors used the Marine Engine Plant Simulator (MEPS) at Kobe University. We carried out the engine watch keeping experiments using this simulator, and the trainees' activity recorded during the experiments. This simulator consists of an engine room and engine control room. During the experiment, the trainee is in the engine control room, and the research staff is in the engine room. We observe how the trainee communicates with the staff in the engine room during the watch keeping experiment. 4th year Kobe University under graduate students volunteered as trainees for our experiments. They had previous experience of the engine plant simulator in the course of their education

2.1 OUTLINE OF EXPERIMENT

The trainees perform the two experiments. One experiment takes 6 minutes. In this experiment, the simulator expresses the engine condition of an ocean-going voyage. We manipulate the simulator to run an alarm every 1 minute during the experiment.

Two types of abnormal alarms are used for the experiments; one is a strainer clogging in the main engine lubricating oil system, and the other a high-level alarm of the bilge tank. The two types of alarms occur twice in the experiment. The order of the alarm is different in the two experiments. In the first experiment, we activate the two types of alarm alternately. In the second experiment, we activate abnormal alarms of the main engine lubricating oil consecutively.

2.2 ABNORMAL ALARM

The high level condition of the bilge tank and the main engine lubricating oil strainer clogging are abnormal events (malfunction); they are set as scenarios in the simulator system. In general, the main engine lubricating oil directly affects the running of the engine. The main engine lubricating oil pressure anomaly actuates a main engine protective device that shuts down the engine. If the trainee is aware of the importance of this alarm, the trainee's level of stress should show on increase.

2.3 RECORD OF ACTIVITY

We record the trainee's behaviour using a video camera. By installing a camera beside the engine control console, the camera can record the trainee's as they operate the buttons and their interactions with the monitors on the console. The camera start recording from the experiment setup, and continues until the experiment is over.

2.4 HEART RATE MEASUREMENT

A heart rate sensor (polar, Inc., RX-800) is affixed to the centre of the trainee's chest. A detection signal from the sensor is sent to a wristwatch type receiver by radio wave. By connecting the receiver to the PC, Heart rate data is available after the experiment. The wristwatch type receiver displays the estimated heart rate of the trainee in real time.

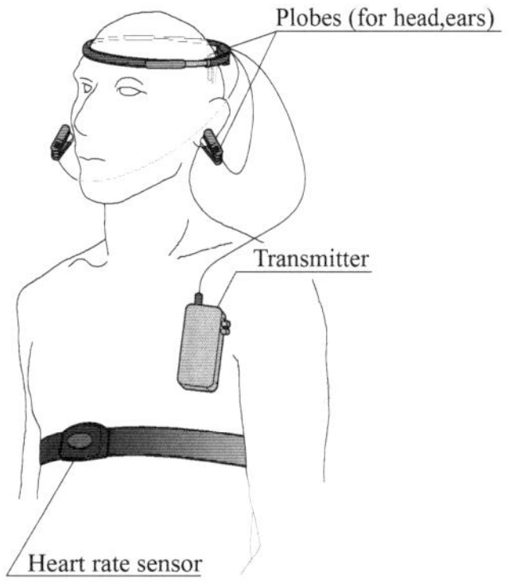

Figure 1 Overview of the measurement equipment

2.5 EEG MEASUREMENT

The portable EEG (Digital Medic, Inc, Muse Brain system) was used to measure the trainee's brain waves in this experiment. This EEG measures brain waves by placing an electrode on the back of the head and in both ear lobes. The detected signal is sent to the receiver by radio wave. The receiver recording the received signal is always connected to the PC.

By analyzing the frequency of the received signal, alpha wave (8-13Hz), beta wave (20-25Hz) and theta wave (4-7Hz) can be determined. In general, a relaxed state is presumed when the component ratio of the alpha wave is large, if the ratio of the beta wave is large to be in a stressed state is presumed.

2.6 NASA-TASK LOAD INDEX

We used a NASA-TLX [4] as our subjective stress evaluation method. The trainee answered questions on the survey sheet after each experiment ended.

2.7 N-BACK TASK WORK

The trainee solved the n-back task by PC prior to the experiment. n-back task [5] is a task of remembering an alphabet and the order of appearance displayed in random order by the PC software. The task difficulty is increased by the number of back times. As the trainee answered the n-back task, we measured the heart rate and EEG.

2.8 MAHALANOBIS GENERALIZED DISTANCE

The Mahalanobis generalized distance [6] [7] is calculated from the biological reactions measured of the trainee. The measurement values of biological reactions used as test data. We calculate the similarity between the training and test data. Mahalanobis Generalized distance uses multiple evaluation items, and calculates the spatial distance between the study data group and the test data. This formula shows the calculation of the Mahalanobis generalized distance

$$D^2 = \left(x_{i1}-\mu_1, x_{i2}-\mu_2, \cdots x_{ij}-\mu_j\right) \Sigma^{-1} \begin{pmatrix} x_{i1}-\mu_1 \\ x_{i2}-\mu_2 \\ \vdots \\ x_{ij}-\mu_j \end{pmatrix}$$

$$\Sigma = \begin{bmatrix} S_{v_1}^2 & S_{v_1v_2} & \cdots & S_{v_1v_j} \\ S_{v_2v_1} & S_{v_2}^2 & \cdots & S_{v_2vj} \\ \vdots & \vdots & \ddots & \\ S_{v_jv_1} & S_{v_jv_2} & \cdots & S_{v_j}^2 \end{bmatrix}$$

x: *test data*
μ : *average of data group*
i : *number of test data*
j : *number of variables*

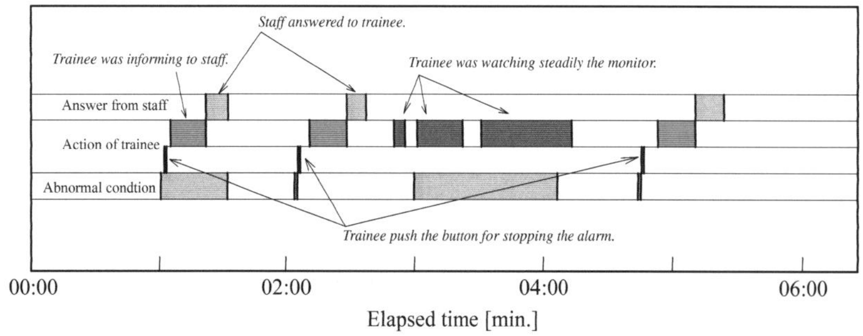

Figure 2 Overview of an experiment

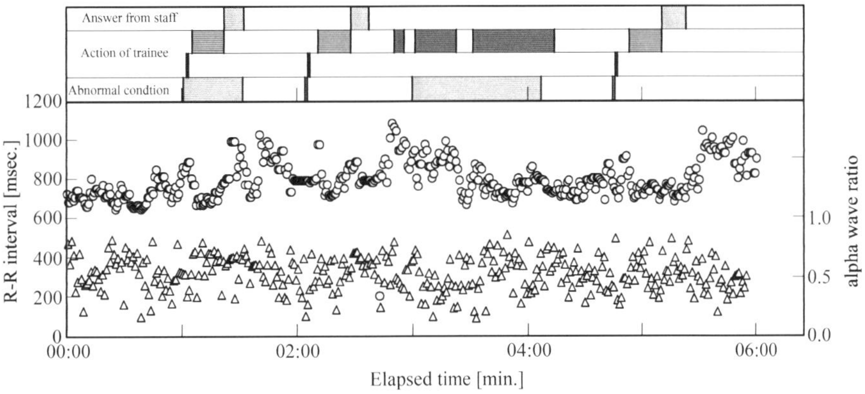

Figure 3.1 Result of measured biological reactions during experiment 1

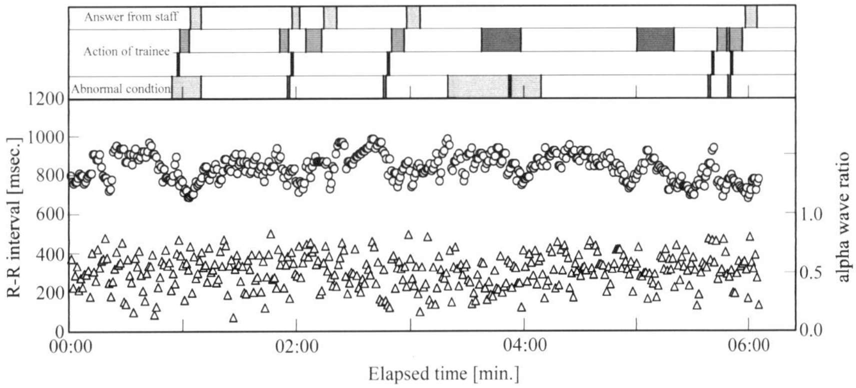

Figure 3.2 Result of measured biological reactions during experiment 2

3. RESULTS AND DISCUSSION

In this report, we focus to the result of analyzing the typical trainee's behaviour. Figure 2 shows an overview of an experiment. The lower line on the figure shows time blocks that abnormal conditions occurred in engine simulator. The middle part of the figure indicates the type of trainee behaviour. The upper part of figure shows block of talking time, our experimental staff in engine room who cooperated with trainee during experiments.

The trainee was almost always standing in front of the engine control console during the experiment, sometimes bending forward, or moving one step to the left and right. The trainees were only required to inform the staff of what was displayed on the engine monitor. As shown in Figures, a trainee was inform to engine room staff at almost every alarm timing.

The engine monitor has an alarm function during an abnormal condition. And, the content of abnormal is display on monitor. When the alarm sounds, trainee will press the "buzzer stop" button because stopping the buzzer sound. Until trainee will press the "confirm" button, the alarm contents are continuously displayed on a small window on the monitor.

If trainee were not pressed the "confirm" button, even then the same abnormality occurs later, the buzzer not sound. Between 3 minutes to 5 minutes of elapsed time, the alarm is not generated by this reason.

3.1 MEASURED HEART BEAT AND EEG

The measured biological reactions are shown in time sequence diagram of Figure 3.1 and 3.2. Open circles indicate the trainees' R-R interval of heart beat. Open triangles are indicates the trainees' alpha wave ratio.

3.2 SCORE OF NASA-TLX

Figure 4 shows the results of measured R-R interval and alpha wave ratio of a trainee solving the n-back task. Open circles are indicates the measured data during the solving for 0-back task. Right grey circles are indicates the measured data during the solving for 1-back task. As can be seen in figure 4, R-R interval of 1-back task work is wider than 0-back task works. In addition, dark grey circles are indicates the data during the 2-back task works. The figure illustrates that R-R interval of 2-back task works shorter than the data of 1-back task works.

To evaluates the experiments 1 and 2 using the NASA-TLX. Then, a trainee evaluated each experiment as follows. Score of experiment 1 is 8.00; score of experiment 2 is 7.53.

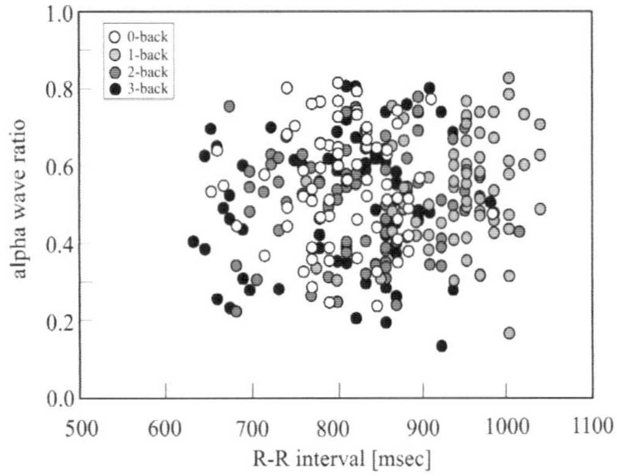

Figure 4 Results of measured R-R interval and alpha wave ratio (during the n-back task)

An experiment takes more long time than n-back task work. The experiment score of NASA-TLX might show as the average evaluation of task load in experience time. On the limited data available, the score of NASA-TLX may overlook the typical response by trainee according to mental state at each experiment time.

Table1 Score of NASA-TLX of a trainee

task/elements	mental	physical	temporal	performance	effort	frustration	TOTAL
0-back	3	3	3	4	3	6	4.27
1-back	3	3	4	12	6	10	7.87
2-back	13	6	6	16	13	14	12.73
3-back	15	4	11	14	17	17	14.13
weight	3	1	2	4	0	5	

The similarity degree is expressed with a numerical value by Mahalanobis generalized distance. By compare the distance between test data and each reference group, the test data were discriminated to each task levels. Figure 5.1 and Figure 5.2 shows the estimated task levels of a trainee during the experiments.

From the determination result, the mental work load as 3-back task level can be seen that many in the early time of the experimental 1. During the experiment, the mental load seems increased when the alarm occurred and their response. The trainee completed the communication with engine room staff, the mental work load level tends to decrease in most of the experiment time.

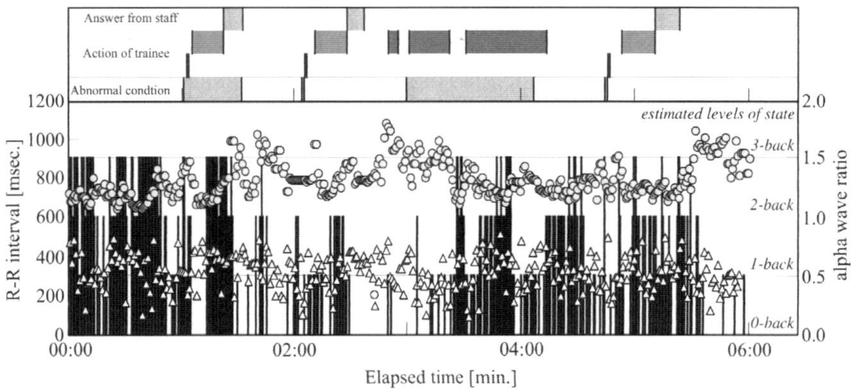

Figure 5.1 Result of the estimate to the task level during experiment 1

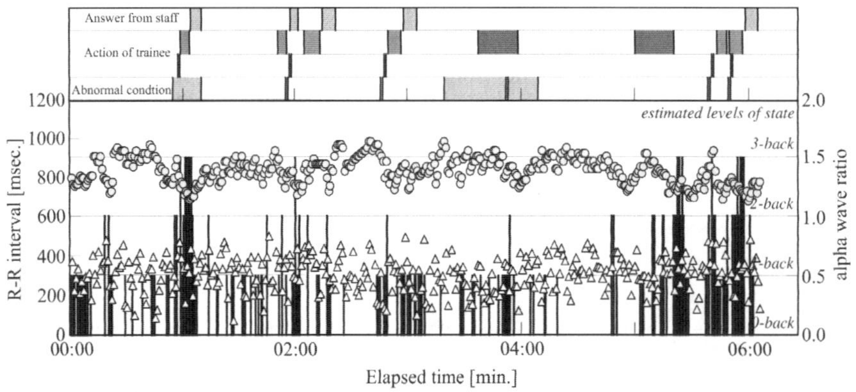

Figure 5.2 Result of the estimate to the task level during experiment 2

4. CONCLUSION

The following conclusions were derived from the results and discussion.

1. The mental state of a trainee is determined by the degree of similarity between the states of the n-back task.
2. The mental workload of a trainee has increased at an alarm occur time and its corresponding time.
3. The mental workload of a trainee tends to decrease by communication with engine room staff.

However, it is necessary to confirm the accuracy of mental load level. We are considering it is necessary to verify the accuracy of estimated mental load by each alarm time as future work.

5. ACKNOWLEDGEMENTS

This work was supported by JSPS KAKENHI. Grant Number:25420864.

6. REFERENCES

1. JAMES REASON, Human error, Cambridge; New York: Cambridge University Press, pp.206-207, 1990

2. A. UCHINO, Measuring of Human Factors by Ship-handling Simulator - Study Circumstances of Workload and Fatigue -, Navigation No.183, Japan Institute of navigation, pp.25-35 [in Japanese], 2012

3. A. UCHINO, H. KOBAYASHI, Human Error Caused by Workload Factor in Ship-handling, No.126, Japan Institute of navigation, pp.65-72 [in Japanese], 2012

4. Development of NASA-TLX (Task Load Index): Results of empirical and theoretical research, SG Hart, LE Staveland - Advances in psychology, Elsevier, 1988

5. T.S. BRAVER, J.D. COHEN, L.E. NYSTROM, J. JONIDES, E.E. SMITH, D.C. NOLL A parametric study of prefrontal cortex involvement in human working memory NeuroImage, 5 (1), pp. 49–62, 1997

6. YANAI, H., Hand Book of Multivariate Analysis, Gendai-Sugakusya, pp. 272. [in Japanese], 1986

7. A.P. DEMPSTER, Element of Continuous Multivariate Analysis, Addison-Wesley Publishing Company, pp.203-209, 1969

7. AUTHORS BIOGRAPHY

Takashi Miwa, lecturer of Kobe University, Speciality is Marine Engineering, Ph.D. in Engineering (Kobe Univ., 2012).

Yanbin Wu is a Ph.D. candidate majoring in Maritime Management in Kobe University. He is focusing on the research of reliable and effective engine-room management method.

Makoto Uchida is a professor working for Graduate School of Maritime Sciences, Kobe University. He obtained a Ph.D. in Osaka University, 1998. His main research area is ship propulsion engineering and marine engineering management.

ULTIMATE STRENGTH PERFORMANCE OF COMMERCIAL SHIPS ON ARCTIC OCEAN

D.K. Park[1], **Y.C. Ha**[1], **B.J. Kim**[1], **J.K. Seo**[1] and **J.K. Paik**[1,2], [1]The Korea Ship and Offshore Research Institute (The Lloyd's Register Foundation Research Centre of Excellence), Pusan National University, Busan, Korea. [2]Department of Mechanical Engineering, University College London, London, UK.

SUMMARY

Generally, global strength analysis (longitudinal strength analysis or hull girder strength analysis) and the local buckling analysis (stiffened panel strength analysis) are performed to identify the structural capacity and safety of ships in the early design stage. The ultimate strength performances of commercial ships on the Arctic Ocean considering the low temperature environment are dealt in this paper. Three types of ship, i.e., oil tanker, bulk carrier and container ship with four different sizes (totally 12 ships) are considered in four low Arctic temperatures and room temperature. ALPS/HULL and ALPS/ULSAP are used for the global strength analysis of hull girders and stiffened panels. The obtained results are summarized in terms of temperature, vessel type, vessel size, loading type and other effects. The important insights and outcomes are documented.

NOMENCLATURE

a	Length of the stiffener between transverse frames (mm)
B	Ship breadth (m)
C.O.V. or CV	Coefficient of variation
D	Ship depth (m)
DBTT	Ductile-brittle transition temperature (°C)
HBM	Horizontal bending moment (GNm)
HT	High tensile steel
K	Ship deadweight (K)
L	Ship length (m)
Longi.	Longitudinal
M_{hu}	Ultimate horizontal bending moment (GNm)
M_u	Ultimate bending moment (GNm)
$M_{u-R.T.}$	Ultimate bending moment at room temperature (GNm)
$M_{u-others}$	Ultimate bending moment at low temperature (GNm)
M_{vu}	Ultimate vertical bending moment (GNm)
MS	Mild steel
NSR	Northern Sea Routes
P	Plate
R.T.	Room temperature
S	Stiffener
t	Plate thickness (mm)
TEU	Twenty-foot equivalent units
Trans.	Transversal
ULS	Ultimate limit state
USGS	United States Geological Survey
VBM	Vertical bending moment (GNm)
w_{oc}	Column-type initial distortions of the longitudinal stiffeners (mm)
w_{opl}	Maximum plate initial deflection (mm)
w_{os}	Sideways initial distortion of the longitudinal stiffeners (mm)
β	Plate slenderness ratio
σ_c	Combination stress (MPa)
$\sigma_{c-others}$	Combination stress at low temperature (MPa)
$\sigma_{c-R.T.}$	Combination stress at room temperature (MPa)
σ_{xu}	Ultimate axial stress in the x direction (MPa)
$\sigma_{Yeq(R.T.)}$	Equivalent yield stress at room temperature (MPa)
σ_{yu}	Ultimate axial stress in the y direction (MPa)
σ_Y	Yield strength (MPa)
$\sigma_{Y-R.T.}$	Yield strength at room temperature (MPa)
℄	Centre line

1. INTRODUCTION

The global warming has brought about many undesired changes such as the various environmental problems including rising sea level, storms, floods, drought, losing of biodiversity and destruction of ecosystems. Therefore, industry stakeholders have introduced legislation to limit these impacts. A number of efforts are now being made to keep a development-free environment or called green growth [1-3].

On the other hand, the decreasing polar ice extent has opened the Northern Sea Routes (NSR) as an advantage of global warming. Shipping on NSR has more cost benefits compared to other routes. It is reported that the Northern Route may reduce not only the distance (almost 40%) but also shipping period (around 35%). For example, the transportation of cargoes from Busan port (Korea) to Rotterdam port (Netherlands) across the Northern Sea Route takes only 14 days in a distance of 12,700km. But, through the Southern Sea Route, it takes 24 days in a distance of 20,100km, as shown in Figure 1 [4]. Moreover, in 2008 the United States Geological Survey (USGS) estimated that the Arctic might contain 13% of the world's undiscovered oil and 30% of its undiscovered gas. Of these hydrocarbon resources, 84% are believed to be offshore, mostly in waters less than

500m deep [5]. These two main benefits have led to the increasing interests in the Arctic operating structures.

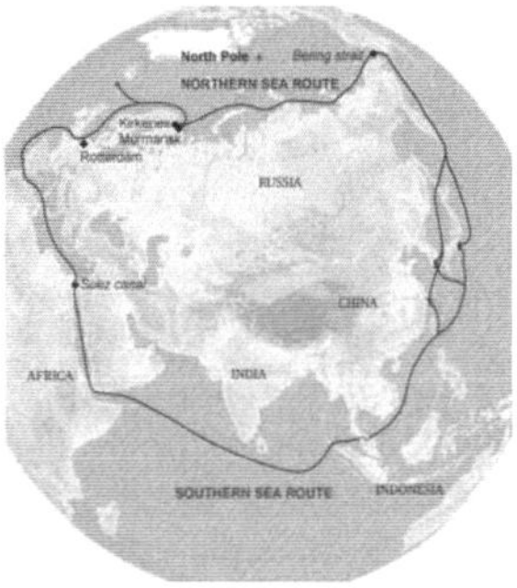

Figure 1: Configuration of Northern Sea Route and Southern sea route

However, ice-free navigation through the NSR is time-limited for about three months per year and requires appropriately strengthened hulls or ice breaker support for safe passage [6, 7]. Therefore, additional construction cost will be charged for NSR-going vessels due to different environmental conditions including low temperature and ice load.

Therefore, to avoid such cost, one of the good alternative ideas is using normal sea-going vessel in Arctic Sea Route during its opening season. In this study, the ultimate strength performance of normal sea-going commercial ships, which are assumed to be able to operate in Arctic sea route, is investigated. Two main objects are namely ultimate hull girder strength behaviour and ultimate stiffened panel strength behaviour.

Arctic sea conditions and the effects on material properties are presented in Section 2. In addition, a method for applying the material properties under low temperature to the ship structures is covered in the Section. Based on this assumption, applied examples of twelve representatives of normal sea-going commercial ships are addressed in Section 3. Ultimate strength analysis of both hull girders and stiffened panels are also illustrated in the section. After that, Arctic sea effects on ultimate strength performance of all structures are analysed and concluded in Sections 3 to 5.

2. NORTHEN SEA CONDITION

2.1 EXPLANATION OF ARCTIC ENVIRONMENT CONDITION

The NSR provides us a decided advantage in the shipping as shown in Figure 1. On the other hand, it also demands us a conquest of the harsh environment. The Arctic's climate is characterised by cold winters and cool summers. Average winter temperatures can be as low as −40°C and the coldest recorded temperature is approximately −68°C.

In the present study, to reflect the Arctic temperature condition, 6 temperature conditions are selected, namely room temperature (R.T., 20°C), 0, −20, −40, −60 and −80°C. In this regards, assumptions in this study are summarised as follows:

- New ice-class ship building is a heavy burden to the economy and time when it is performed only for the NSR voyage purpose.
- Non-ice class ships can sail though the ice-free NSR.
- The 12 normal (non-ice class) commercial ships are adopted for the purpose of the research. It is also assumed that global warming may continue the reduction of Arctic ice extent. .

The details of analysis results for twelve vessels are covered in applied example Section.

2.2 MATERIAL PROPERTIES AT LOW TEMPERATURE

A series of tensile coupon tests were performed to investigate the material properties at low temperatures on metallic materials for ships [8]. In Figures 2(a) to 2(c), the test results showed increase of strength and irregular fracture strain tendency as the temperature decreased, as shown in. The dramatic increase was observed to on Grade A mild steel than other grades of mild steel and it was relatively dull on high tensile steels. In the present study, the results on Grade A and Grade AH were used to estimate the yield strength of mild steel (MS), 32 Grade high tensile steel (HT32) and 36 Grade high tensile steel (HT36) at targeted low temperatures.

Test results set up two quadratic functions that compute the ratios of yield strength increase to temperature for each material. The estimated yield strengths of different grades of carbon steel at low temperatures with the quadratic functions are shown in below a list. The estimation is based on the minimum yield strength requirements of materials [9]. Yield strength at 0°C showed no difference with the R.T. on materials and increased with a same ratio with the test results. Next, the estimated material properties were used in ultimate strength analysis.

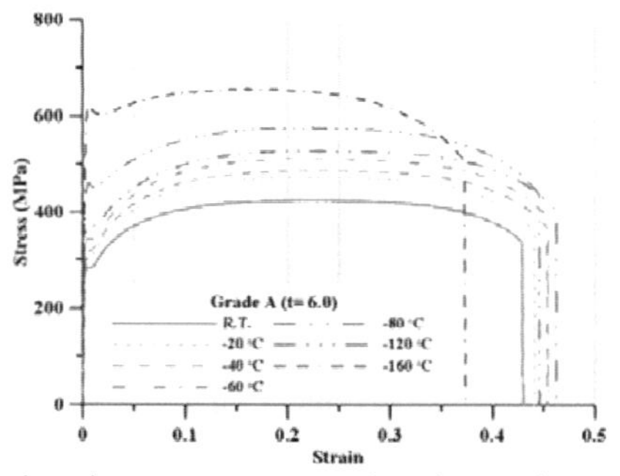

(a) Engineering stress versus engineering strain curves of Grade A mild steel

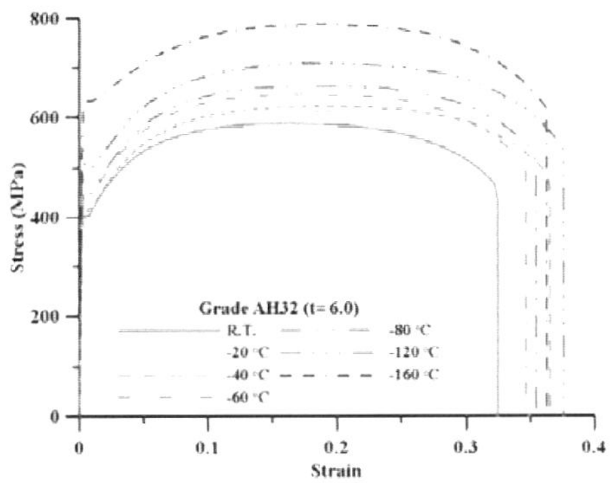

(b) Engineering stress versus engineering strain curves of Grade AH32 high tensile steel

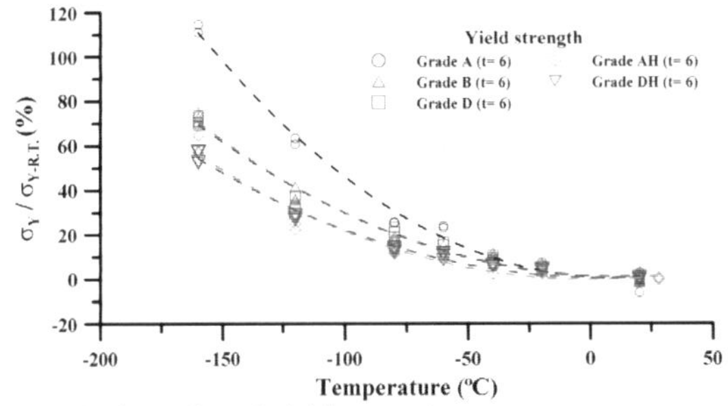

(c) The ratios of yield strength to temperature on different grades carbon steel

Figure 2: Material properties at low temperatures

- Yield strength on different materials at target temperatures

Material \ Temperature (°C)		-80	-60	-40	-20	0	20
MS24	σ_Y (MPa)	309.4	280.2	258.2	243.3	235.0	235.0
	σ_Y increase (%)	31.7	19.3	9.9	3.5	0.0	0.0
HT32	σ_Y (MPa)	357.4	337.8	323.8	315.3	315.0	315.0
	σ_Y increase (%)	13.5	7.2	2.8	0.1	0.0	0.0
HT36	σ_Y (MPa)	402.8	380.7	364.9	355.4	355.0	355.0
	σ_Y increase (%)	13.5	7.2	2.8	0.1	0.0	0.0

2.3 APPLICATION OF MATERIAL PROPERTIES TO VESSELS

Ship structures are subjected to various types of load, which may be grouped according to their characteristics in time: static loads, low-frequency dynamic loads, high-frequency dynamic loads and impact loads. In ship structural analysis and design, the most common loads are the static and low-frequency dynamic loads, the latter usually being treated as static or quasi-static loads.

A ship hull in an intact condition will sustain applied loads smaller than the design loads, and in normal seagoing and approved cargo loading conditions it will not suffer any structural damage such as buckling and collapse. However, the loads acting on the ship hull are uncertain both because of the nature of rough seas and because of possibly unusual loading/unloading of cargo. Therefore, present study did not considered dynamic hardening and DBTT (Ductile-Brittle Transient Temperature) which are important variables for structural analysis under high-frequency dynamic loads and impact loads.

Kim et al. [10] performed a study on low temperature effects on ultimate hull girder strength of commercial ships. But, they applied uniform temperature distribution on whole section of hull girders. In practice, temperature distributions are founded as shown in Figure 3.
In this regards, following assumptions are applied in this study.

- Five types of Arctic temperatures (R.T. (or 0°C), −20, −40, −60 and −80°C) are applied to the hull girders that are exposed to air.
- In the case of Northern Sea Route voyage, unfrozen seawater condition is assumed and its temperatures are applied to 0°C.
- Seawater temperature (0°C) is applied to the hull girder structures below sea level.
- Temperatures of inner side shell are determined by cargo type.

Different temperatures are applied to the each vessel based on the above assumptions.

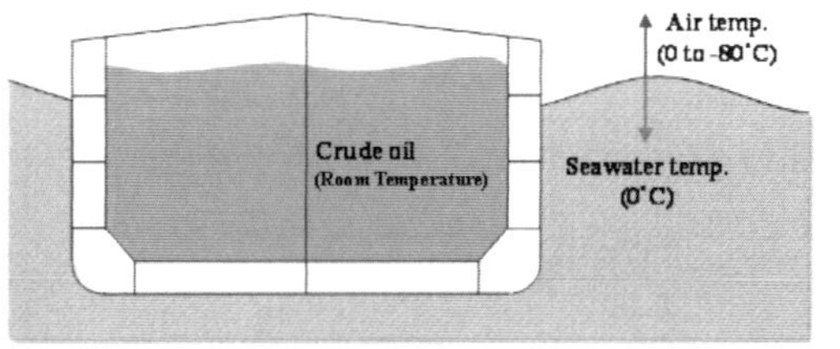

Figure 3: Explanation of temperature distribution of Arctic sea-going vessel

3. TARGET STRUCTURES AND TEMPERATURE MAPPINGS

Three types of ships namely oil tanker, bulk carrier and container ship are selected as the most representative vessels that occupy up to 60 to 70% of whole operating vessels in several recent decades [11]. Four types of different size vessel per each ship type are selected. Twelve types of target structures namely 4 oil tankers (Panamax, Aframax, Suezmax and VLCC), 4 container ships (3500, 5000, 7500 and 13000 TEU) and 4 bulk carriers (Handysize, Supramax, Kamsarmax and Capesize) are considered in order to investigate the low temperature effects on hull girder structures. Principal dimensions with scantling draught data, which are the most important factor to consider the temperature distribution on outer side member, are presented in Figures 4 to 6. Two types of different size are presented for each target structures.

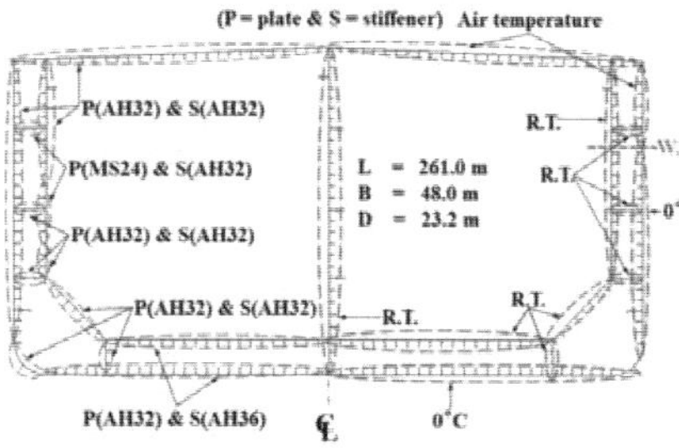

Figure 4: Midship section of Suezmax oil tanker with applied temperature and material properties

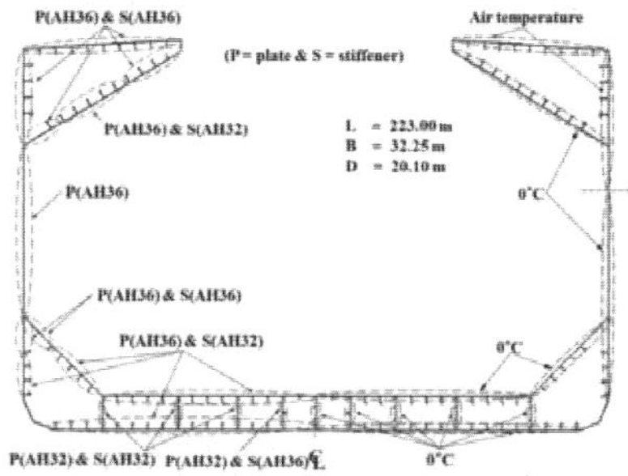

Figure 5: Midship section of 82K Kamsamax bulk carrier with applied temperature and material properties

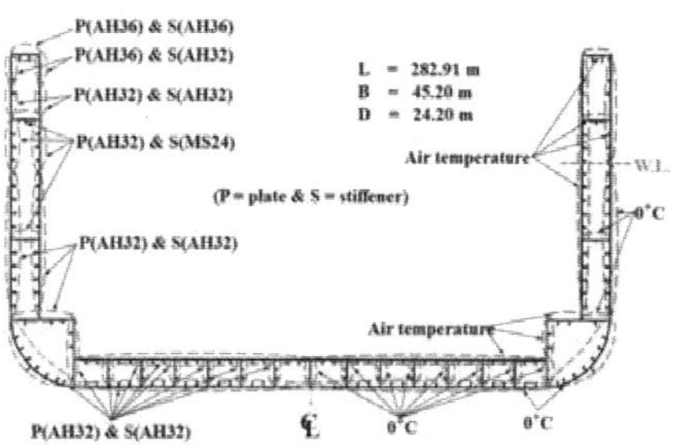

Figure 6: Midship section of 7500 TEU container ship with applied temperature and material properties

In the case of oil tanker, R.T. is applied to inner side members based on the oil heating, which is generally applied during operation. In the case of bulk carrier, 0°C is applied to the inner side members based on the assumptions that inner side temperature is not lower than room temperature and the hatch cover. In practice, the same material properties are applied on inner side member for oil tankers and bulk carriers as there is no difference in terms of material properties between R.T. and 0°C from the test results. In the case of container ships, ambient temperatures are applied to the deck and inner bottom members because this vessel has open type structures.

3.1 ULTIMATE STRENGTH ANALYSIS OF HULL GIRDERS

Global strength assessment for longitudinal strength and local strength assessment for buckling strength are carried out at the initiatory ship design stage. In this context, hull girder structure and stiffened panel structure are considered for the ultimate strength examination. There are several numerical methods to evaluate ultimate longitudinal strength performance of ships and ship-shaped offshore structure. With regard to this, numerous comparison studies have been performed to investigate their accuracy and efficiency subjected to corrosion [12, 13] and grounding [14, 15].

The ALPS/HULL [16] progressive hull collapse analysis program is adopted for the ultimate hull girder strength analysis and ALPS/ULSAP [17] buckling strength analysis for stiffened panel structures is used in this study. The details of ALPS/HULL program may be referred in a previous study [18] by Paik and Thayamballi.

Structural initial imperfection of plate and stiffener members is considered as in Eq. (1).

$$w_{opl} = 0.1\beta^2 t,\ w_{oc} = w_{os} = 0.0015a \qquad (1)$$

where,

a = the length of the stiffener (between transverse frames),

t = plate thickness,

w_{opl} = the maximum plate initial deflection,

w_{oc} = the column-type initial distortions of the longitudinal stiffeners,

w_{os} = the sideways initial distortion of the longitudinal stiffeners,

β = plate slenderness ratio.

Generally, vertical bending moments (VBM) are mostly applied in ships and ship-shaped offshore structures. In this study, horizontal bending moments (HBM) are also considered for the purpose of the research. Figures7, 9 and 11 present analysis results of ultimate hull girder strength under VBM and HBM. Five air temperatures namely room temperature (R.T.), −20, −40, −60 and −80°C are applied for evaluating the ultimate strength performance of target structures. As mentioned earlier, in the case of 0°C, material tensile coupon test results are same with R.T. Therefore, only room temperature case is analysed.

To draw a general tendency, the obtained ultimate strength results are analysed by applying the consistent variables such as vessel size or carrying capacity. For example, ship

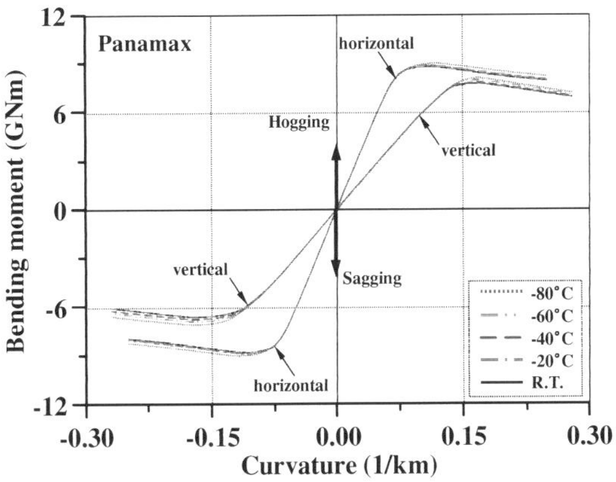

(a) Panamax

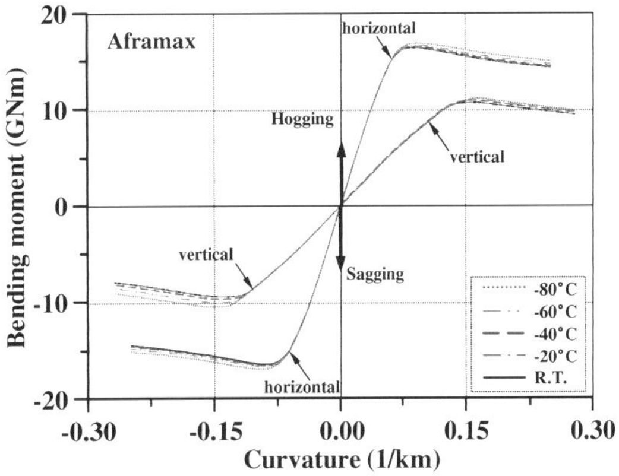

(b) Aframa

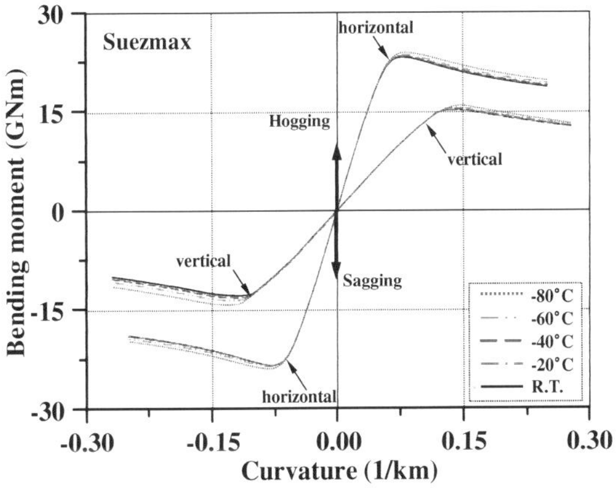

(c) Suezmax

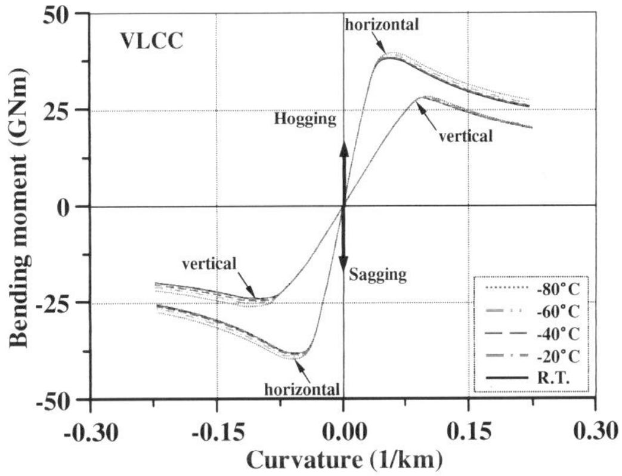

(d) VLCC

Figure 7: Ultimate hull girder strength analysis for oil tankers

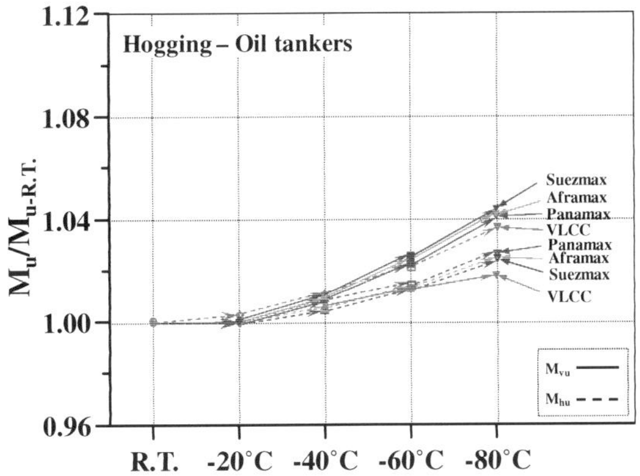

(a) In hogging

(b)

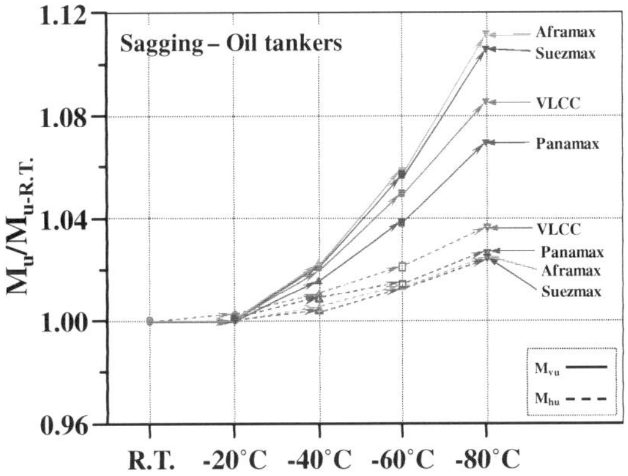

(b) In sagging

Figure 8: Deviation of ultimate hull girder strength of oil tankers with temperature
(M_{vu}= ultimate vertical bending moment, M_{hu}= ultimate horizontal bending moment)

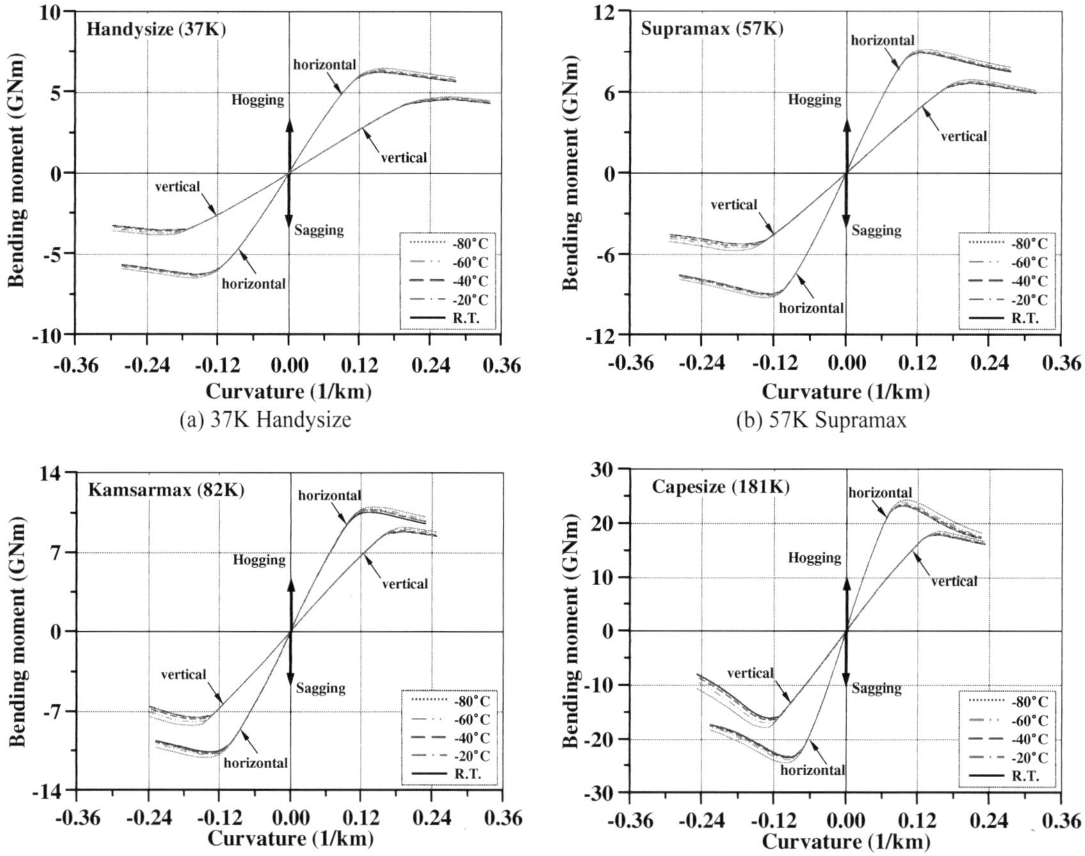

(a) 37K Handysize

(b) 57K Supramax

(c) 82K Kamsarmax

(d) 181K Capesize

Figure 9 Ultimate hull girder strength analysis for bulk carriers

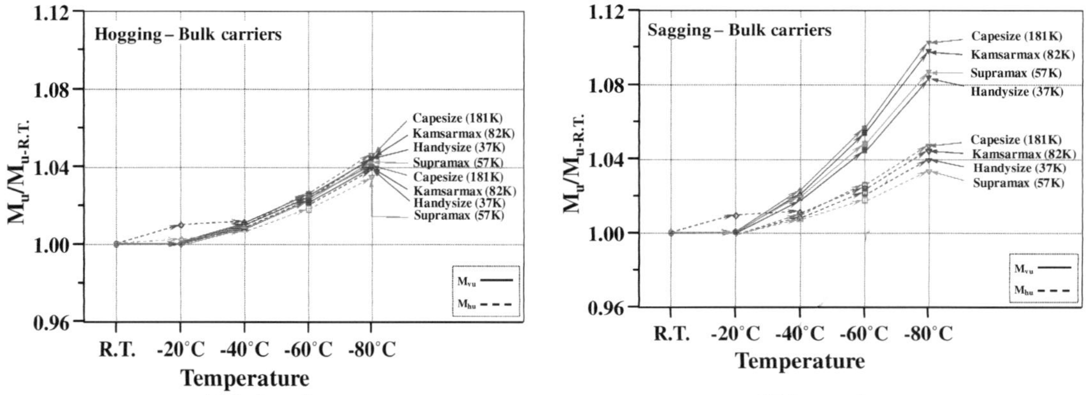

(a) In hogging

(b) In sagging

Figure 10: Deviation of ultimate hull girder strength of bulk carriers with temperature (M_{vu}=ultimate vertical bending moment, M_{hu}=ultimate horizontal bending moment)

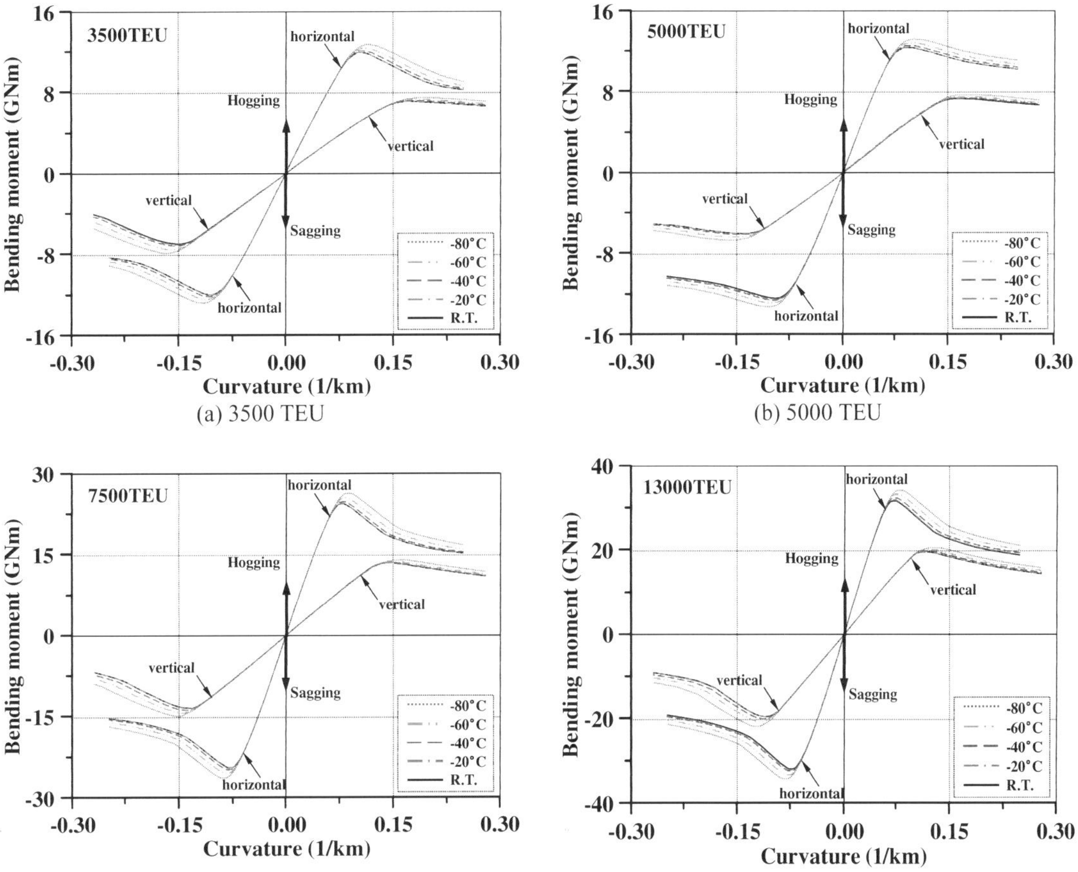

(a) 3500 TEU

(b) 5000 TEU

(c) 7500 TEU

(d) 13000 TEU

Figure 11: Ultimate hull girder strength analysis for container ships

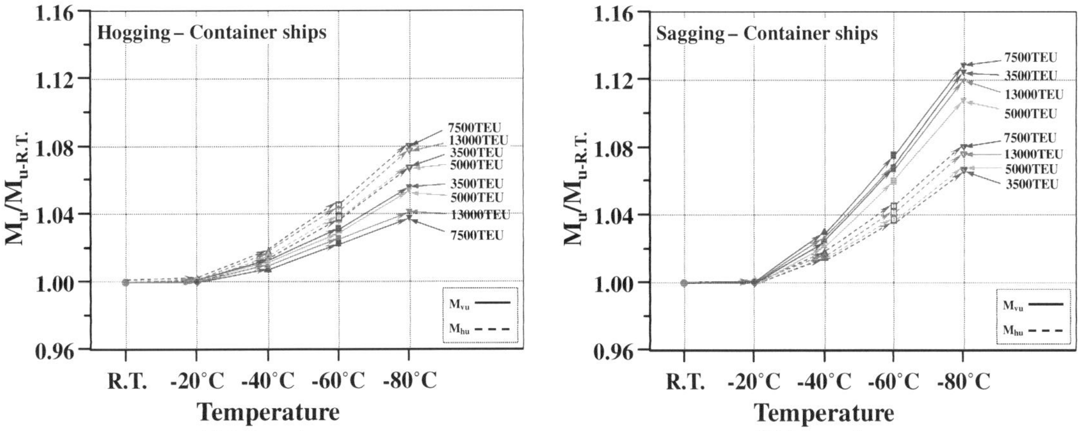

(a) In hogging

(b) In sagging

Figure 12: Deviation of ultimate hull girder strength of container ships with temperature (M_{vu}= ultimate vertical bending moment, M_{hu}= ultimate horizontal bending moment)

length (L), deadweight (K) and shipping capacity of container box (twenty-foot equivalent units, TEU) are applied to the target ships as shown in Figures 8, 10 and 12. Structural members above sea water level are exposed to Arctic ambient temperature, leading to the increase in the strength of materials. The deck part members are applied with compression in sagging. Therefore, the ultimate strength increase in sagging is greater than in hogging. This is observed in all the cases.

The ultimate hull girder strength of oil tankers and container ships increases as temperature decreases. But, it has an irregular tendency with ship capacity, while bulk carrier shows a proportional tendency to temperature decrease and capacity increase in sagging with VBM.

3.2 ULTIMATE STRENGTH ANALYSIS OF STIFFENED PANELS

Generally, stiffened panels of deck and outer bottom parts are targeted for evaluating the local strength assessment at beginning state of ship design. In fact, axial forces such as tension (above the neutral axis of structural members under hogging moment and below the neutral axis of structural members under sagging moment) and compression force (below the neutral axis of structural members under hogging moment and above the neutral axis of structural members under sagging moment) are applied at mid-ship section when the bending moments are applied. The maximum axial forces are applied to the farthest hull girder structural member from neutral axis of structure, i.e., deck and bottom structural members. In the present study, deck stiffened panel, which is exposed to Arctic ambient, is selected to investigate the low temperature effect on the ultimate strength of stiffened panels. Examples of deck stiffened models are shown in Figure 13.

The applied loads shall be defined as four types of loading combinations (case I to case IV) as shown in Figure 14. Next, Figures15, 17 and 19 present the ultimate strength analysis results.

It is observed that the results of ultimate strength on oil tankers and bulk carriers are varied with the temperature decreases are slightly changed when the loading conditions are case III and case IV compared to the container ships. In these cases, scantlings of deck stiffened panels for container ships are significantly thicker than other ships. These thicker member have caused the fluctuation of ultimate stiffened panel strength of container ships.

Figures 16, 18 and 20 show the summarised results in terms of mean values and coefficient of variation (C.O.V. or CV) depending on temperatures variation and vessel size. Generally, ship size has no effect on ultimate strength at low temperature on oil tankers and bulk carriers, but has effect on ultimate strength at low temperature on container ships. Low temperature has an effect on larger container ships than smaller ships. It is also found that the ultimate strength variation is proportional to the temperature decrease.

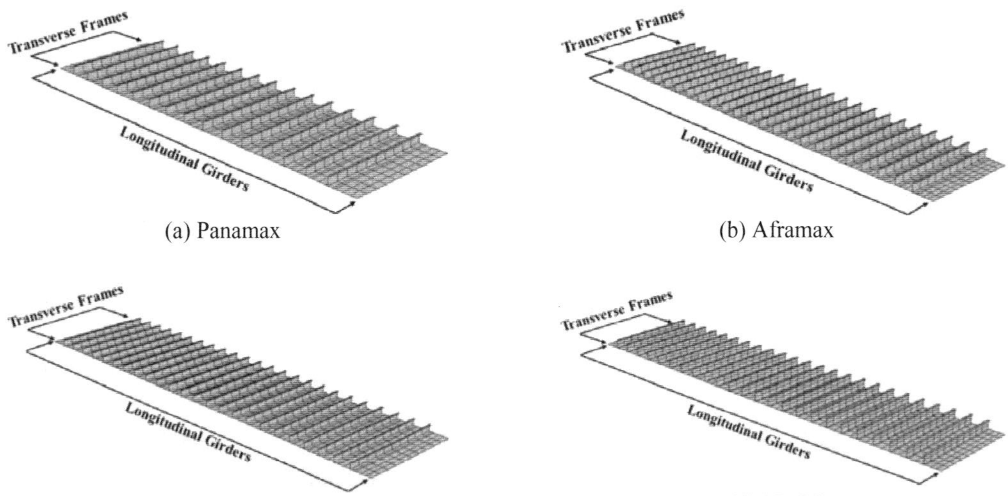

(a) Panamax (b) Aframax (c) Suezmax (d) VLCC

Figure 13: Deck stiffened panel model for oil tankers

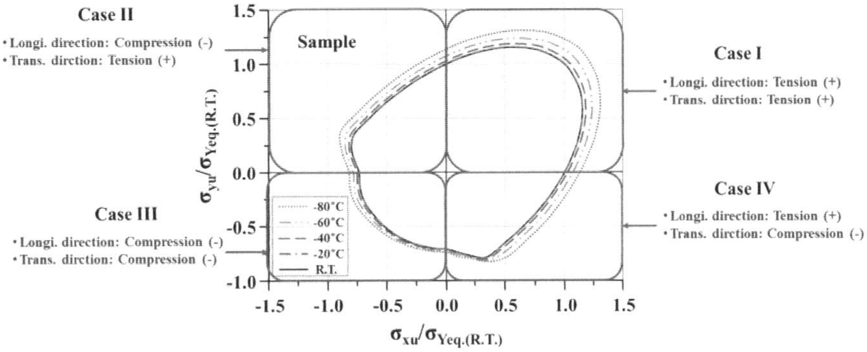

Figure 14: Definition of analysis case of ultimate stiffened panel

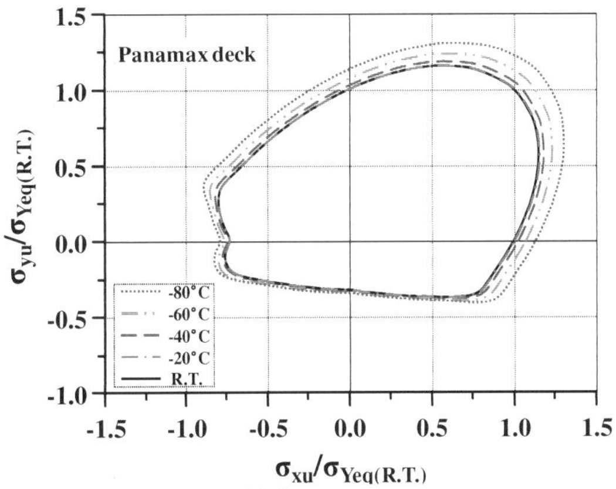

(a) Panamax

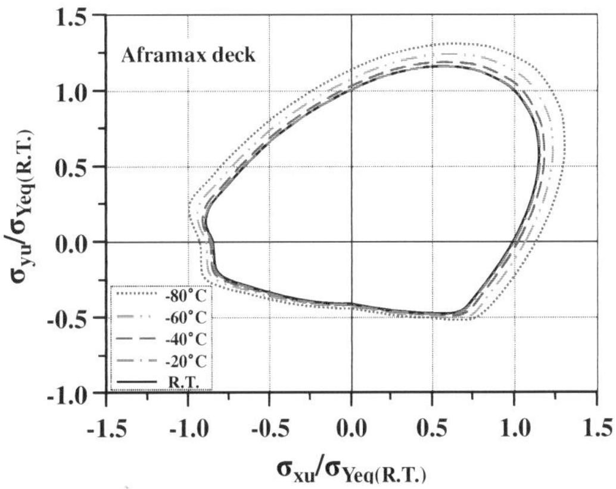

(b) Aframax

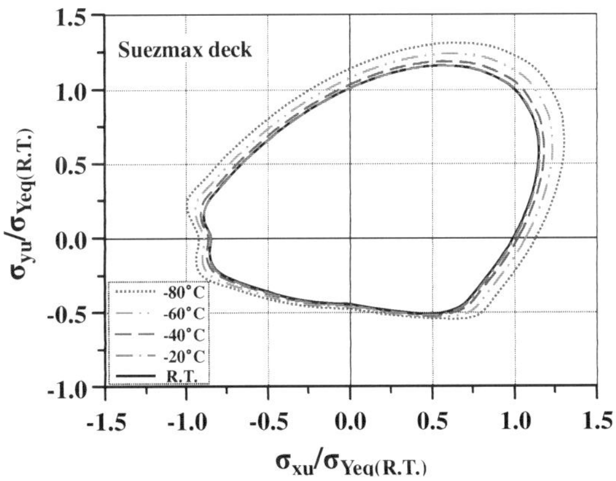

(c) Suezmax

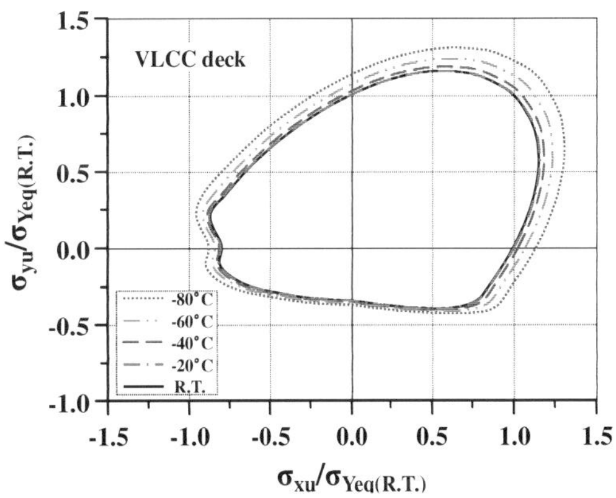

(d) VLCC

Figure 15: Ultimate strength analysis of deck stiffened panel for oil tankers

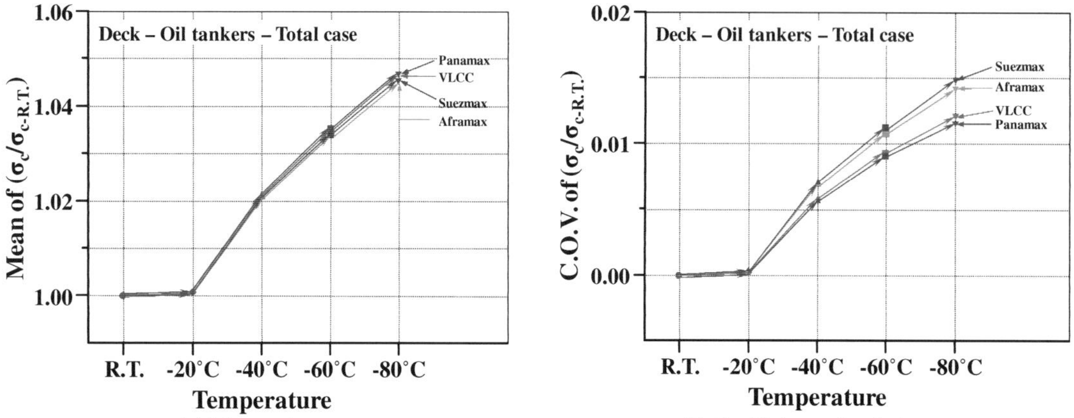

(a) Mean values

(b) Coefficient of variations

Figure 16: Deviation of ultimate stiffened panel strength of oil tankers with temperature

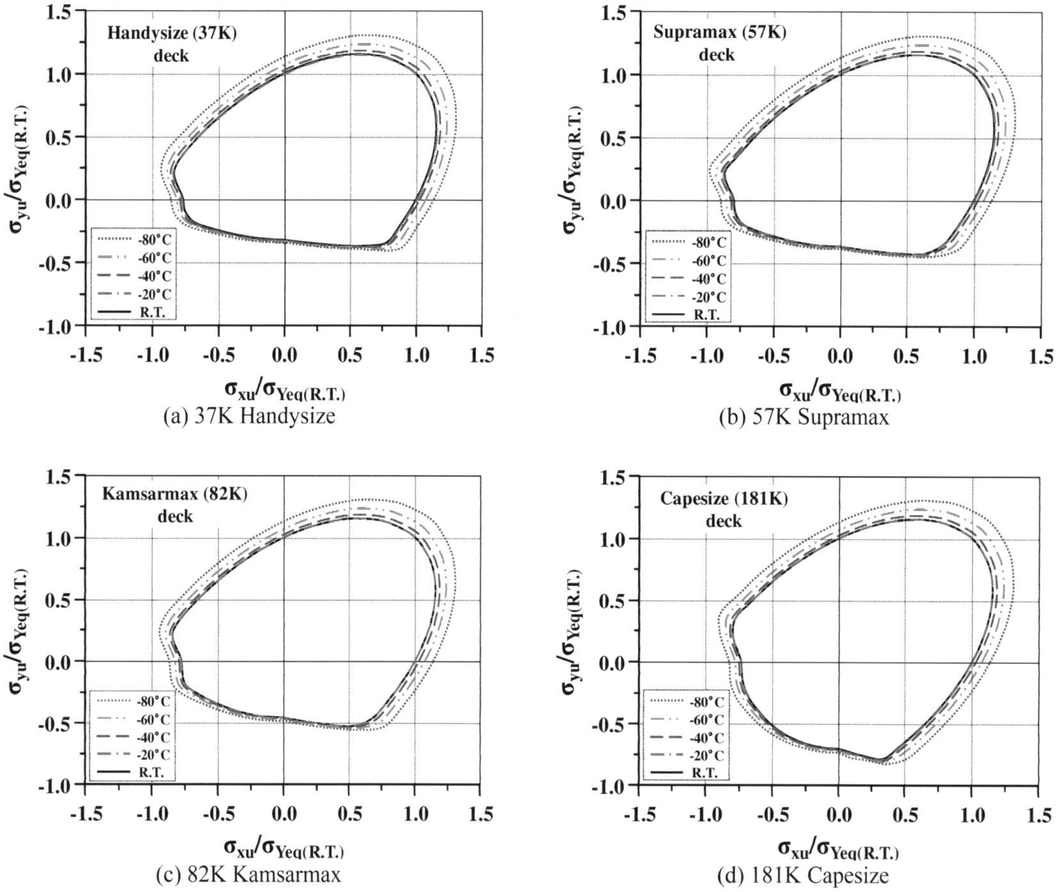

(a) 37K Handysize

(b) 57K Supramax

(c) 82K Kamsarmax

(d) 181K Capesize

Figure 17: Ultimate strength analysis of deck stiffened panel for bulk carriers

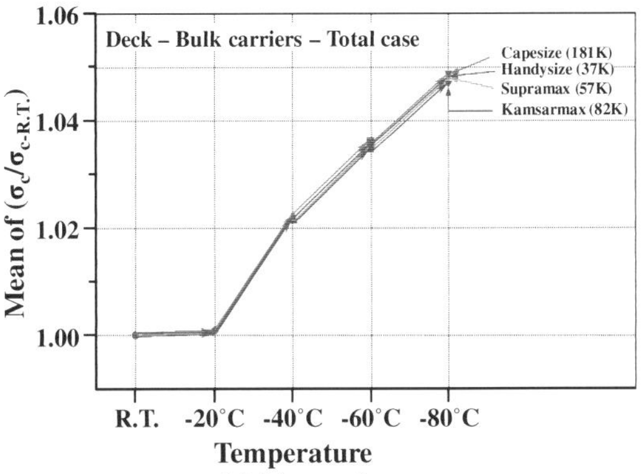

(a) Mean values

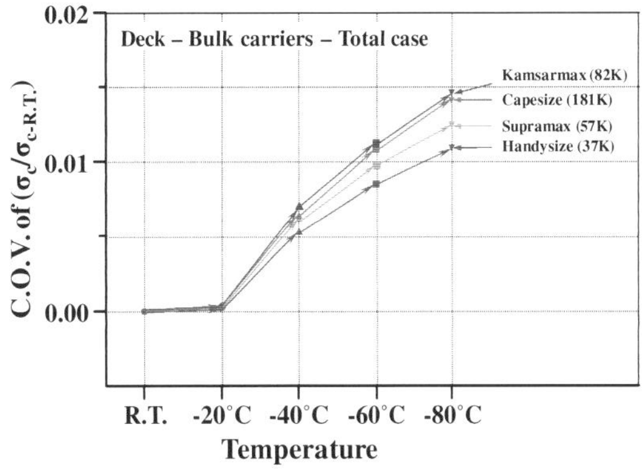

(b) Coefficient of variations

Figure 18: Deviation of ultimate stiffened panel strength of bulk carriers with temperature

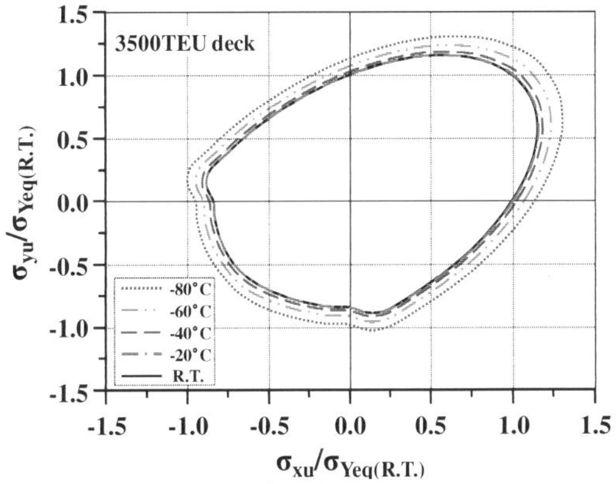

(a) 3,500 TEU

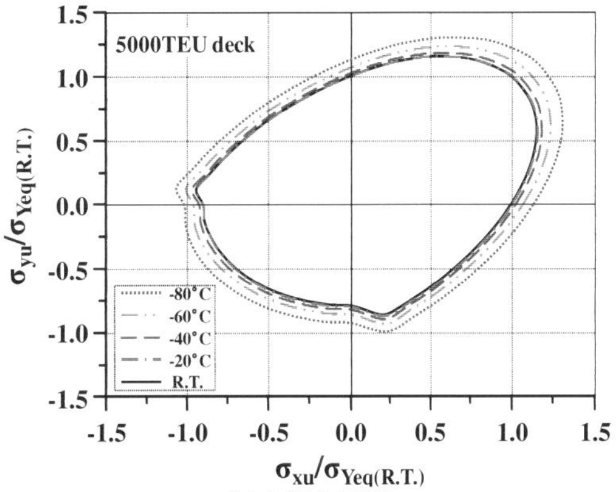

(b) 5,000 TEU

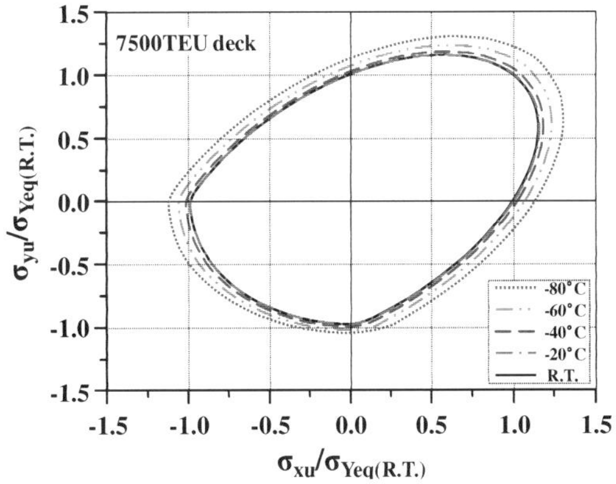

(c) 7,500 TEU

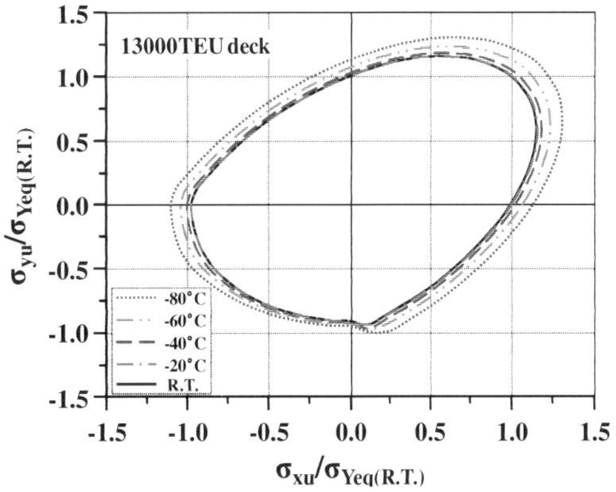

(c) 13,000 TEU

(d)

Figure 19: Ultimate strength analysis of deck stiffened panel for container ships

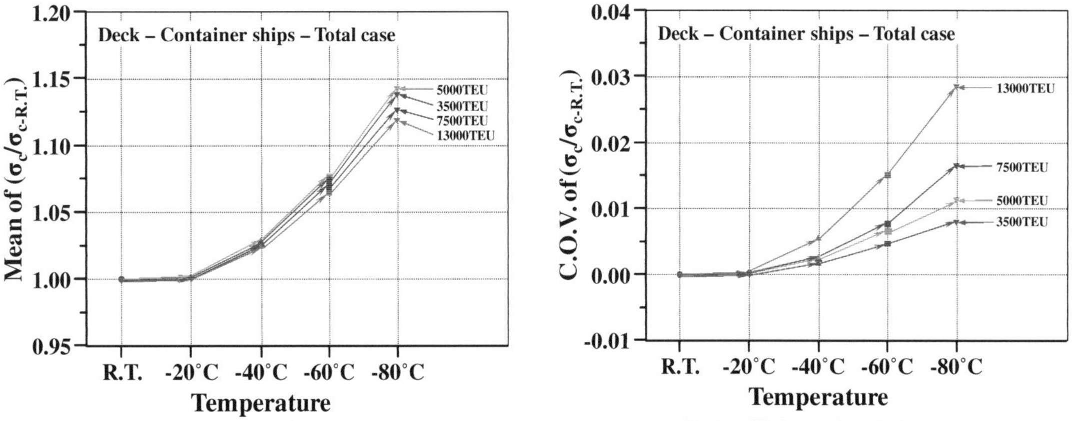

(a) Mean values (b) Coefficient of variations

Figure 20: Deviation of ultimate stiffened panel strength of container ships with temperature

4. DISCUSSION

The ultimate strength of structures at low temperature increased less than the increase of yield strength. Less increase of the ultimate strengths for hull girder strength and stiffened panel at low temperature are also observed in previous studies [19, 20]. It attributes to a temperature mapping method and a progressive buckling. Different temperature mapping on different structure members relevant to a real operating situation affects it, i.e., higher temperature (lower strength) is also applied on structures with a low temperature environment as mentioned in Section 3.1. Moreover, a progressive buckling of structural members under increase load occurs. The most highly compressed member will collapse earlier and the stiffness of the overall hull will decrease gradually. As loads continually increase, buckling and collapse of more structural members will occur progressively until the ultimate limit state (ULS) is reached [18]. In other words, the low temperature environment affected on the ultimate strength for hull girder strength and stiffened panel, but it was not the only variable in the ultimate strength estimation.

4.1 HULL GIRDERS

Figures 22 to 24 show the variation of ultimate hull girder strength at different temperatures. Mean and C.O.V. are calculated with temperature variation as well. In case of vertical bending moment, as the temperature decreases, it is observed that the mean values are 0.1 to 6.4% for oil tankers, 0.1 to 6.7% for bulk carriers and 0.0 to 8.2% for container ships. On the other hand, as the temperature decreases under horizontal bending moment, the mean values are 0.1 to 2.8% for oil tankers, 0.3 to 4.1% for bulk carriers and 0.1 to 7.2% for container ships. Mean value increases higher under vertical bending moment than the mean value under horizontal bending moment for ships.

There are differences in hogging and sagging bending moment results under vertical bending moment between oil tankers and bulk carriers, as shown in Figure 24(a). These differences are shown in Figure 22(a) for oil tankers and Figure 23(a) for bulk carriers. These differences could be due to the difference in terms of mid-ship shape and composition of materials of each structural member. On the other hand, hogging and sagging bending moment results under horizontal bending moment are similar for each type of ship.

The C.O.V. increases as the temperature decreases and it is sensitive under vertical bending moment. The increment largely varies in the order from the most sensitive to the least sensitive as follows: container ship, oil tanker and bulk carrier.

The effects of low temperature on each vessel size (i.e., length, shipping capacity and deadweight) are presented in Figures 25 to 27. There are no significant differences covering the ship size of each type of vessel except for Panamax class oil tanker case in sagging. This difference is caused by ratio of ship breadth (B) to depth (D). The B/D ratios for other three oil tankers are 2.00 to 2.09, but the B/D ratio for Panamax class double hull oil tanker is 1.56.

From Figures 22 to 27, it could be said that the ultimate hull girder strength is more sensitive to low temperature under vertical bending moment than under horizontal bending moment, except for container ships under hogging moment. Moreover, the low temperature has less effect on ship size, except for Panamax class oil tanker in sagging. The obtained results could be summarised as the following order and as shown in Figure 21.

Ultimate hull girder strength at low temperature:

Oil tanker < Bulk carrier < Container ship

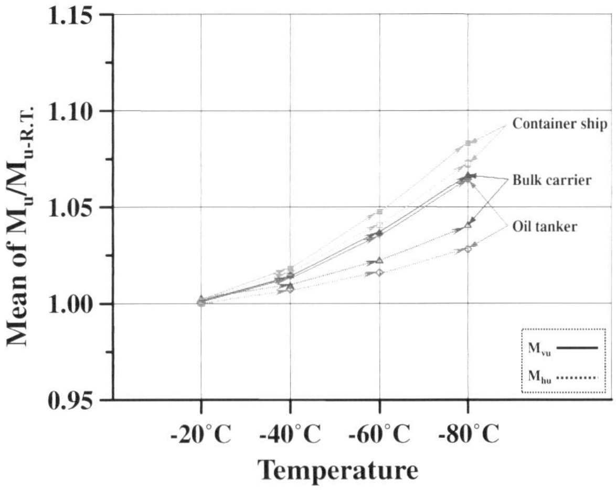

Figure 21: Comparison of mean value for commercial ships in terms of ultimate longitudinal strength

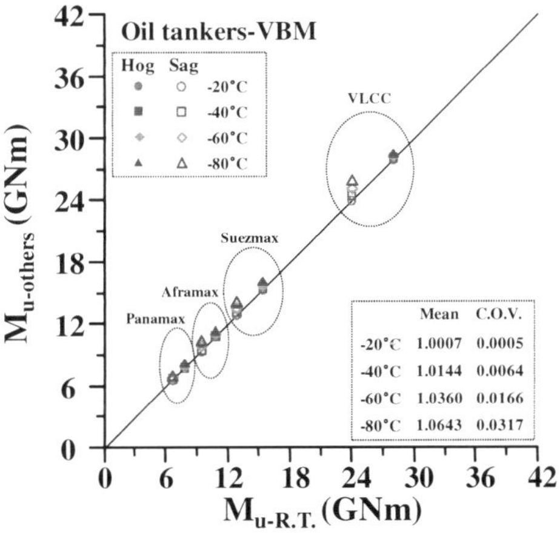

(a) Oil tankers – Vertical bending moment

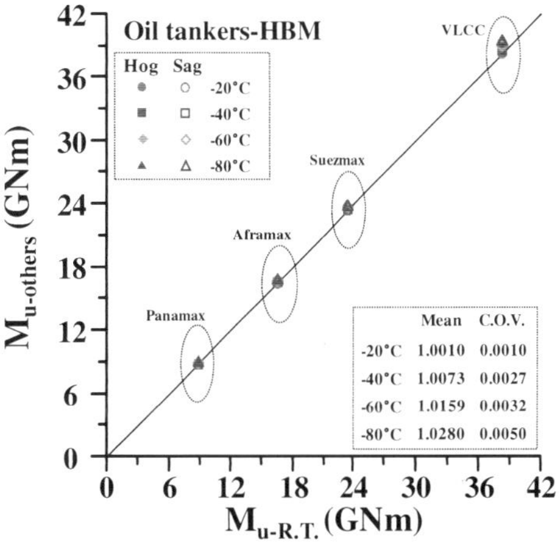

(b) Oil tankers – Horizontal bending moment

Figure 22: Effect of temperature variation on ultimate hull girder strength of oil tankers

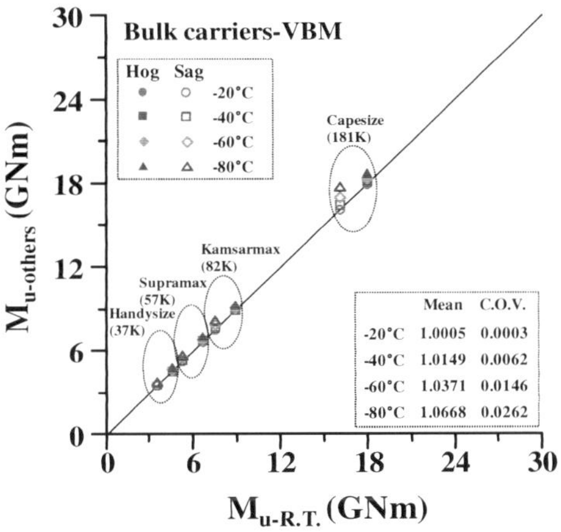

(a) Bulk carriers – vertical bending moment

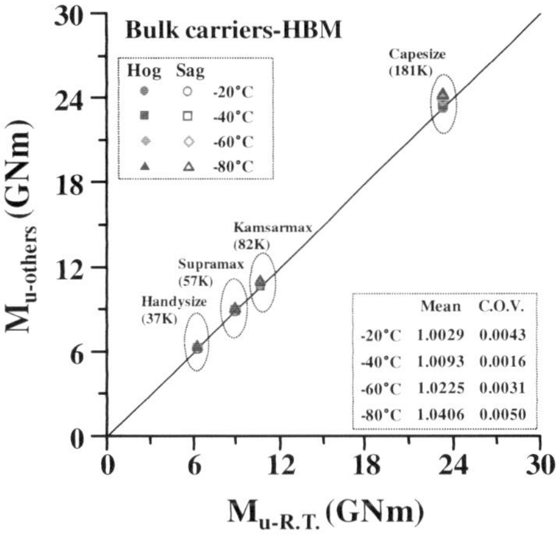

(b) Bulk carriers - horizontal bending moment

Figure 23: Effect of temperature variation on ultimate hull girder strength of bulk carriers

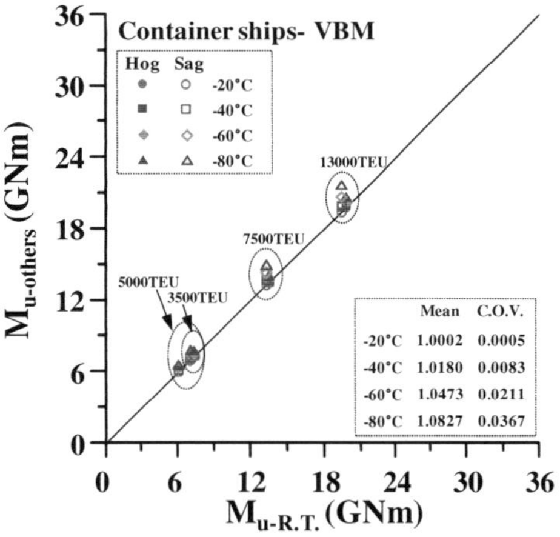

(a) Container ships – vertical bending momen
t

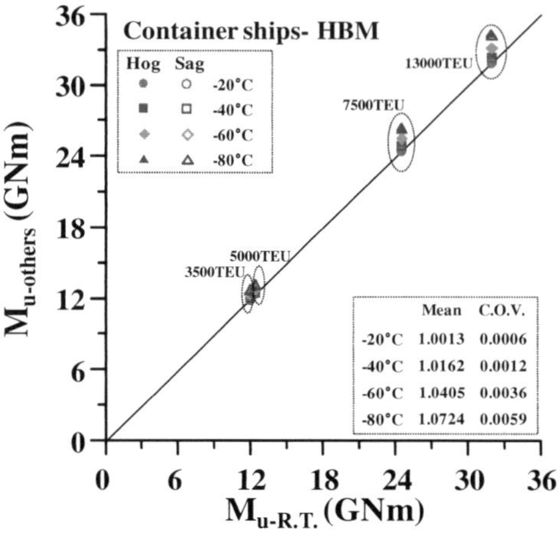

(b) Container ships – horizontal bending moment

Figure 24: Effect of temperature variation on ultimate hull girder strength of container ships

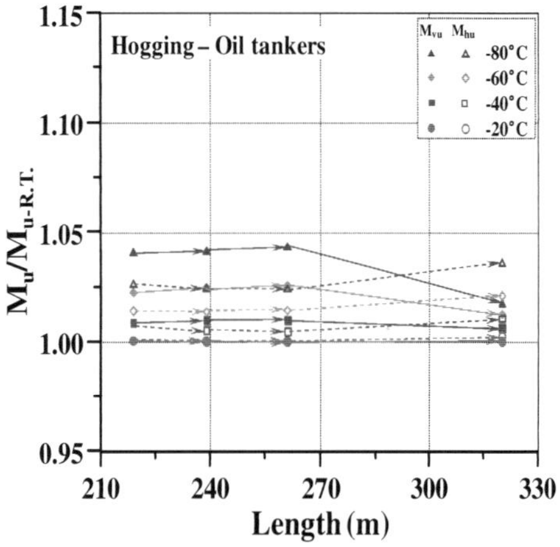

(a) In hogging

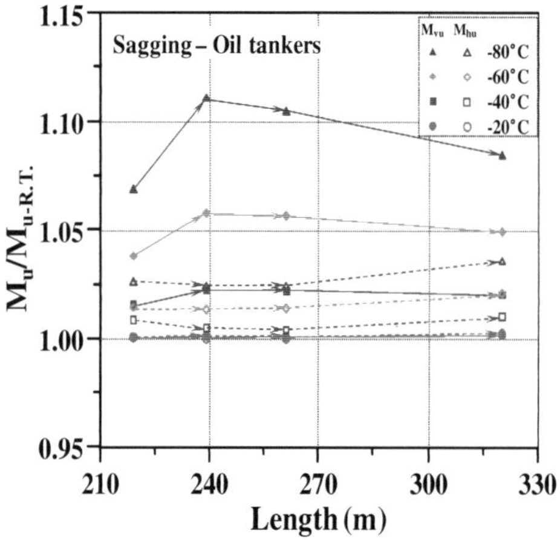

(b) In sagging

Figure 25: Trend of ultimate hull girder strength of oil tankers with vessel length

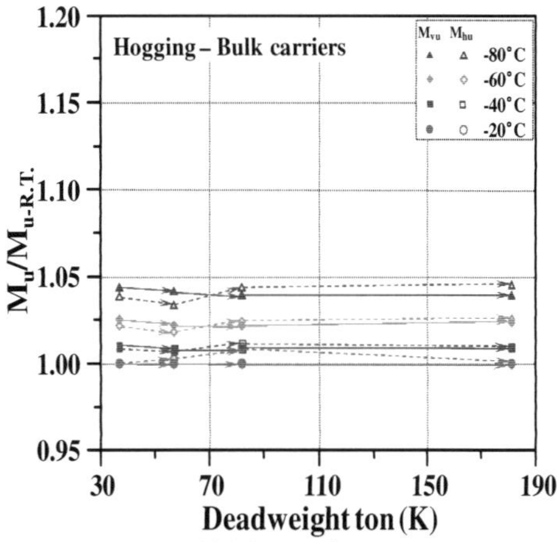

(a) Mean values

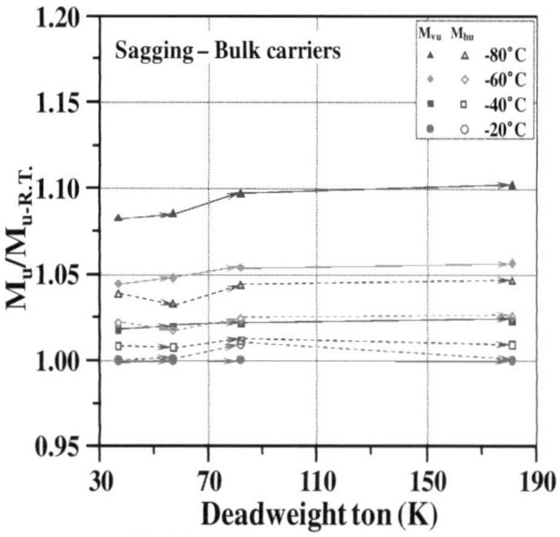

(b) Coefficient of variations

Figure 26: Trend of ultimate hull girder strength of bulk carriers with deadweight

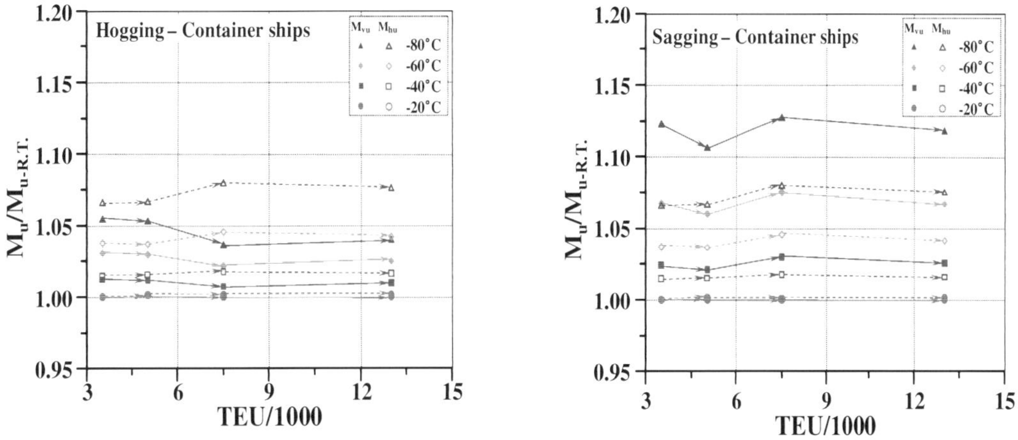

(a) In hogging (b) In sagging

Figure 27: Trend of ultimate hull girder strength of container ships with shipping capacity (TEU)

4.2 STIFFENED PANELS

Figures 28 to 30 show the variation of ultimate deck stiffened panel strength at different temperatures. Mean and C.O.V. values are calculated with temperature variation from the results. The mean values of ultimate deck stiffened panel strength in case I are same for the three types of ship, i.e., 0.1 to 13.5% for oil tankers, bulk carriers and container ships. The increase ratio is similar to that of material yield strength at low temperature for high tensile steel, as shown in the previous list as the deck stiffened panels are made of high tensile steel, as described in Figures 4 to 6.

The mean values in case III are at least 0.1 to 7.7% for oil tankers, 0.3 to 8.2% for bulk carriers and 0.1 to 12.2% for container ships. In other words, the low temperature effects are more sensitive for container ships than oil tankers and bulk carriers. Thick deck stiffened panels for container ships affected the ultimate deck stiffened panel strength.

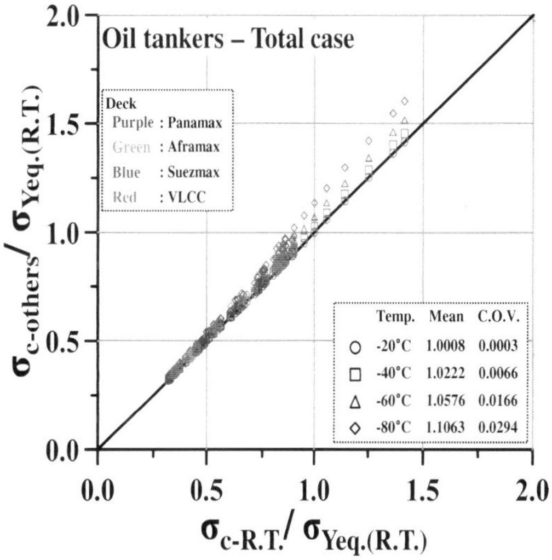

Figure 28: Effect of temperature variation on ultimate deck stiffened panel strength of oil tankers

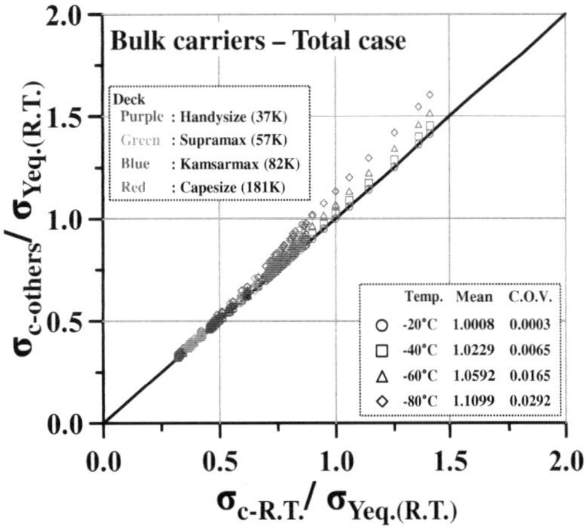

Figure 29: Effect of temperature variation on ultimate deck stiffened panel strength of bulk carriers

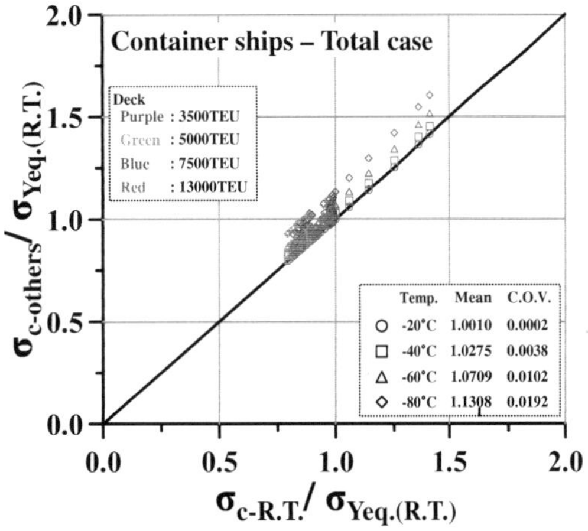

Figure 30: Effect of temperature variation on ultimate deck stiffened panel strength of container ships

Ultimate deck stiffened panel strength is more sensitive for oil tankers and bulk carriers when combination stress to equivalent stress increase in Figures 28 and 29. On the other hand, the ultimate deck stiffened panel strength is similar for container ships in all cases, as shown in Figure 30. The obtained results could be summarised as the following order:

Ultimate deck stiffened panel strength at low temperature:
Oil tanker < Bulk carrier < Container ship

5. CONCLUDING REMARKS

In the present study, operability studies of normal (non-ice class) vessels in Northern Sea Route (Arctic sea) are dealt. In total, 12 vessels, i.e., three types of general vessels namely oil tankers, bulk carriers and container ships are targeted. For each vessel type, four representative sizes are considered in order to investigate the general tendency of ultimate strength performance. A series of analyses for target vessels about global (hull girder) and local (stiffened panel) strength are performed with a consideration of low temperature effect.

The temperature distributions are assumed by general common sense as shown in Figure 3, and those distributions are applied to the vessels as shown in Figures 4 to 6. The obtained results are compared from the perspective of temperature, vessel size and vessel type.

The obtained results can be summarised as follows:

- Strength of materials increases as temperature decreases for each material. In consideration of Arctic sea navigation, the yield strength increases to 31.7% on Grade A (not allowed for ice class ships but generally used for normal ships) and 13.5% on high tensile steel (Grade AH and DH) at 80°C than those at room temperature.
- Temperature distribution for operation in Arctic Sea needs a practical service condition. In this study, reliable temperature distributions are applied according to the vessel types.
- The ultimate hull girder strength and the ultimate deck stiffened panel strength of three types of ships namely oil carrier, bulk carrier and container ship increase as the temperature decreases. Generally, the low temperature has greater effects on the ultimate hull girder strength under sagging than hogging and under vertical moment than horizontal bending moment. The ultimate hull girder strength increases at −80°C, 6.4% for oil tankers, 6.7% for bulk carriers, and 8.3% for container ships under VBM and 2.8% for oil tankers, 4.1% for bulk carriers, and 7.2% for container ships under HBM. This tendency to increase from material strength increases in quadratic function mode as like material strength did in material tests.
- In cases of the ultimate deck stiffened panel strength, the low temperature has greater effects than others on loading case I as combination stress increases. Moreover, larger ships tend to be sensitive to low temperature on container ships, while ship size has no definite effect on oil tankers and bulk carriers.
- The ultimate strength variation at low temperature with temperature distribution assumptions in this study is as the following order:
  Oil tanker < Bulk carrier < Container ship

The documented results will be useful to analyse the operability of various normal sea-going (non-ice class) ships for the voyage of Northern Sea Route (Arctic sea) from a point of ultimate strength performance.

Further studies on ice class ships and more practical temperature distribution methods with low temperature effect should be considered.

6. ACKNOWLEDGEMENTS

This research was supported by the Leading Foreign Research Institute Recruitment Programme through the National Research Foundation of Korea (NRF) funded by the Ministry of Science, ICT & Future Planning (MSIP) (Grant No.: 2014040731).

In addition, this research was supported by the Basic Science Research Programme through the National Research Foundation of Korea (NRF) funded by the Ministry of Education, Science and Technology (Grant No.: 2013R1A6A3A01057671).

This research was also supported by Lloyd's Register Foundation. The Lloyd's Register Foundation supports the advancement of engineering-related education, and funds research and development that enhances safety of life at sea, on land and in the air.

7. REFERENCES

1. IMO, Prevention of Air Pollution from Ships, MEPC 57/4/10, London, 2008.

2. IMO, Prevention of Air Pollution from Ships, Second IMO GHG study, MEPC 59/INF.10, London, 2009.

3. IMO, Prevention of Air Pollution from Ships, Second IMO GHG study, MEPC 61/INF.10, London, 2010.

4. Kwon, Y.I., 'Creation of conditions to attract transit cargoes to the far eastern ports', Far East Investment Congress (The 6th Pacific Economic Congress), Vladivostok, Russia, September, 2013.

5. DNV, Arctic Resource Development- Risks and Responsible Management, A Joint Report from FNI and DNV Prepared for the ONS Summit 2012, Høvik, Norway, 2012.

6. WIKIPEDIA, 'Arctic',http://en.wikipedia.org/wiki/Arctic, 2015.

7. WIKIPEDIA, 'Northern Sea Route', http://en.wikipedia.org/wiki/Northern_Sea_Route, 2015.

8. PARK, D.K., LEE, B.J., HAN, H.W., KIM, B.J., SEO, J.K. and PAIK, J.K., 'Material Properties and Crashworthiness of ASTM A131 Steel Plated Structures at Low Temperature: An Experimental and Numerical Study', The 12th International Symposium on Practical Design of Ships and Other Floating Structures(PRADS 2013), Changwon, Korea, October, 2013.

9. ASTM Standard A 131/A 131M–08, 'Standard Specification for Structural Steel for Ships', American Society for Testing and Materials (ASTM), New York, 2008.

10. KIM, D.K., PARK, D.K., SEO, J.K., PAIK, J.K. and KIM, B.J., 'Effects of low temperature on mechanical properties of steel and ultimate hull girder strength of commercial ship', Journal of Korean Institute of Metals and Materials, 50(6), 427-432, 2012.

11. ISF. International Shipping Federation (www.marisec.org).

12. KIM, D.K., PARK, D.K., KIM, J.H., KIM, S.J., KIM, B.J., SEO, J.K. and PAIK, J.K., 'Effect of corrosion on the ultimate strength of double hull oil tankers - Part I: stiffened panels', Structural Engineering and Mechanics, 42(4), 507-530, 2012.

13. KIM, D.K., PARK, D.K., PARK, D.H., KIM, H.B., KIM, B.J., SEO, J.K. and PAIK, J.K., 'Effect of corrosion on the ultimate strength of double hull oil tankers - Part II: hull girders', Structural Engineering and Mechanics, 42(4), 531-549, 2012.

14. KIM, D.K., PEDERSEN, P.T., PAIK, J.K., KIM, H.B., ZHANG, X.M. and KIM, M.S., 'Safety guidelines of ultimate hull girder strength for grounded container ships', Safety Science, 59, 46-54, 2013.

15. PAIK, J.K., KIM, D.K., PARK, D.H., KIM, H.B. and KIM, M.S., 'A new method for assessing the safety of ships damaged by grounding', International Journal of Maritime Engineering, 154(A1), 1-20, 2012.

16. ALPS/HULL, A Computer Program for Progressive Collapse Analysis of Ship Hulls, Proteus Engineering, Stevensville, MD, USA (www.maestromarine.com), 2012.

17. ALPS/ULSAP, A Computer Program for Ultimate Limit State Assessment for Stiffened Panels, Proteus Engineering, Stevensville, MD, USA (www.maestromarine.com), 2012.

18. PAIK, J.K., THAYAMBALLI, A.K., Ultimate Limit State Design of Steel-Plated Structures, Wiley, Chichester, UK, 2003.

19. KIM, M.S. and CHENG, Y.F., 'Limit states rules: Past experiences and future scenarios', The 11th International Marine Design Conference (IMDC 2012), Glasgow, UK, June, 2012.

20. KIM, Y.S., PARK, D.K., KIM, S.J., LEE, D.H., KIM, B.J., HA, Y.C., SEO, J.K. and PAIK, J.K., 'Ultimate Strength Assessment of Ship Stiffened Panel under Artic Conditions', Journal of the Society of Naval Architects of Korea, 51(4), 283-290, 2014.

8. AUTHORS BIOGRAPHY

Dae Kyeom Park is a postdoctoral researcher, at The Korea Ship and Offshore Research Institute (The Lloyd's Register Foundation Research Centre of Excellence), Pusan National University.

Yeon Chul Ha is an assistant professor at The Korea Ship and Offshore Research Institute (The Lloyd's Register Foundation Research Centre of Excellence), Pusan National University.

Bong Ju Kim and **Jung Kwan Seo** are associate professors at The Korea Ship and Offshore Research Institute (The Lloyd's Register Foundation Research Centre of Excellence), Pusan National University.

Professor Jeom Kee Paik is a Professor of the Department of Naval Architecture and Ocean Engineering and serves as president of the Korea Ship and Offshore Research Institute (KOSORI) and as director of the Lloyd's Register Foundation Research Centre of Excellence at Pusan National University (PNU) in Korea. In 2015, he also joined University College London's Department of Mechanical Engineering as a professor and faculty member. He is a Fellow and Vice President of SNAME, USA and a Fellow and Council Member of RINA, UK. He was awarded William Froude Medal from RINA in 2015, the David W. Taylor Medal from SNAME in 2013, and a Doctor Honoris Causa from the University of Liege in Belgium in 2012. His research interests are safety engineering and design of ships and offshore structures in extreme and accidental conditions associated with risk assessment and management.

PREDICTIONS OF ICE SCOUR LOADS AND RATE EFFECT EVALUATION FROM SMALL SCALE 1g TESTS

S Arnau and **A Ivanović,** University of Aberdeen, UK

SUMMARY

Ice scouring the seabed has been investigated by means of reduced scale physical model testing. The aim of this study is focused on the prediction of ice scour loads on a cohesionless seabed in saturated conditions. The influence of different soil parameters such as permeability and relative density have been studied over a range of scouring depths, widths and drifting velocities of an iceberg in the laboratory under 1g conditions. The effect of the drifting speed of an iceberg on the drag force was found to be rather important: an iceberg scouring the seabed with a mean drifting speed of 0.1 m/s can generate scour loads twice as big as the static loads. The methods used today to predict ice scour loads only consider the static loads, assuming drained conditions, and they should be revised as unrealistic loads can be obtained.

1. INTRODUCTION

Marine structures such as pipelines are at risk of being damaged when in contact with scouring ice features. The ice scour process begins when a floating ice mass with a deep keel is driven into shallow water by environmental forces such as winds, currents and waves. After contact with the seabed the ice continues to scour until the initial kinetic energy and the work done by driving forces is dissipated through work done on the seabed, or until the soil resistance exceeds the strength of the ice keel. In this last scenario, the ice feature can continue to scour the seabed but it is likely that the part of the ice feature in contact with the seabed will experience substantial modifications in shape which is known as ice keel degradation. An ice feature also has the ability to pitch and heave as indicated in Figure 1. In either case, the drag force associated with an ice feature scouring the seabed has to be determined and the accuracy of the model chosen to predict the maximum scouring depth of an ice feature will be dictated partly by the accuracy of the predicted scouring load. It is also known that during the movement of an iceberg the soil below will displace horizontally and vertically, known as sub-gouge deformation, as reported previously by [1] (Figure 1).

A number of studies have been undertaken to investigate the ice scouring process either through small scale, large scale or centrifuge tests and a compilation of these can be found in [2]. For this study, small rigid indentors have been used to simulate icebergs. A few ice scour analytical models exist in literature which attempt to explain the resistance forces associated with the movement of the iceberg while in contact with the seabed. These analytical models are used later on to obtain a gouge depth distribution based on the strength of the iceberg keel, energy dissipation, iceberg geometry and soil properties; the most common models referenced are those developed by [1], PRISE model [3], [4] and [5] where the scouring event is treated as a drained event (F_{static}).
[6] and [7] attempted to provide useful empirical equations to estimate the scour forces in undrained conditions (F_{static} + $F_{dynamic}$) from small scale laboratory tests. Multiple linear regression analysis from laboratory results were performed in order to gain further insight into the measured force's dependence on iceberg velocity (v), scouring depth (D) and scour width (B). The regression analysis yielded the following relationship represented in Eq. (1), where the scour load (F) is a function of the scour depth (D), scour width (B), iceberg velocity (v) and the mean grain size D_{50}.

$$F = 13.2v + 33.4D + 11.6B - 0.6D_{50} - 2.4 \qquad (1)$$

The rate effect occurs because the process of sand cutting is accompanied by a tendency for the particles of sand to move apart and for the void space between particles to increase. The increase of the void space is filled by additional pore fluid.

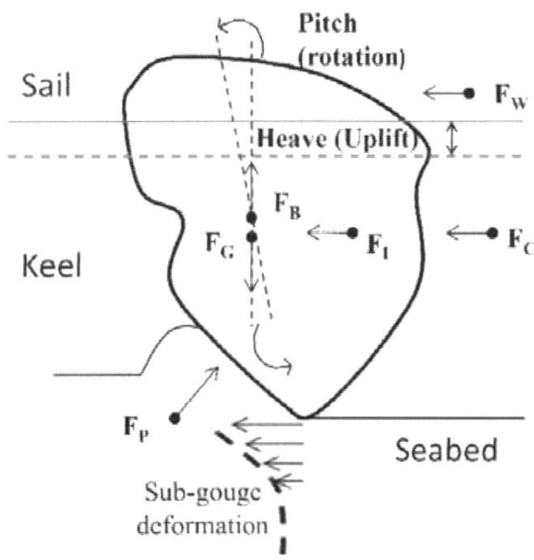

Figure 1: General schematic representation of the ice gouging process. F_W: wind, F_C: water current, F_P: seabed resistance, F_G: weight, F_B: ice buoyancy, F_I: inertial force.

The pore fluid flows from the surrounding soil governed by a pressure gradient which finally increases the effective stresses in the soil [8]. The work carried out by [9] on the performance of pipeline ploughs, where the rate effect for this particular component is studied, has been the starting point to analyse the rate effect during ice scouring (see Eq.

2). It is thought that because Eq. (2) was obtained from the performance of pipeline ploughs where no kinematic wedges were observed in the failure mechanism of the soil, the constant C will differ from the original study. This is the reason why the present study is concentrated in changing the geometry of the object being dragged (rake angle α, scour depth D, scour width B) rather than test with different soils, which was the focus of [9].

$$\frac{F_{Dynamic}}{D^3} = C \times \frac{vD[\Delta e/(1+e_0)]}{C_V} \quad (2)$$

where $F_{dynamic}$ is the total measured drag force minus F_{static} ; D is the scour depth (associated with the drainage path length); v is the velocity of the object penetrating the soil; e_0 is the initial void ratio and Δe is the void ratio increase from initial conditions to e_{crit} (associated with the dilation potential S); C_V can be calculated with Eq. (3) as reported by [10]; and C is a constant and is thought to be a function of the object's geometry penetrating the soil and also a function of the shear band configuration.

$$C_V = \frac{E_0 k}{\gamma_w} \quad (3)$$

where E_0 is the one-dimensional stiffness modulus; k is the permeability; γ_w is the unit weight of water.

2. EXPERIMENTAL SETUP

The scaled models of an iceberg keel were tested using a 4.8 m long and 0.8 m wide and 0.3 m deep sand channel, a facility based at the University of Aberdeen where the drag force and the associated soil disturbance were measured (Figure 2). In order to measure the drag force, 'full' models were towed along the full channel width (Figure 2b) and a custom-made load cell, with a maximum load capacity of 1000 N, was used to obtain the reaction force in the vertical and horizontal direction. In order to monitor shearing zones of the sand and sub-gouge deformation (not reported here) as the iceberg model is moving, the channel is divided into two parts by means of a plexiglass wall (Figure 2a). Particle Image Velocimetry technique (PIV) is then used to study the sand deformation pattern. The software utilised to perform the PIV analysis was GeoPIV [11]. Sand 'pluviation' was used before each test in dry conditions to ensure a uniform relative density across the full channel lengths. The sand characteristics are indicated in Table 1.

Three different models widths (B) of 150 mm, 250 mm and 500 mm were used in dry and saturated conditions to study how the drag force changes with the width of the object being dragged. The rake angle α of the three models is 150° (see Figure 4). A set of tests related to the former were undertaken at fixed penetration (fixed vertical displacement) depths ranging from 1 cm to 4 cm. The choice of this range was limited by real scouring conditions as a ratio between the scour width B and scour depth D (e.g. aspect ratio B/D) is usually around 30 [12] and for the thickness of the shear band which poses a limitation for very shallow tests [13]. Additionally, two more scale models (150 mm wide) with a rake angle α of 120° and 90° were also used to make the force-prediction model more robust.

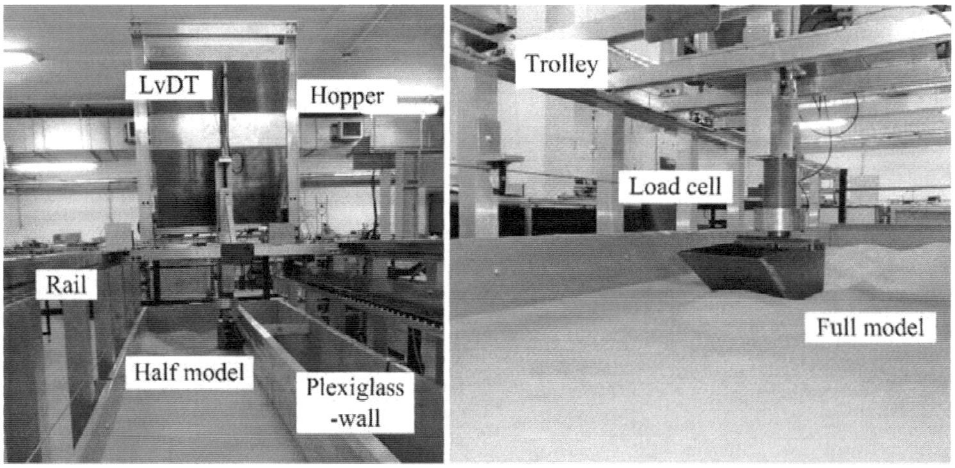

a b

Figure 2: Test facility for ice scouring tests. (a) Half model and PIV set-up. (b) Full model set-up.

When water was introduced in the experiments (saturated conditions), the seabed preparation method was varied for a vibration device allowing the samples to be at relative densities ranging from 45% up to 90% in order to test at different permeability (k) and dilation potential (S). Additionally, differential pore water pressure transducers were buried in the seabed at different depths to capture the pore water pressure reduction near the dilating shear band (see Figure 4). The carriage where all the electronic devices were mounted on and used to attach the iceberg model can travel at different speeds ranging from 0.27 to 0.01 m/s. All the instrumentation was monitored using a USB data acquisition system with data sampling at 0.1 s intervals.

3. EXPERIMENTAL RESULTS

A typical behaviour of an iceberg scale model scouring a seabed can be described as follow: after the initial contact with the seabed, forces increase steadily as the sand is being accumulated in front of the model. This transient process ended when the sand cleared towards the sides equals the sand being pushed in front of the model, then the force become steady and starts to fluctuate as the successive shear bands are being formed and overridden by the next shear band. The successive creation of shear bands were studied in plane strain conditions using PIV where kinematic wedges were observed within the soil failure pattern. A schematic representation of the soil failure mechanism is presented in Figure 4.

3.1 RESULTS AND OBSERVATIONS FROM DIRECT SHEAR BOX TEST

Direct shear box tests were used to obtain several soil parameters used later on in this study: the angle of internal friction at low and high normal stresses and the critical state line associated with e_{crit} was obtained for a range of normal effective stresses and initial void ratio e_0.

The methodology used to obtain the critical state line from a shear box test, see Eq. (4) and Figure 3, is based on that described in [9]. Two sets of data are represented in Figure 3, one corresponding to an initial void ratio $e_0 = 0.54$ (dense) and the other at $e_0 = 0.76$ (loose). The assumed thickness of the shear band ($28d_{50}$) was selected as it caused the values of e_{crit} from test at different e_0 to converge. This slightly differs from the assumed thickness proposed by [9] although it is still in the range of $20\text{-}40d_{50}$, where $40d_{50}$ is twice the thickness of the shear band.

$$e_{crit} = e_0 + \frac{\Delta H}{28 d_{50}}\left(1 + e_0\right) \tag{4}$$

where e_{crit} is the void ratio at which no further change in volume and stresses occur; e_0 is the initial void ratio at what the sample is prepared; ΔH is the increase in height of the sample; and d_{50} is the particle size diameter. A constant critical friction angle φ_{crit} of 32° was obtained for a range of normal effective stresses and initial void ratio e_0. The peak friction angle φ_{peak} was found to vary with stresses as indicated by [14]. At very low effective stresses high levels of dilation were found as represented in Figure 3.

3.2 ANALYTICAL PREDICTION OF ICE SCOUR LOADS (F_{static})

From the tests in dry conditions it was possible to study the static, non-rate-dependent, component of the scour force. In this case, the scour depth and width and the sand accumulated in front of the iceberg model as it moves, combined with the angle of internal friction were found to be the key parameters that define F_{static}. The theory of kinematic boundary wedges described by [15] has been implemented by other researchers to estimate the ice scour loads in drained conditions (e.g. [1] [3] [5]). The same methodology was used in this study to calculate the scour loads in drained conditions (F_{static}) and a detailed explanation of how to compute F_{static} for a range of iceberg geometries and soil properties can be found in [13].

Figure 4 represents the geometry of the failure mechanism created in front of an iceberg model with the frontal face inclined at an angle α = 150°. As can be observed, a kinematic wedge is now represented in front of the frontal face of the model and a perfectly rough interface is created with a rake angle of (90°-φ). The same type of failure mechanism can be created for the scenario where α = 90° and α = 150°. From this type of soil failure mechanism with kinematic wedges derives that the basal shear component Fr (the frictional force generated at the base of the kinematic wedge) is the main contributor of the gouge force (F).

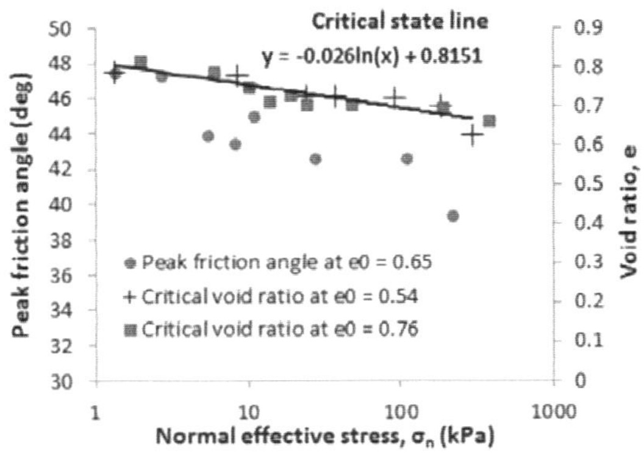

Figure 3: Direct shear box test results. Increase in friction angle with reduction of the normal effective stress and representation of the critical state line.

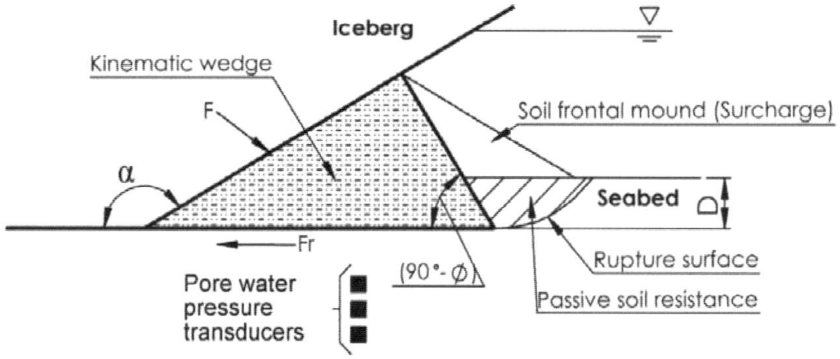

Figure 4: Assumed soil failure mechanism for sand in front of an iceberg model.

3.3 NORMALIZATION OF THE DYNAMIC COMPONENT OF THE SCOUR FORCE ($F_{dynamic}$)

Figure 5 to Figure 8 represent the influence of parameters such as scour depth *D*, scour width *B*, drifting speed *v*, permeability *k* and initial void ratio e_0 on $F_{dynamic}$. The same tests carried out in dry conditions were repeated in undrained conditions (sand saturated with water) so that $F_{dynamic}$ can be finally obtained as indicated in Eq. (2). Figure 5 shows the influence of the scour velocity *v* (or drifting speed) on the scour force for different scour depth in dense sand. The first finding is observed in Figure 6 which is constructed using the slope of the rate effect *dF/dv* defined by the trend lines in Figure 5. Figure 6 shows that for a rigid metal box with inclined frontal face, the increase in force is rather linear with the scour depth *D*. Tests were also carried out at different scour widths *B* and it was found that the rate effect increase linearly with ($B^{1.5}$) as represented in Figure 7. This differs from Eq. [1] which indicates that scour loads increase linearly with the object's width (*B*). A possible explanation for this disagreement could be that the aspect ratios (*B/D*) used to calibrate Eq. (1) were close to 1 (or even smaller) whereas aspect ratios used for this present study were always kept above 5 to mimic real ice scour scenarios. Also, as reported by [16] this type of low aspect ratios can lead to a soil failure mechanisms which are not fully developed. Additionally, Eq. (1) was derived from laboratory studies where only two different scour widths were utilized (100 mm and 200 mm) whereas for the present study three different widths (150, 250 and 500 mm) were used. This disagreement in how the scour force varies with the object's width is not very important at small scales but when forces are extrapolated to real scour scenarios the error also increases exponentially.

Figure 8 represents all the data obtained from the tests at different towing speeds, scour depths, scour widths, initial void ratios and rake angles. As indicated by [9] the horizontal axis has been normalised using the dimensionless number (vDS/C_V).

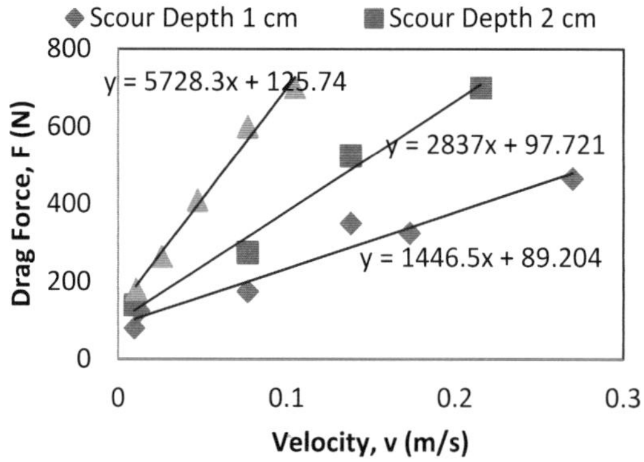

Figure 5: Scour force (F) measured in saturated conditions for a 0.152 m model width and different scour depths (D) and towing speeds (v).

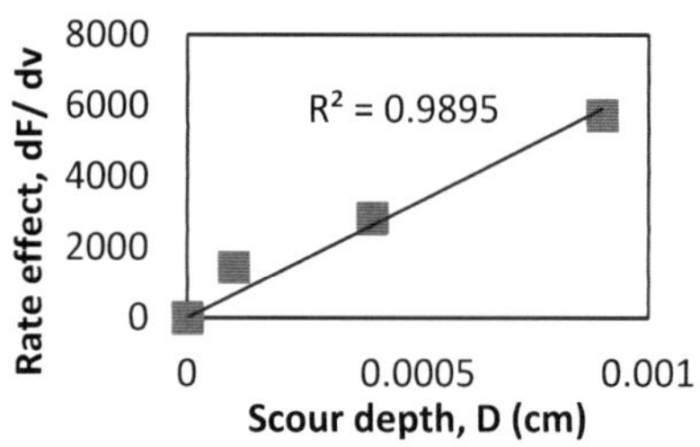

Figure 6: Rate effect (dF/dv) vs scour depth (D).

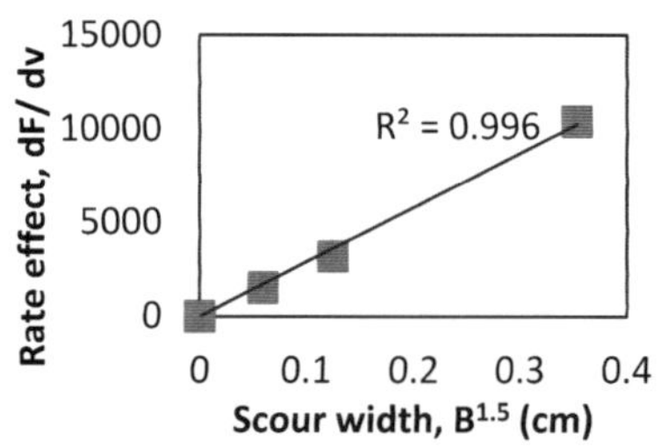

Figure 7: Rate effect (dF/dv) vs scour width (B).

There is also a need to compare tests not only carried out at different scour depths (D) but also at different scour widths (B) and this is why the force on the vertical axis in Figure 8 has been normalised using $B^{1.5}$. As mentioned earlier on, the constant C in Eq. (2) is thought to be a function of the object's geometry. This is observed in Figure 8, where tests with three different rake angles (α) are represented. Although there is no much difference for α = 120° and 150° a change on the slopes of the three regression lines is observed, especially for α = 90°.

3.4 SCALING UP THE RATE EFFECT ($F_{dynamic}$)

When scaling up the rate effect by means of the dimensionless number (vDS/Cv) some soil parameters need to be determined at the same stress level as the prototype. For instance, Figure 3 can be used to determine e_{crit} at higher stress level which in turn will reduce the dilation potential S. The one-dimensional soil stiffness E_0 is also stress dependent and its value will differ from the model scale. For a 15 m wide iceberg with a rake angle α = 150° scouring a cohesionless seabed at 2.1 m depth and at two different drifting speeds of 0.03 and 0.1 m/s, (vDS/Cv) works out to be 0.2 and 0.42 respectively. According to Eq. (5), which was extracted from Figure 8, the $F_{dynamic}$ for this two drifting speeds is 12.1 MN and 25.4 MN respectively. F_{static} for this particular scouring scenario is 26MN. Then, the total scour force will be 38.1 MN for an iceberg scouring the seabed at 0.03 m/s and 51.4 MN if the drifting speed is increased to 0.1m/s.

There are few well documented centrifuge test results in saturated sand conditions which were undertaken during PRISE program [17] and the results are plotted in Figure 9. For this set of tests, two drifting speeds were utilized: 0.03 m/s and 0.1m/s. Furthermore, the soil utilized was also very similar to the one used for this present study and the soil properties for the two different studies are found in Table 1. Figure 9 compares the scaled up scouring force obtained during PRISE studies and those calculated by means of Eq. (5) for two scour width B of 15 m and 30 m at a scale prototype over a range of scour depth (D).

$$\frac{F_{total} - F_{static}}{B^{1.5}} = 104 \cdot 10^4 \times \frac{vDS}{C_V}$$

4. CONCLUSIONS

Reduced scale models of icebergs were used in a large sand channel with carefully prepared sand beds to investigate the rate effects of an iceberg scouring a cohesionless seabed at different drifting speeds. The results were compared with existing methods for rate effect evaluation during ploughing and the right modification to these methods have been presented here to describe the rate effect during ice scouring.

It was found that the rate effect increases linearly with D (scour depth) and $B^{1.5}$ (scour width), this differs for ploughing and is thought to be consequence of a different geometry and shear band configuration. Three different rake angles were investigated (150°, 120° and 90°) which in turn affect the $F_{dynamic}$ prediction and pore water pressure reduction. Figure 8 shows that when α = 150° the rate effect increases more rapidly than for the case when α = 90°.

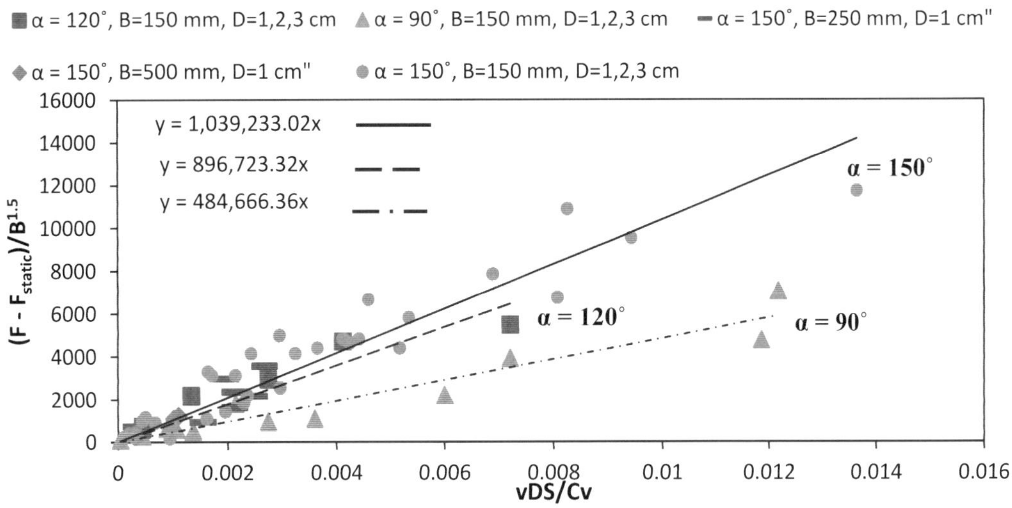

Figure 8: Normalized drag force against vDS/C_v which is a dimensionless number grouping different towing speeds (v), scour depths (D), initial void ratios e0. C_v is a function of permeability (k), one dimensional stiffnes modulus (E_0) and the soil unit weight (γ).

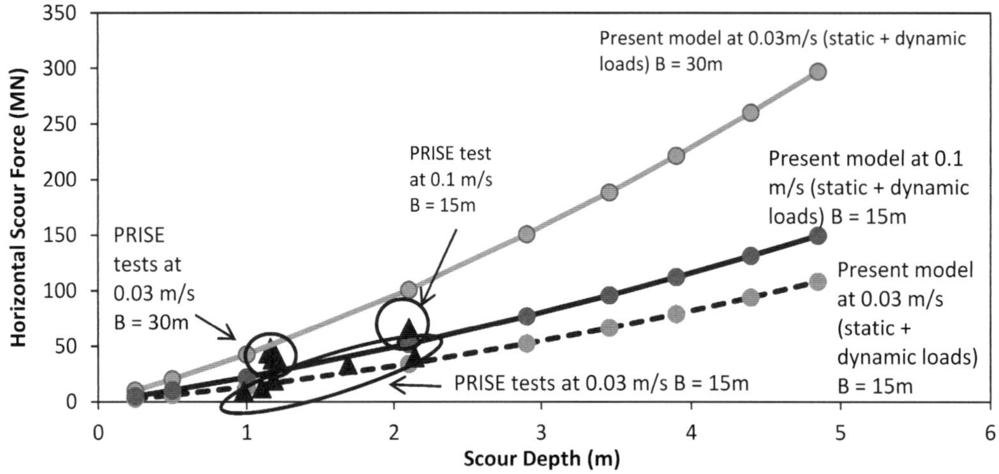

Figure 9: Comparison of measured (PRISE study) and calculated scouring forces in saturated conditions for sand at two drifting speeds, 0.03 m/s and 0.1 m/s, and two scouring widths (B), 15 m and 30m, for a range of scouring depths (D).

Table 1: Summary of the material properties

Parameter	For this study	PRISE study
Mean grain size (D_{50})	0.177 mm	0.140 mm
Effective grain size (D_{10})	0.126 mm	0.100 mm
Specific gravity	2.66	2.66
Uniformity coefficient	1.43	1.50
Maximum Unit weight (γ)	17.2 kN/m^3	17.4 kN/m^3
Minimum Unit weight (γ)	14.47 kN/m^3	13.8 kN/m^3
Critic friction angle (φ_{critc}) Peak friction angle (φ_{peak}) (at prototype scale) Peak friction angle (φ_{peak}) (at model scale)	32° 38° 50°	32° 36.5° -
Repose angle (β)	30°	-
e_{min} e_{max} Potential dilation (S) (at prototype scale) (dense)	0.52 0.77 0.16	-
E_0 Loose (at low stresses) E_0 Dense (at low stresses) E_0 Loose (at 100kPa normal stress) E_0 Dense (at 100 kPa normal stress)	6000 kPa 9000 kPa 10000 kPa 13000 kPa	-
k Loose k Dense	$1.2 \cdot 10^{-4}$ m/s $6 \cdot 10^{-5}$ m/s	-

In the same way as proposed by [9] an empirical relationship, Eq. (5), has been presented here to determine the dynamic component of the scouring force. Figure 8 represents the rate effect for different soil parameters, which are stress dependent, and also for different scour widths (associated with the scale of an iceberg). Thus, Eq. (5) can be used to estimate scouring forces at prototype scale. If Eq. (1) is used to compute the scouring loads at prototype scale the computed forces are between one and two orders of magnitude smaller and therefore unrealistic. A possible explanation for this disagreement is that the scouring loads in Eq. [1] increase linearly with the scouring width whereas in this study the scouring loads were found to increase proportionally to $B^{1.5}$.

Finally, current models to predict scour loads in a cohesionless seabed only take into account the static component and assume drained conditions. This implies that if such models are used to compute scour forces on a seabed with rather low friction angle but high permeability, the resultant forces could be underestimated.

5. REFERENCES

1. BEEN, K., KOSAR, K., HACHEY, J., ROGER, B.T. & PALMER, A.C., 'Ice scour models', *Proceedings of the 9th International Conference on Offshore Mechanics and Artic Engineering, Houston, Volume 5*, pp. 179, 1990b.

2. BARRETTE, P., 'Offshore pipeline protection against seabed gouging by ice: An overview', *Cold Regions Science and Technology,* vol. 69, no. 1, pp. 3-20, 2011.

3. WALTER, D.J. & PHILLIPS, R., 'PRISE- Force models for drained and undrained steady state ice

scouring', *Contract report, C-Core Publication 98-C33*, 1998.

4. KIOKA, S., ABE, S., SAKAI, H. & SAEKI, H., 'Mechanical model of ice scour', *Proceedings of the 15th international conference on port and ocean engineering under artic conditions (POAC), Helsinki.*, pp. 537, 1999.

5. CROASDALE, K., COMFORT, G. & BEEN, K., 'Investigation of ice limits to ice gouging', *Proceedings of the 18th International Conference on Port and Ocean Engineering under Artic Conditions (POAC), Potsdam*, pp. 23, 2005.

6. BARKER, A. & TIMCO, G., 'Laboratory experiments of ice scour processes: rigid ice indentor', *Cold Regions Science and Technology,* vol. 35, pp. 195-206, 2002.

7. BARKER, A. & TIMCO, G., 'Laboratory experiments of ice scour processes: buoyant ice model', *Cold Regions Science and Technology,* vol. 36, pp. 103-114, 2003.

8. PALMER, A.C., 'Speed effects in cutting and ploughing', *Geotechnique,* vol. 49, no. 3, pp. 285-294, 1999.

9. LAUDER, K.D., BROWN, M.J., BRANSBY, M.F. & GOODING, S., 'Variation of tow force with velocity during offshore ploughing in granular materials', *Canadian Geotechnical Journal,* vol. 49, pp. 1244-1255, 2012.

10. FINNIE, I.M.S., 'Performance of shallow foundations in calcareous soils', *Ph.D. thesis*, University of Western Australia, Perth, Australia, 1993.

11. WHITE, D.J., TAKE. W.A. & BOLTON, M.D., 'Soil deformation measurement using particle image velocimetry (PIV) and photogrammetry', *Geotechnique,* vol. 53, no. 7, pp. 619-631, 2003.

12. KING, T., PHILLIPS, R., BARRETT, J. & SONNICHSEN, G., 'Probabilistic pipeline burial analysis for protection against ice scour', *Cold Regions Science and Technology,* vol. 59, no. 1, pp. 58-64, 2009.

13. ARNAU, S. & IVANOVIC, A., 'Predictions of ice scour loads from small scale tests under 1g conditions', *Proceedings of the third international symposium on frontiers in offshore geotechnics (ISFOG) , Oslo.*, ed. V. Mayer, , 10-12 June, pp. 933, 2015.

14. BOLTON, M.D., 'The strength and dilatancy of sands', *Geotechnique,* vol. 36, no. 1, pp. 65-78, 1986.

15. HETTIARATCHI, D.R.P. & REECE, A.R., 'Boundary wedges in two-dimensional passive soil failure', *Geotechnique,* vol. 25, no. 2, pp. 197-222, 1975.

16. HETTIARATCHI, D.R.P. & REECE, A.R., 'Symmetrical three-dimensional soil failure', *Journal of Terramechanics,* vol. 4, no. 3, pp. 45-67, 1967.

17. WINSOW, R. & PARSON, G., 'Pressure ridge ice scour experiment phase 3c, Extreme ice scour event - Modeling and interpretation, Milestone 2: Centrifuge test PRSA02 data report', *Contract report for BP Alaska, Chevron Petroleum Technology, Exxon Production Research Company, Petro-Canada and Union Texas Petroleum, C-CORE Publication 97-C6.* 1997

6. AUTHORS BIOGRAPHY

Sergi Arnau, PhD candidate at Aberdeen University, UK. Responsible for the project related to ice scouring.

Ana Ivanović, PhD, Senior Lecturer at the University of Aberdeen with extensive research in soil structure interaction with applications ranging from icebergs, fishing gear elements and ground anchorages.

PRESSURE-IMPULSE DIAGRAM FOR PREDICTION OF STRUCTURAL DAMAGE TO THE LNG FPSO TANKS UNDER SLOSHING IMPACT LOADS

SE Lee, BJ Kim, JK Seo, YC Ha, JK Paik*, The Korea Ship and Offshore Research Institute (The Lloyd's Register Foundation Research Centre of Excellence) at Pusan National University, Korea
T Matsumoto, ClassNK, Japan
SH Byeon, STX Offshore and Shipbuilding Co., Ltd, Korea
(*Also, Department of Mechanical Engineering, University College London, London, UK)

SUMMARY

Sloshing impact loads can cause severe structural damage of cargo tanks in LNG-FPSOs (FLNGs). Sloshing studies can be classified into two groups, namely hydrodynamics- and structural mechanics-related studies. The present study is a sequel to the authors' previous papers [1-2] while it is mainly related to the structural mechanics. In the present paper, the new method for probabilistic sloshing assessment previously developed by the authors is briefly addressed. The nonlinear impact structural response characteristics under sloshing impact loads are examined by the nonlinear finite element ANSYS/LS-DYNA method. An ISO-damage curve representing the pressure-impulse diagram is derived for self-supporting prismatic shape LNG cargo containment system of a hypothetical FLNG. The insights developed from the present work can be useful for damage-tolerant design of cargo tanks in FLNGs.

NOMENCLATURE

C	Cowper-Symonds coefficient
D	Maximum deformation level (mm)
I_a	Impulsive asymptote (bar·s)
I_o	Impulse (bar·s)
K	Spring stiffness (N/mm)
T	Natural period (s)
P_a	Quasi-static asymptote (bar)
P_o	Peak pressure (bar)
q	Cowper-Symonds coefficient
t_{dec}	Decay time of an impact action (ms)
t_{dur}	Duration time of an impact action (ms)
t_{dur}'	Idealized duration time of an impact action (ms)
t_o	Rise time of an impact action (ms)
x_{max}	Maximum dynamic displacement (mm)
α	Constant of a fitting equation
β	Constant of a fitting equation
$\dot{\varepsilon}$	Strain rate (mm/s)
ε_f	Fracture strain under a static load (mm)
ε_{fd}	Fracture strain under a dynamic load (mm)
ν	Poisson's ratio
ρ	Density of a material (kg/m^3)
σ_Y	Yield stress under a static load (N/mm^2)
σ_{Yd}	Yield stress under a dynamic load (N/mm^2)
ω	Natural circular frequency (rad/s)

1. INTRODUCTION

Tank sloshing in a liquefied natural gas floating production storage offloading unit (LNG-FPSO or FLNG) is unavoidable due to continuous loading and unloading processes together with the effect of winds, waves and currents. Often, repetitive sloshing impact loads can cause severe structural damage of LNG cargo containment systems (CCS) and their sub-structures.

Sloshing studies can be classified into two groups, hydrodynamics-related studies that aim to identify the impact pressure profile over time and structural mechanics-related studies that aim to calculate the dynamic structural damage. To quantify structural damage under sloshing impact loads, a number of researches have been extensively conducted over the last 50 years. As a result, some of classification societies, shipyards and 3rd part authorities proposed their own assessment procedure[3-13]. It was, however, recognized that sloshing impact loads obtaining from current procedures might not represent a whole set of credible sloshing scenarios since it is conducted under the specific environmental conditions or a limited range of tank resonant periods.

To improve the sloshing scenario sampling, the authors previously proposed the fully probabilistic design procedure from how to select credible scenario probabilistically to how to determine a design sloshing load[1] and the nonlinear dynamic structural analysis under actual and design sloshing loads was followed[2]. As a sequel, the present study applies a new approach utilizing pressure-impulse (P-I) diagram for assessing a damage extent of cargo tanks in FLNG. P-I diagram method typically had been used to the field of animal and human injuries, and structural damage subjected to blast loads[14-28] by utilizing a single degree of freedom (SDOF) model[20-25] and a finite element method (FEM)[24-28]. In the present paper, the authors try to examine a feasibility of P-I method to quantify structural damage of cargo tanks in FLNG.

The aims of this study are to briefly address the developed probabilistic design approach and to derive the P-I diagram for the maximum deformation level based on the numerical results of a nonlinear FEM (NLFEM). Using ANSYS/LS-DYNA method, the nonlinear impact structural response characteristics under sloshing impact loads are examined. Through a series of parametric studies, the P-I diagram is finally derived.

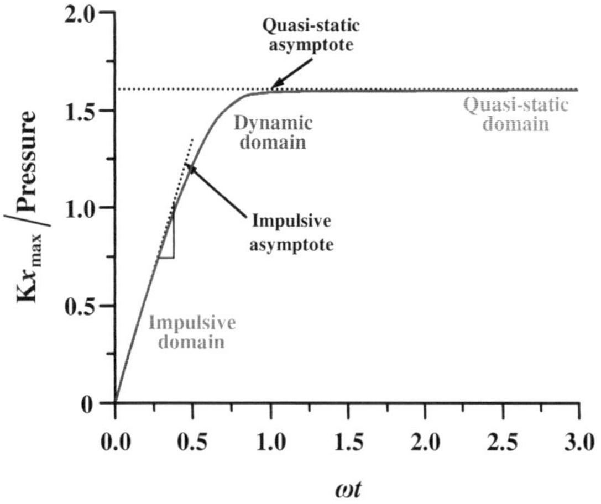

Figure 1: Typical response spectrum.

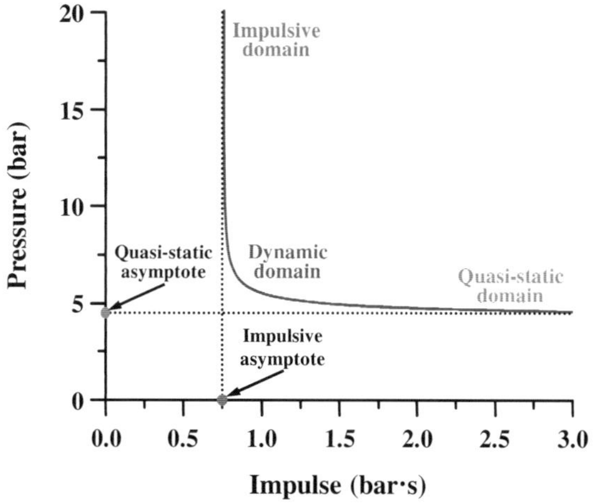

Figure 2: Typical pressure-impulse diagram.

2. CHARACTERISTICS OF P-I DIAGRAM

To the structural designers, either maximum deflection or stress at the end of the loading states is a primary object for assessing loading capacity of a target structure. On the other hand, in structural design of dynamic system, time-histories of the dynamic response are of significant. For practical design purposes, a structural behaviour subjected to an impact-pressure action can be idealized within three domains depending on a ratio of a duration time of the impact action, t_{dur} to a natural period of a structure, T as follow[29]:

- Quasi-static domain: $t_{dur}/T \geq 3$
- Dynamic domain: $0.3 \leq t_{dur}/T < 3$
- Impulsive domain: $t_{dur}/T \leq 0.3$

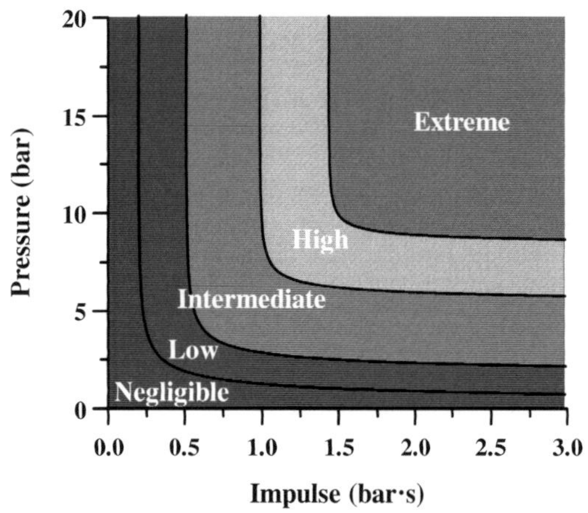

Figure 3: Damage level of P-I diagram.

By plots known as a response spectra illustrating a maximum peak response, x_{max} to abovementioned three domains, the structural behaviour under given loading can be simplified as shown in Fig. 1. There are various forms of response spectra plots describing the relationships between the maximum value of a response parameter and a characteristic of the dynamic system. Among them, the P-I diagram as a representative of a response spectrum is widely used for the structural damage assessment as shown in Fig. 2.

Compared to the typical response spectrum plot, benefits of the P-I diagram are easily differentiating behaviour domains, in a form of impulsive and quasi-static asymptotes. For estimating structural safety of the dynamic system under combined pressure-impulse actions, the left and below a basis level indicate no specified damage, while those to the right and above mean damage in excess of it. As shown in Fig. 3, if the dynamic system shows multiple damage-levels in the P-I diagram, each of them represents damage severity from negligible to extreme according to a damage criteria. In this regards, the P-I diagram method can be an efficient approach to quantify structural damage levels and will systematically help structural engineers identify the safety and structural capacity for an expected impact pressure action.

3. PROBABILISTIC DESIGN PROCEDURE

As shown in Fig. 4, this chapter briefly summarizes the developed probabilistic design procedure including how to determine the design sloshing load, to examine nonlinear structural response, and to derive the P-I diagram of the FLNG CCS.

3.1 DESIGN SLOSHING LOADS

This section addresses the *Design Stage I* which was presented in the authors' previous study[1] aiming to determine nominal values for tank sloshing loads in structural design of FLNG. The applicability of the method was demonstrated using an applied example of a hypothetical FLNG operating in a natural gas site off a

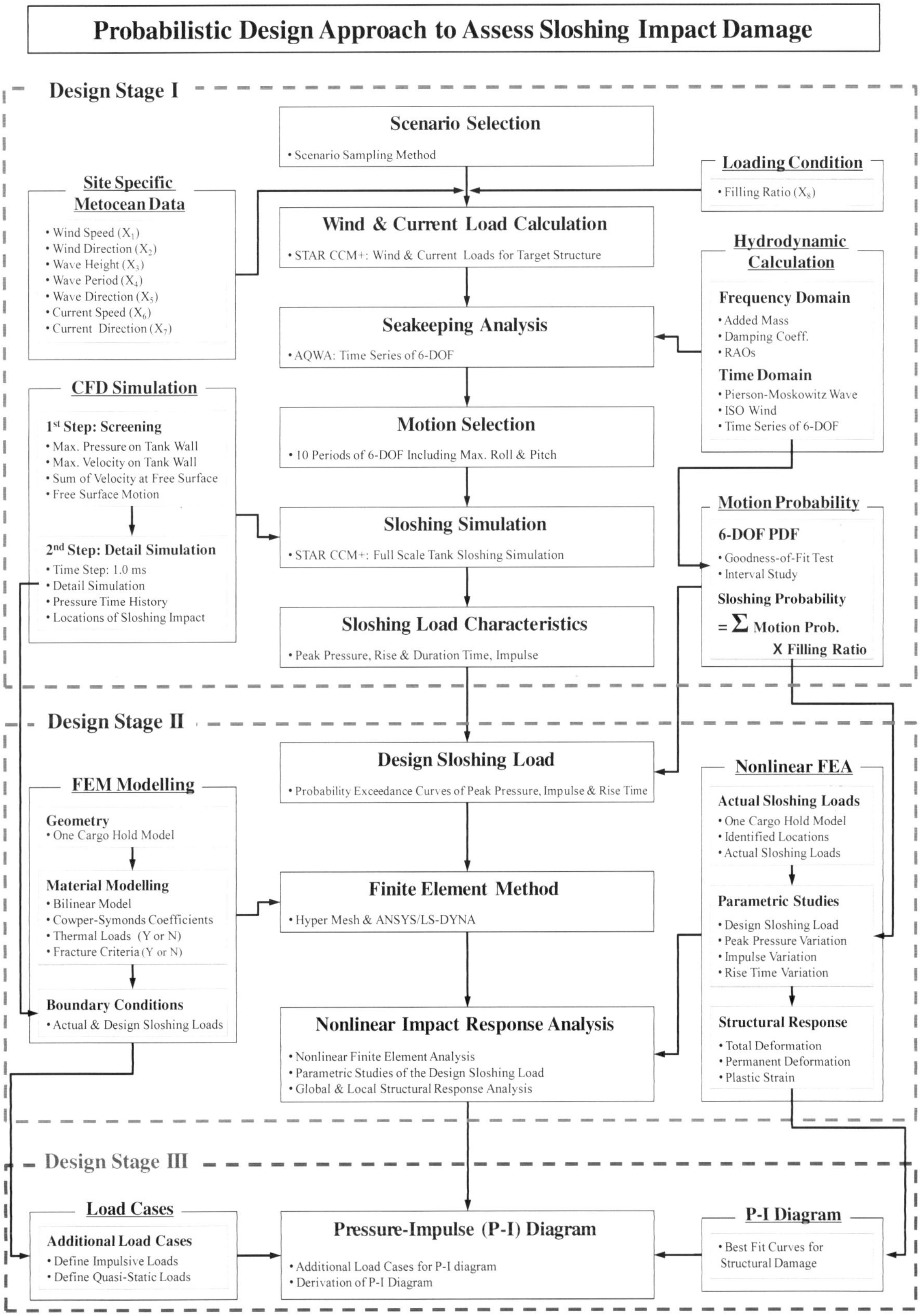

Figure 4: Procedure of a probabilistic design approach to assess structural damage under sloshing impact loads.

hypothetical oceanic region. As a result, sloshing probability of exceedance curves for peak pressure, impulse and rise time were suggested[1]. Proposed procedures are as follows:

- Credible sloshing scenario selection by a probabilistic sampling method
- Wind and current loads using computational fluid dynamics (CFD) for all directions
- Seakeeping analysis in frequency- and time-domain taking into account winds, waves and currents using a 3D panel method
- Select a range of motion for sloshing simulation
- CFD modelling of sloshing simulation in full scale
- Convergence studies
- Sloshing simulation (1st and 2nd Steps[1])
- Goodness-of-fit test and interval studies for each motion DOF
- Estimate the best probability of density function (PDF) for each motion DOF by Anderson-Darling method
- Define sloshing probability
- Sloshing load characteristics for peak pressure, impulse and rise time
- Derivation of sloshing probability exceedance curves
- Define a design sloshing load by an acceptance criteria

3.2 NONLINEAR STRUCTURAL RESPONSE

This section explains the *Design Stage II* which is presented in the authors' previous study[2] aiming to develop a NLFEM model and to examine the dynamic structural characteristics under actual and design sloshing loads obtained from the previous research[1]. Through this study, NLFEM model considering nonlinear dynamic material properties, strain rate effects, impact load profile effects and local structural effects was developed. As a result of the parametric studies for the design sloshing load, a location where the most severe structural damage is expected was found[2]. Proposed procedures are as follows:

- NLFEM modelling (geometry, material modelling and loading condition)
- Mesh convergence test
- Nonlinear structural response analysis under actual sloshing loads
- Parametric studies for a design sloshing load (peak pressure, impulse and rise time)
- Dynamic structural response (plastic strain, and maximum and permanent deformations)
- Identify the weakest location

3.3 DERIVATION OF P-I DIAGRAM

This section primarily introduces the *Design Stage III* aiming to derive the P-I diagram based on the parametric studies for the range of the actual sloshing load profiles. Through this study, the structural capacity of the LNG CCS could be identified and the derived P-I diagram could be used to evaluate the expected structural damage under sloshing loads. Further procedures are as follows:

- Identification of the range of actual sloshing impact profiles
- NLFEM series analyses for the weakest location
- Quantifying a damage extent
- Derivation of the P-I diagram for maximum deformation
- Estimation of impulsive and quasi-static asymptotes
- Evaluate the structural capacity and safety

4. FINITE ELEMENT MODELLING

This chapter states the developed NLFEM model for nonlinear dynamic structural analysis.

4.1 GEOMETRY

A hypothetical barge-type FLNG with the SPB tank was developed as shown in Fig. 5. The target structure had the following dimensions: length 411 m, breadth 80 m, depth 40 m, draft 15.1 m and total capacity 628 K. The details of the tank arrangement and a schematic drawing of a mid-ship section were presented in the authors' previous study[1].

4.2 MATERIAL MODEL

In the dynamic structural response analysis of NLFEM, material properties are major factors since yield stress and fracture strain vary along with the dynamic effects, namely strain rate effects. The Cowper–Symonds model is usually used to evaluate strain rate effects[30-31].

$$\frac{\sigma_{Yd}}{\sigma_Y} = 1.0 + \left(\frac{\dot{\varepsilon}}{C}\right)^{1/q} \qquad (1)$$

$$\frac{\varepsilon_{fd}}{\varepsilon_f} = \left[1.0 + \left(\frac{\dot{\varepsilon}}{C}\right)^{1/q}\right]^{-1} \qquad (2)$$

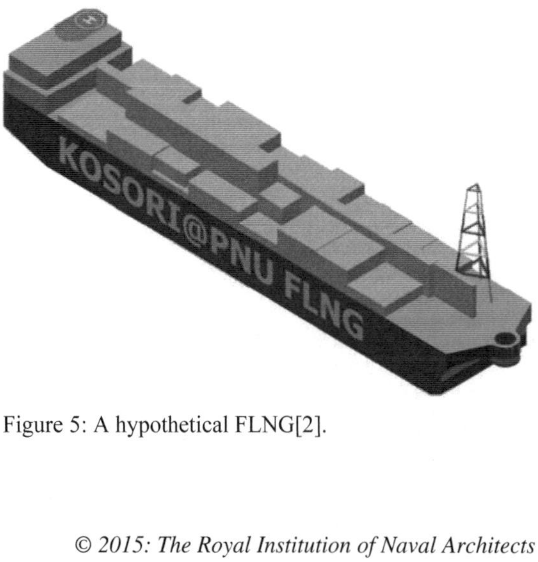

Figure 5: A hypothetical FLNG[2].

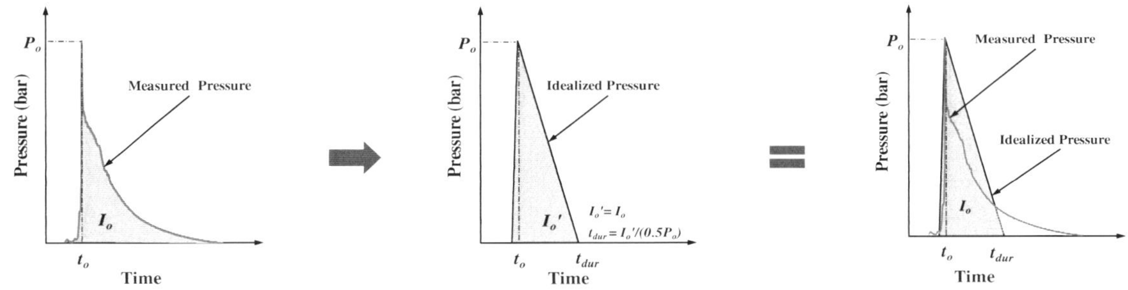

Figure 6: Idealisation of measured sloshing impact pressure characteristics applied in the dynamic structural analysis[2].

where σ_Y and ε_f are the yield stress and the fracture strain under static load, σ_{Yd} and ε_{fd} are the yield stress and the fracture strain under dynamic load, $\dot{\varepsilon}$ is the strain rate, and C and q are the experimentally determined Cowper-Symonds coefficients.

MAT_PIECEWISE_LINEAR_PLASTICITY in ANSYS/ LS-DYNA is adopted in order to take into account the elasto-plastic behaviour and strain rate. This element characterises not only the nonlinear stress-strain relationships of the material, but also the yield stress with factors C and q [32]. The materials in the tank and hull structures are aluminium and mild steel, respectively. The SPB tank and its supporting structures are assumed to be constructed of aluminium alloy at a temperature of -163°C and the hull structures of mild steel at room temperature. The effect of thermal loads within cargo tanks is not considered because this study is focusing on the dynamic structural behaviour. Applied material properties obtained by the material coupon test are shown in Table 1. It is noted that the insulation system is not included in this study.

4.3 LOADING AND BOUNDARY CONDITIONS

Characterisation of the sloshing impact load is one of challenging issues. The actual sloshing loads can be idealised as impulsive loading, characterised by the peak pressure, impulse and rise time, as shown in Fig. 6. In the parametric analyses, each sloshing load was idealized and their results for the idealized duration time, t_{dur}' were summarised in Table 2.

To simplify the NLFEM computations, the boundary conditions are symmetrical around the centre plane and fixed at both ends of the hull structure which can be found in the authors' previous study[2]. The total simulation time is set to be three times the duration time of an applied sloshing load to consider the dynamic stress propagation in the structural model.

4.4 MESH CONVERGENCE STUDY

In authors' previous paper[2], a mesh convergence study was performed for six types of mesh sizes under the

Table 1: Material properties of cargo tanks

Units	Al. at -163℃	Mild steel at RT
ρ	2700	7889
ν	0.33	0.3
σ_Y	133.25	235.0
C	6500	40.4
q	4	5

Table 2: Actual sloshing load profiles and idealized duration time

Scenario No.	t_o	P_o	t_{dec}	I_o	t_{dur}'
Scenario-02	18	15.31	292	0.95	124
Scenario-04	10	1.32	195	0.06	88
Scenario-12	67	7.45	1,388	1.33	356
Scenario-13	18	3.22	269	0.17	108
Scenario-14	35	0.71	643	0.10	279
Scenario-19	27	5.36	578	0.93	345
Scenario-24	22	0.59	769	0.09	316
Scenario-25	34	8.58	966	1.50	350

Table 3: Calculated impulse for parametric studies

P_o	t_{dur}'				
	88	108	124	316	350
0.592	0.026	0.032	0.037	0.094	0.103
1.324	0.058	0.071	0.082	0.209	0.231
3.225	0.142	0.174	0.199	0.510	0.564
5.361	0.236	0.289	0.331	0.847	0.926
8.587	0.378	0.463	0.530	1.357	1.502
15.312	0.674	0.825	0.946	2.420	2.678

Table 4: Calculated duration time for impulsive conditions

P_o	I_o	
	0.250	0.500
5.361	93	187
8.587	58	116
15.312	33	65
20.000	25	50

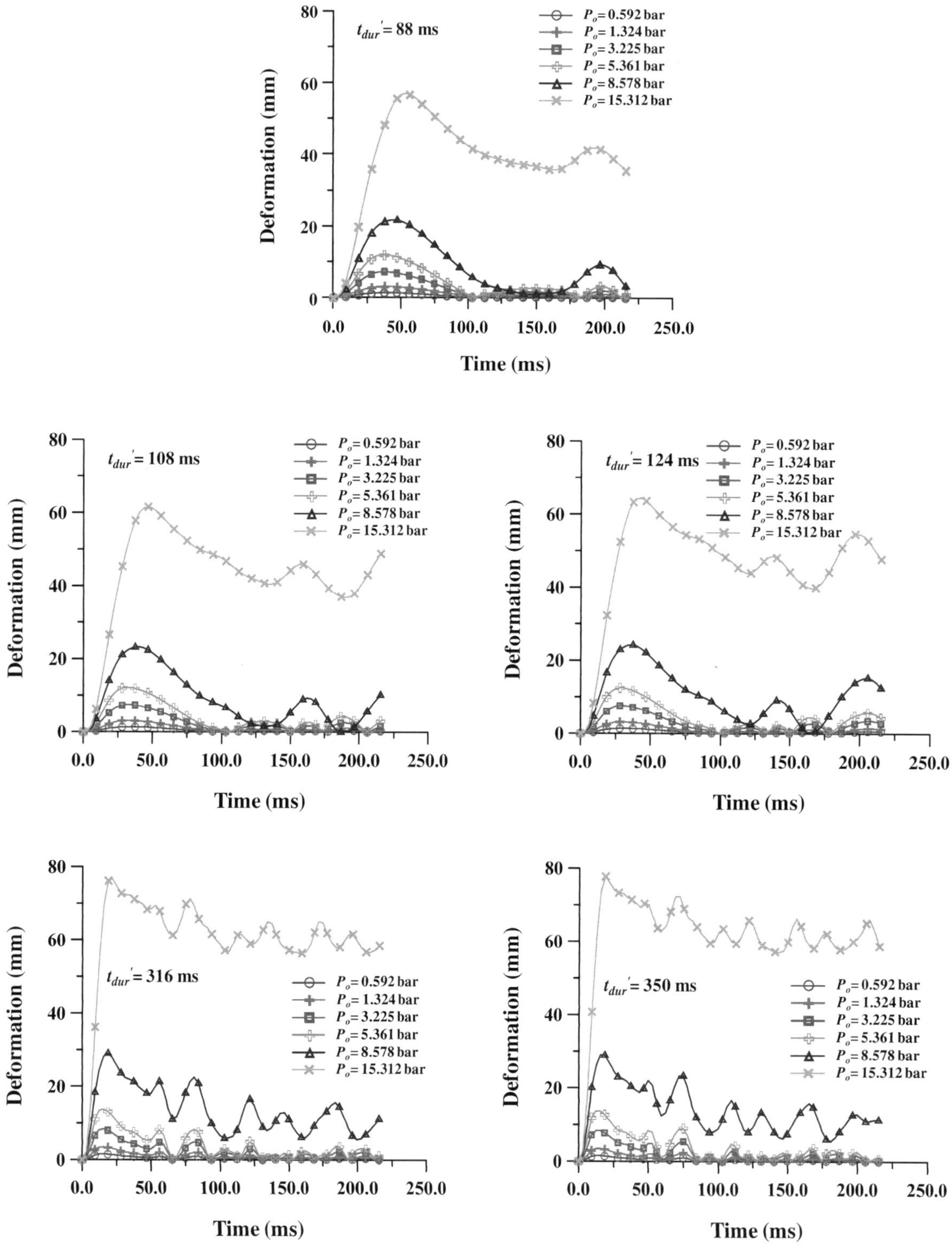

Figure 7: Deformation time-histories of the FLNG CCS under sloshing impact loads.

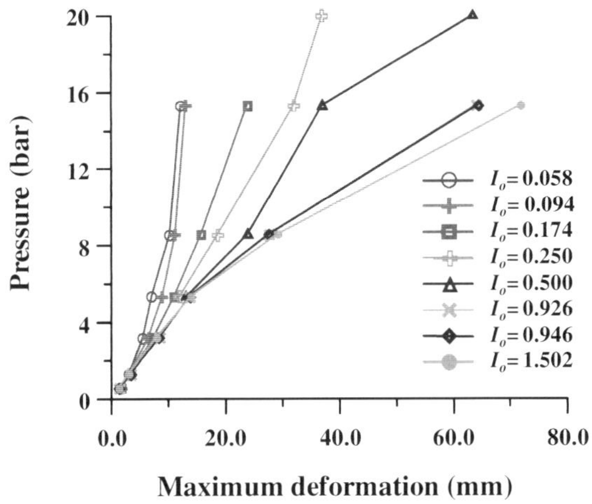

Figure 8: Maximum deformation of the FLNG CCS with varying impact load profiles.

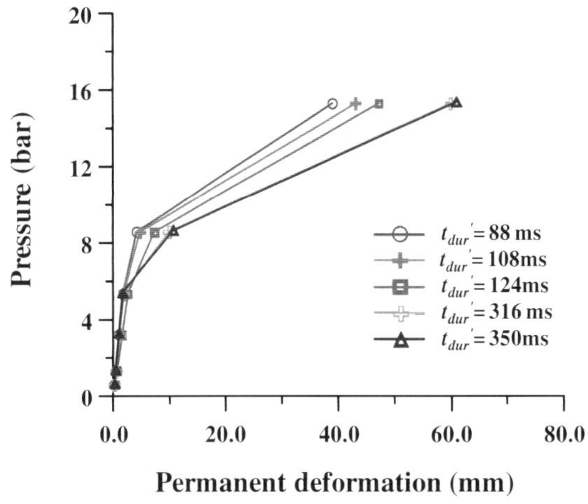

Figure 9: Permanent deformation of the FLNG CCS with varying impact load profiles.

sloshing load. As a result, the element sizes for the SPB tank and the hull structure were determined as 100 and 400 mm, respectively. The number of elements generated and applied in this study is approximately 3.4 million.

5. NUMERICAL RESULTS

5.1 PARAMETERIC STUDIES

To derive the P-I diagram of the FLNG CCS, the peak pressure and idealized duration time from the actual sloshing loads are selected as an input parameter. Thirty cases series analyses are performed. The parameters considered in this part are as follows:

- Peak pressure, P_o : 0.592, 1.324, 3.225, 5.361, 8.587, 15.312 bar
- Idealized duration time, t_{dur}' : 88, 108, 124, 316, 350 ms

The calculated impulse values according to the selected peak pressure and idealized duration time are summarized in Table 3. Furthermore, to capture the damage level near impulsive domain in the P-I diagram, eight more cases of series analyses are performed. The parameters considered in this part are as follows:

- Impulse, I_o : 0.250, 0.500 bar·s
- Peak pressure, P_o : 5.361, 8.587, 15.312, 20.000 bar

The calculated duration time according to the impulse level and peak pressure is presented in the Table 4. The dynamic structural response characteristics of the FLNG CCS are investigated using ANSYS/LS-DYNA[33]. Maximum deformation is defined as the maximum value and permanent deflection is the average value after the duration time.

Fig. 7 describes the deformation time histories for varying the peak pressure and the idealized duration time. Figs. 8 and 9 show maximum and permanent deformations of the FLNG CCS. It is observed that as impulse and the peak pressure increase, the maximum deformation increases and the maximum deformation is more sensitive to the variation of impulse than peak pressure. It is found that the permanent deformation is not severe when sloshing peak pressure is under 6.0 bar. However, the damage level increases significantly over 6 bar of the peak pressure.

5.2 DERIVATION OF THE P-I DIAGRAM

In order to generate the P-I diagram for the FLNG CCS, a series of NLFEM analyses are conducted. The sloshing impact loads such as peak pressure and impulse, corresponding to the LNG CCS damage, will be plotted in the P-I space based on the maximum deformation level. Finally, the ISO-damage curves which are the boundary lines between different damage levels, could be obtained using curve fitting method. The fitted P-I diagram is expressed analytically as follows:

$$(P-P_a)(I-I_a)=\alpha(P_a/2+I_a/2)^{\beta} \tag{3}$$

Table 5: Approximated values of asymptotes and constants

D	P_a	I_a	α	β
6	2.284	0.038	0.083	-5.989
8	2.957	0.031	0.143	-1.131
10	3.516	0.010	0.464	-1.183
15	5.408	0.102	0.221	-0.135
20	6.254	0.096	0.547	-0.150
30	8.334	0.163	0.813	-0.303

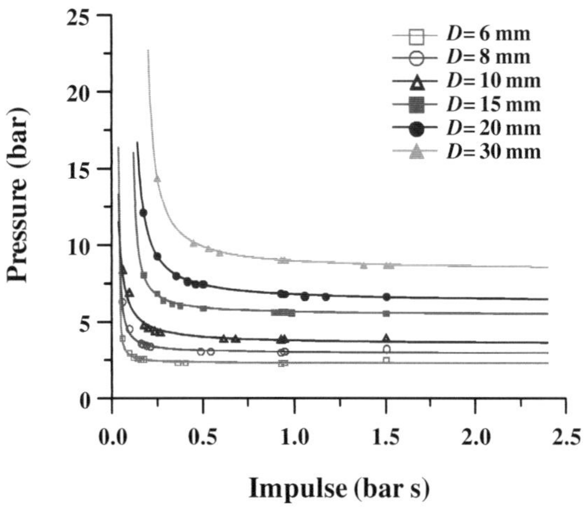

Figure 10: Pressure-impulse diagram of the FLNG CCS according to the damage level.

where P_a is the pressure asymptote for maximum deformation level D, I_a is the impulsive asymptote for maximum deformation level D, α and β are constants.

Fig. 10 illustrates the P-I diagram of the FLNG CCS. Estimated values of asymptotes and constants are summarized in Table 5 along with the maximum deformation level.

6. DISCUSSION

Sloshing impact loads in cargo tanks such as LNGC, FPSO, FLNG, VLCC are one of the challenging issues for structural designers. For over the last 50 years, a number of design procedure have been proposed and applied to the industrial purposes.

Apart from the industrial usage of existing design procedures, it is still controversial issues how to determine the sloshing scenarios and the design sloshing load accurately. To resolve this issue, the authors introduced a novel concept utilizing a probabilistic approach to sample sloshing scenarios and to analyze the sloshing impact load profiles. Previously, the feasibility was fully reviewed with an applied example[1].

As a sequel to the abovementioned research, the authors further developed the nonlinear structural analysis procedure to accurately predict the nonlinear dynamic structural behaviour subjected to the sloshing impact load. As an preliminary step, the NLFEM modelling technique was established[2]. In addition to this, the authors apply new concept, P-I diagram method, for assessing structural damage which is not yet extended in sloshing impact analysis. In the present study, the result of an applied example is presented from Sec. 5.2 and Sec. 5.3 and finally the P-I diagram is derived. It is concluded that the applicability of new approach is fully confirmed.

7. CONCLUSIONS

The aim of the present study is to briefly introduce the developed probabilistic design procedure of sloshing assessment for the FLNG CCS and to derive the P-I diagram based on parametric studies of sloshing impact loads. Based on the results obtained from this study, the following conclusions can be drawn.

(1) A procedure to generate the P-I diagram of the FLNG CCS under sloshing impact loads is newly introduced based on a series of NLFEM analyses. Finally, the P-I diagram of the FLNG CCS is derived for ISO-damage levels.

(2) It is observed that as the effect of impulse and peak pressure is proportional to the maximum and permanent deformations, and the maximum deformation is more sensitive to a variation of impulse than peak pressure.

(3) It is found that regardless of duration time, very similar damage level of the permanent deformation is appeared when the peak pressure is under 6.0 bar.

(4) The probabilistic design approach to assess sloshing impact damage which covers from determining the design sloshing load to derivation of the P-I diagram through nonlinear dynamic structural analysis is completed. The feasibility of the developed procedure has been confirmed by an applied example of a hypothetical FLNG.

(5) It is believed and hoped that the design procedure developed in the present study will be helpful for engineers and designers in the shipbuilding and offshore industries.

Authors will continue the work on sloshing induced impact
loads and design loads, as there is a need to further refine industry standard approach. It is noted that to improve the accuracy of the proposed procedure, the effect of an insulation system and a thermal load in cargo tanks should be considered.

8. ACKNOWLEDGEMENTS

This work was supported by ClassNK Joint R&D for Industry Program, and carried out jointly by Pusan National University, STX Offshore & Shipbuilding Co., Ltd. and ClassNK as “Limit States Design of Type B Independent LNG Cargo Tanks”.

It was also supported by Leading Foreign Research Institute Recruitment Program through the National Research Foundation of Korea (NRF) funded by the

Ministry of Science, ICT & future Planning (MSIP) (Grant No. 2014040731).

This study was undertaken at The Lloyd's Register Foundation Research Centre of Excellence at Pusan National University. Lloyd's Register Foundation (LRF), a UK registered charity and sole shareholder of Lloyd s Register Group, Ltd, invests in science, engineering and technology for public benefit, worldwide.

9. REFERENCES

1. PAIK, J.K., LEE, S.E., KIM, B.J., SEO, J.K., HA, Y.C., MATSUMOTO, T. and BYEON, S.H., 'Toward A Probabilistic Approach to Determine Nominal Values of Tank Sloshing Loads in Structural Design of LNG FPSOs', *Journal of Offshore Mechanics and Arctic Engineering*, Vol. 137(2), 2015.

2. LEE, S.E., KIM, B.J., SEO, J.K., HA, Y.C., MATSUMOTO, T., BYEON, S.H. AND PAIK, J.K., 'Nonlinear Impact Response Analysis of LNG FPSO Cargo Tank Structures under Sloshing Loads', *Ships and Offshore Structures*, 2015.

3. ABS, Guidance Note on Strength Assessment of Membrane-Type LNG Containment Systems Under Sloshing Loads, American Bureau of Shipping, Houston, TX, US, 2006.

4. DNV GL, Sloshing Analysis of LNG Membrane Tanks, DNV classification notes no.30.9, Det Norske Veritas, Oslo, Norway, 2006.

5 LR, Ship Right-Sloshing Assessment Guidance Document for Membrane Tank LNG Operations, Lloyd's Register, London, UK, 2009.

6 BV, Strength Assessment of LNG Membrane Tanks under Sloshing Loads-Guidance Note NI 564 DTR00 E, Bureau Veritas, Paris, France, 2011.

7. HWANG, J.O., CHUN, S.E., JOH, K.H., SAMBOS, P., DE LAUZON, J., WHITE, N., KIM, M.S., PARK, J.B., and LEE, J.M., 'Direct Assessment of Structural Capacity Against Sloshing Using Dynamic Nonlinear FE Analysis', *The 24h International Offshore and Polar Engineering Conference, ISOPE*, Busan, Korea, June 15–20, 2014.

8. ITO, H., SUH, Y.S., CHUN, S.E., SATISH KUMAR, Y.V., HA, M.K., PARK, J.J., YU, H.C., and WANG, B., 'A Direct Assessment Approach for Structural Strength Evaluation of Cargo Containment System Under Sloshing inside LNGC Tanks based on Fluid Structure Interaction', *The 27th International Conference on Offshore Mechanics and Arctic Engineering*, Estoril, Portugal, June 15-20, 2008.

9. NAM, S.K., KIM, W.S., NOH, B.J., SHIN, H.C., and CHOI, I.H., 'The Parametric Study on the Response of Membrane Tanks in a Mark III Type LNG Carrier Using Fully Coupled Hydro-Elastic Model', *International Conference on Ship and Offshore Technology, ICSOT*, Busan, Korea, September 14-15, 2006.

10. PASTOOL, W., OSTVOLD, T.K., BYKLUM, E., and VALSGARD, S., 'Sloshing Load and Response in LNG Carriers for New Designs, New Operations and New Trades', *GasTech 2005*, Bilbao, Spain, March 14-17, 2005.

11. GRACZYK, M., MOAN, M.K., and WU, M.K., 'Extreme Sloshing and Whipping-Induced Pressures and Structural Response in Membrane LNG Tanks', *Ships Offshore Structures*, Vol. 2(3), pp. 201–216, 2007.

12. GRACZYK, M., and MOAN, T., 'A Probabilistic Assessment of Design Sloshing Pressure Time Histories in LNG Tanks', *Ocean Engineering*, Vol. 35(8-9), pp. 834–855, 2008.

13. KUO, J.F., CAMPBELL, R.B., DING, Z., HOIE, S.M., RINEHART, A.J., SANDSTORM, R.E., YUNG, T.W., GREER, M.N. and DANACZKO, M.A., LNG tank sloshing assessment methodology: The new generation, *International Journal of Offshore and Polar Engineering, ISOPE,* Vol. 19(4), pp. 241-253, 2009.

14. SMITH, P.D. and HETHERINGTON, J.G., 'Blast and Ballistic Loading of Structures', *Butterworth Heinemann*, London, UK, 1994.

15. JARRETT, D.E., 'Derivation of British Explosives Safety Distances', *Annals of New York Academy of Science*, 152(1), pp. 18-35, 1968.

16. BAKER, W.E., COX, P.A., WESTINE, P.S., KULESZ, J.J. and STREHLOW, R.A., 'Explosion Hazards and Evaluation', *Elsevier Scientific Pub. Co.*, New York, US, 1983.

17. MAYS G.C., SMITH, P.D., 'Blast Effects on Buildings: Design of Buildings to Optimize Resistance to Blast Loading', *Thomas Telford Services Ltd*, 1995.

18. MERRIFIELD, R., 'Simplified Calculations of Blast Induced Injuries and Damage', *Report no.*

37, Health and Safety Executive Specialist Inspector, 1993.

19. LI, Q.M. and MENG, H., 'Pressure-Impulse Diagram based on Dimensional Analysis and Single-Degree-of-Freedom Model', *Journal of Engineering Mechanics*, Vol. 128(1), pp. 87–92, 2002.

20. LI, Q.M. and MENG, H., 'Pulse Loading Shape Effects on Pressure-Impulse Diagram of Elastic Plastic Structures', *International Journal of Mechanical Sciences*, Vol. 44, pp. 1985-1998, 2002.

21. FALLAH, A.S. and LOUCA, L.A., 'Pressure–Impulse Diagrams for Elastic–Plastic Hardening and Softening Single-Degree-of-Freedom Models subjected to Blast Loading', *International Journal of Impact Engineering*, Vol. 34, pp. 823-842, 2007.

22. MA, G.W., SSHI, H.J. and SHU, D.W., ' P-I Diagram Method for Combined Failure Modes if Rigid-Plastic Beams', *International Journal of Impact Engineering*, Vol. 34, pp. 1081-1094, 2007.

23. PARLIN, N.J., DAVIDS, W.G., NAGY, E. and CUMMINS, T., 'Dynamic Response of Lightweight Wood-based Flexible Wall Panels to Blast and Impulse Loading', *Construction and Building Materials*, Vol. 50, pp. 237-245, 2014

24. LAN, S.R. and CRAWFIRD, J.C., 'Evaluation of the Blast Resistance of Metal Deck Proofs', *Proceedings of the fifth Asia-Pacific conference on shock & impact loads on structures*, Changsha, Hunan, China, November 12-14, 2003.

25. SCHNEIDER, P., 'Predicting Damage of Slender Cylindrical Steel Shells under Pressure Wave Load', *Journal of Loss Prevention in the Process Industries*, Vol. 11(3), pp. 223-228, 1998.

26. WESEVICH, J.W. and OSWALD, C.J., 'Empirical based Concrete Masonry Pressure–Impulse Diagrams for Varying Degrees of Damage', *Structures Congress 2005*, New York, US, 2005.

27. SHI, Y., HAO, H. and LI, Z.H., 'Numerical Derivation of Pressure-Impulse Diagrams for Prediction of RC Column Damage to Blast Loads', *International Journal of Impact Engineering*, Vol. 35, pp. 1213-1227, 2008.

28. SOHN, J.M., KIM, S.J., KIM, B.H. and PAIK, J.K., 'Nonlinear Structural Consequence Analysis of FPSO Topside Blast Walls', *Ocean Engineering*, Vol. 60, 2013.

29. NORSOK, Actions and action effects, Norwegian Standards, 1999.

30. JONES, N., Structural impact, *Cambridge University Press*, Cambridge, UK, 1970.

31. PAIK, J.K. and THAYAMBALLI, A.K., Ultimate limit state design of steel-plated structure, *John Wiley & Sons*, Chichester, UK, 2003.

32. ANSYS/LS-DYNA, User's manual (version 15.0), ANSYS Inc., Pennsylvania, US, 2014.

10. AUTHORS BIOGRAPHY

Dr. Sang Eui Lee is a postdoctoral research fellow in The Korea Ship and Offshore Research Institute (The Lloyd's Register Foundation Research Centre of Excellence) at Pusan National University in Korea. His research interests are in the area of fluid structure interactions under impact loads including sloshing, slamming and green waters.

Dr. Bong Ju Kim is an associate professor of the Korea Ship and Offshore Research Institute (The Lloyd's Register Foundation Research Centre of Excellence) at Pusan National University in Korea. His research interests are in the area of nonlinear structural mechanics and design associated with limit states and risk-based approaches. He also leads the Korean government funded project to build large-scale test facilities for hydrocarbon explosions and fires. He has been a member of the technical committee III.1 Ultimate Strength of International Ship and Offshore Structures Congress.

Dr. Jung Kwan Seo is an associate professor of the Korea Ship and Offshore Research Institute (The Lloyd's Register Foundation Research Centre of Excellence) at Pusan National University in Korea. He serves as director of Subsea Research Centre of the KOSORI. He has been a member of the specialist committee V.8 Risers and Pipelines of International Ship and Offshore Structures Congress (ISSC).

Dr. Yeon Chul Ha is an assistant professor in the Korea Ship and Offshore Research Institute (The Lloyd's Register Foundation Research Centre of Excellence) at Pusan National University in Korea. He is a member of SNAME. Ha serves as director of the Marine and Information Technology Research Centre of KOSORI. Before he joined the KOSORI in 2010, he served as director of R&D Center of WORLDENG, Co. Ltd. His research interests include information technology for shipbuilding and offshore industry.

Professor Jeom Kee Paik is a Professor of the Department of Naval Architecture and Ocean Engineering and serves as president of the Korea Ship and Offshore Research Institute (KOSORI) and as director of the Lloyd's Register Foundation Research Centre of Excellence at Pusan National University (PNU) in Korea. In 2015, he also joined University College London's Department of Mechanical Engineering as a professor and faculty member. He is a Fellow and Vice President of SNAME, USA and a Fellow and Council Member of RINA, UK. He was awarded William Froude Medal from RINA in 2015, the David W. Taylor Medal from SNAME in 2013, and a Doctor Honoris Causa from the University of Liege in Belgium in 2012. His research interests are safety engineering and design of ships and offshore structures in extreme and accidental conditions associated with risk assessment and management.

Mr. Toshiyuki Matsumoto is an operating officer and general manager at the department of Research Institute in ClassNK, Chiba, Japan.

Mr. Su Hwan Byeon is a senior manager at Shipbuilding Technology Division, Structure Technology Team in STX Offshore & Shipbuilding, Jinhae, Korea. He commenced his career in 2000 as an assistant engineer and has been involved in hull key-plan design for 15 years.

DETERMINATION OF FIRE ACCIDENTAL LOADS FOR ALUMINIUM SAFETY HELIDECKS

SJ Kim, JB Park, BJ Kim and JK Seo, The Korea Ship and Offshore Research Institute (The Lloyd's Register Foundation Research Centre of Excellence), Pusan National University, Korea
JK Paik, The Korea Ship and Offshore Research Institute (The Lloyd's Register Foundation Research Centre of Excellence), Pusan National University, Korea, and Department of Mechanical Engineering, University College London, London, UK
WH Shin, Research Institute Team, N.K Co., Ltd., Korea
JS Park, Central Research Institute, Samsung Heavy Industries Co., Ltd., Korea

SUMMARY

Helicopters are essential in supporting offshore oil and gas activities around the world. The helideck structure must satisfy the safety requirements associated with various environmental and accidental loads. There have been a number of fire accidents offshore due to helicopter collision (take-off and/or landing) in recent decades. To prevent further accidents, a substantial amount of effort has been directed toward the management of fire in the safety design of offshore helidecks. The aims of this study are to introduce and apply a procedure for quantitative risk assessment and management of fires by defining the fire loads with an applied example. The frequency of helicopter accidents in the Gulf of Mexico and the North Sea are considered, and design accidental levels for both regions are suggested. The proposed procedures for determining design fire loads can be efficiently applied in offshore helideck development projects, and the application includes the assessment of design fire loads and the quantification of the effects of risk control options, such as optimization of the helideck pancake profile, the location and the number of water deluge systems.

NOMENCLATURE

F_{Fire}	Fire frequency
$F_{Release}$	Leak frequency
$P_{Ignition}$	Ignition probability.
α	Scale factor of Weibull distribution
β	Shape factor of Weibull distribution
μ	Mean of normal distribution
σ	Standard deviation of normal distribution

1. INTRODUCTION

Helicopters are the primary modes of transport for offshore platforms. They carry people, equipment and necessities to offshore installations. Unlike other transport, such as ships, helicopters take off and land on offshore installations. The offshore structures can therefore be damaged by helicopter accidents.

Many helicopter accidents (referred to as heli-accidents in this paper) have been reported over the last ten years [1]. Figure 1 shows examples of heli-accidents on offshore platforms.

Figure 1: Heli-accidents on West Navion (left, 2001) and Araon (right, 2013)

The frequency of heli-accidents is lower than for other events (explosions, fires, collisions or dropped objects), but they do also have the potential risk factor of fires, as combustibles, in the form of highly volatile fuel in fuel tanks, are always present in helicopters.

If a heli-accident occurs on a helideck, fuel can leak from the tank. Released fuel can cause a pool fire, which can structurally damage the helideck.

API [2] and DNV [3] recommend that fire safety facilities such as fire extinguishers, water sprays, fire resistant equipment, etc. are installed to prevent such structural damage.

There are, however, only recommendations for fire suppression facilities in the rules and standards, and the guidelines for structural safety designs to combat fires and the definition of design fire loads are weak.

In particular, a helideck made from aluminium is sensitive to temperature and heat flux, compared with other materials such as carbon steel, stainless steel, nickel alloy, etc. A structural design for aluminium helidecks that specifically addresses the risk of fires is therefore necessary and should be developed.

There has been extensive research into the fire risk assessment and management of topside structures on offshore installations, but for helidecks this has been neglected. The topside structures and helidecks on offshore platforms are totally different [4-10].

The aims of this study are to (i) introduce and (ii) apply a procedure for quantitative fire risk assessment and management for helidecks on offshore installations, and to (iii) define the design fire loads for helicopter collisions on helidecks, with applied examples.

In the present study, credible fire scenarios are selected in the procedure, and a reasonable accident acceptance level for helidecks is suggested, to help define design fire loads to ensure the structural safety of aluminium helidecks in case of fires.

2. A PROCEDURE FOR QUANTITATIVE FIRE RISK ASSESSMENT AND MANAGEMENT

Unlike the prescriptive procedures associated with rules, standards or qualitative risk assessment and management [9], the procedure for quantitative risk assessment and management (QRA&M) for fires on offshore platforms proposed by Paik et al. [10] efficiently evaluates risks and defines the design loads. Figure 2 illustrates a procedure for the quantitative risk assessment and management of offshore installations in case of fires. The procedure can be applied to the risk assessment of helidecks, and is used in this study.

The procedure for QRA&M in fire accidents, as shown in Figure 2, can be divided into eight steps, described in detail in Table 1.

Table 1: Details of the procedure for quantitative fire risk assessment and management

Steps	Descriptions
1	Investigation of metocean data and functional requirements of target structure
2	Selection of credible fire scenarios with metocean and operation-related parameters using a probabilistic approach
3	Obtaining fire loads (temperature, heat flux, etc.) of each fire scenario by experiment and/or computational fluid dynamics (CFD)
4	Calculation of fire accident frequencies of fire scenarios using historical data
5	Generation of fire (temperature, heat flux, flame length, etc.) exceedance curves with fire loads and frequencies, and determination of design fire loads
6	Conducting structural consequence response analysis by the nonlinear finite element method and/or experiment, and investigating the consequences of fire accidents
7	Calculation of fire risk [risk = frequency x consequence]
8	Decision making (redesign of the structure or adoption of risk control options)

In this study, steps 1-5 (not the entire procedure) are carried out to obtain a fire exceedance curve, which can define the design fire loads of helidecks.

The partial procedure carried out applies a (i) probabilistic method to select scenarios, (ii) computational fluid dynamics (CFD) to obtain fire loads and (iii) historical data to calculate the fire frequency.

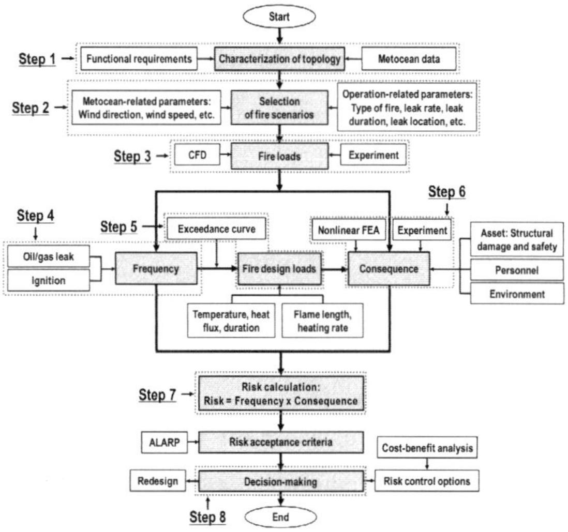

Figure 2: A procedure for quantitative risk assessment and management for fires

3. SELECTION OF CREDIBLE FIRE SCENARIOS ON HELIDECKS

When selecting credible fire accident scenarios, all possible parameters that can have an effect on fire loads should be considered, such as temperature, heal flux, flame length, etc. Fires can also be affected by wind (direction and speed) and leak (rate, direction, duration and position).

In the case of a jet fire, the scenarios should take into account the leak rate, direction, duration and position associated with the release. The characteristics of a pool fire, however, mean that only the operation parameters of leak amount and position are required.

In this study, the four parameters (i) wind direction, (ii) wind speed, (iii) leak amount and (iv) leak position are considered when selecting the scenarios, as fires caused by heli-accidents are similar to pool fires.

3.1 WIND DIRECTION AND SPEED

Unlike ships that travel around the world, offshore platforms operate in a fixed, specific location. The metocean characteristics of the specific ocean can therefore be used for the risk assessment of offshore helidecks.

Figure 3 and Table 2 present the wind direction (wind rose) and the metocean data, which in this applied example are the probability of wind direction and speed in the ocean off West Africa.

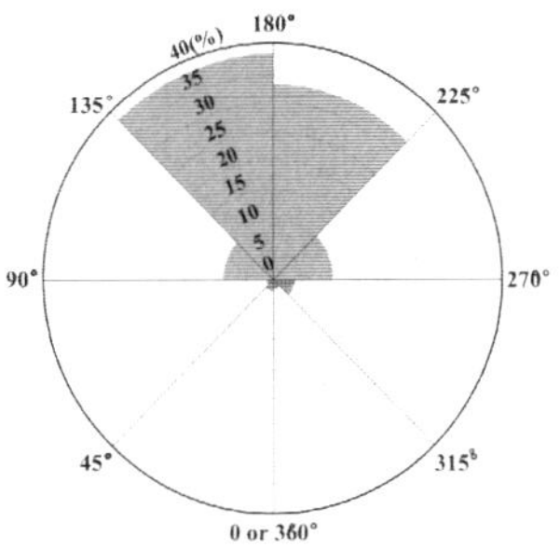

Figure 3: Probability of wind direction (wind rose) in the ocean of West Africa from the south

Table 2: Probability of wind direction and speed in the ocean off West Africa

Wind direction (°)	Probability (%)	Wind speed (m/s)	Probability (%)
0-45	2.137	0-2	9.508
45-90	1.501	2-4	33.831
90-135	8.101	4-6	40.270
135-180	38.775	6-8	14.870
180-225	33.628	8-10	1.446
225-270	10.572	10-12	0.065
270-315	3.420	12-14	0.010
315-360	1.871	14+	0.005
Total	100.005	Total	100.005

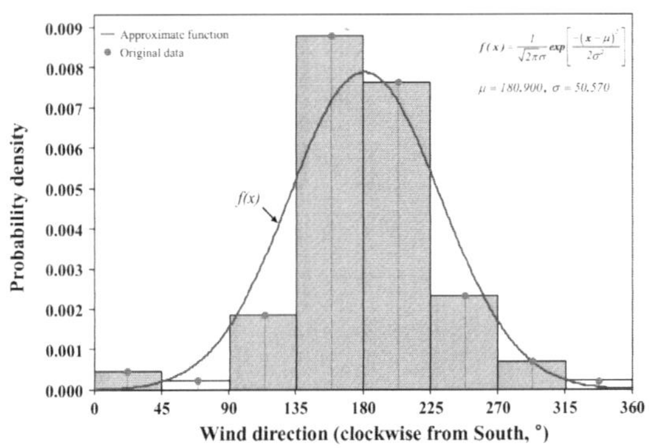

Figure 4: Probability density distribution of wind direction

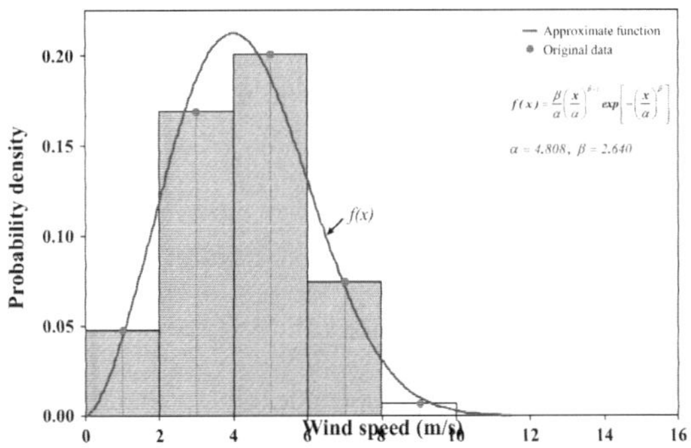

Figure 5: Probability density distribution of wind spe ed

As the original (raw) data are discontinuous, they must be converted to continuous functions so that a probabilistic approach can be used. Best-fit analysis should be conducted before the probability density function (PDF) is defined.

Figures 4 and 5 show the PDFs of wind direction and speed. The best-fitted function for wind direction is normal distribution, and for speed is Weibull distribution.

3.2 LEAK AMOUNT

Calculating a leak amount is difficult, as helicopter accidents on offshore platforms are extremely rare, and the capacities of helicopter fuel tanks and engines are all different. Several assumptions (type of helicopter and/or fuel, the location of the offshore platforms, the amount of fuel in the tank, etc.) must therefore be made.

Two assumptions are made in this study: first, that one type of fuel is used in the helicopters, and two, that they have a full fuel condition when they depart from land.

The leak amounts given in the study are calculated using the following steps:

1) Definition of helicopter (type of fuel)
2) Investigation of helicopter characteristics (capacity of fuel tank, mileage, etc.)
3) Definition of fuel condition on the land (a full condition is assumed)
4) Calculation of the distance between the land and the offshore installation
5) Calculation of the minimum required fuel from the land to the offshore installation
6) Calculation of the remaining fuel in the helicopter, which will be the leak amount

To obtain the leak amount, a helicopter with JP-8 fuel type, which is generally used in offshore installations, is assumed. The specifications of the helicopter are given in Table 3.

Table 3: Helicopter information used in the examination of offshore installations

Capacity of fuel tank (L)	2877
Mileage (km/L)	0.35
Density of fuel (kg/L)	0.811

To investigate the air-line distance between the land and offshore installations, the ocean regions of Brazil and West Africa, where many platforms are installed, are considered.

Figure 6 shows offshore platforms in the ocean regions of Brazil and West Africa, and the shortest radius from offshore to land, as the shortest distance is generally preferred.

After investigating the distance from the land in Figure 6, the minimum amount of fuel required to the platforms, and the remaining amount of fuel when the helicopter arrives at the platforms can be calculated. These are shown in Table 4. The remaining fuel is then considered as the leak amount.

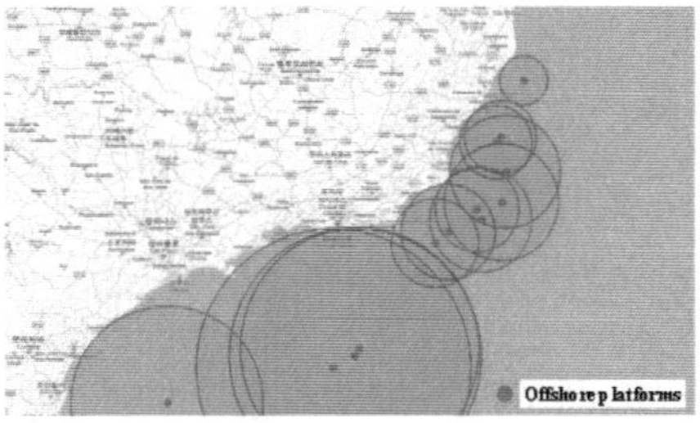

(a) The ocean of Brazil

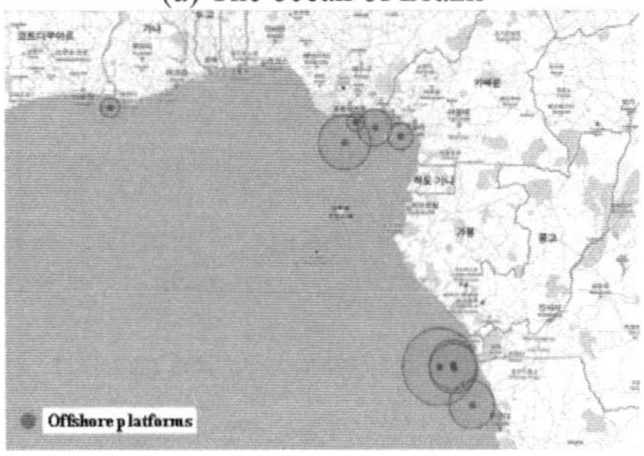

(b) The ocean of West Africa

Figure 6: Location and distance of offshore installations from the nearest land

Table 4: Distance of offshore platforms from the nearest land, the minimum required fuel and the remaining fuel

Offshore No.	Shortest distance from land (km)	Minimum required fuel (L)	Fuel in tank after landing on offshore (L)
1	332.00	948.57	1928.43
2	322.58	921.66	1955.34
3	296.77	847.91	2029.09
4	282.76	807.89	2069.11
5	202.94	579.83	2297.17
6	180.00	514.29	2362.71
7	138.00	394.29	2482.71
8	135.71	387.74	2489.26
9	123.48	352.80	2524.20
10	123.33	352.37	2524.63
11	122.58	350.23	2526.77
12	120.00	342.86	2534.14
13	113.00	322.86	2554.14
14	111.29	317.97	2559.03
15	110.00	314.29	2562.71
16	108.00	308.57	2568.43
17	106.00	302.86	2574.14
18	93.00	265.71	2611.29
19	88.24	252.11	2624.89
20	87.10	248.86	2628.14
21	84.38	241.09	2635.91
22	81.67	233.34	2643.66
23	60.00	171.43	2705.57
24	53.33	152.37	2724.63
25	45.00	128.57	2748.43
26	40.00	114.29	2762.71

Table 5 shows the probability of the leak amount. It is difficult to make a PDF (normal, Weibull distribution function, etc.), so it is re-categorised, as shown in Table 6, to generate the normal distribution function.

Table 5: Probability of practicable leak amount

Leak amount (L)	Leak amount (kg)	Probability (%)
1849.57-1972.87	1500-1600	7.69
1972.87-2096.18	1600-1700	7.69
2096.18-2219.48	1700-1800	0.00
2219.48-2342.79	1800-1900	3.85
2342.79-2466.09	1900-2000	3.85
2466.09-2589.40	2000-2100	42.31
2589.40-2712.70	2100-2200	23.08
2712.70-2836.00	2200-2300	11.54
Total		100.00

Table 6: Re-categorised index of leak amount for probability density distribution

Index	Leak amount (kg)	Probability (%)
0-1	1700-1800	0.00
1-2	1800-1900	3.85
2-3	1500-1600	7.69
3-4	2100-2200	23.08
4-5	2000-2100	42.31
5-6	2200-2300	11.54
6-7	1600-1700	7.69
7-8	1900-2000	3.88
	Total	100.00

Figure 7 illustrates the histogram and PDF of the re-categorised leak amount, used to select fire scenarios.

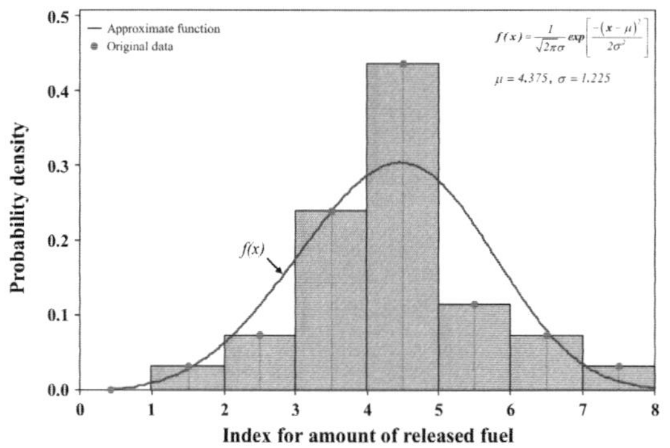

Figure 7: Probability density distribution of index for amount of released fuel

LEAK POSITION

3.3

leak position in the topside structure is relevant to The essing units such as pipes, equipment, vessels, etc., procthe crash areas are considered as leak positions on and helideck. However, this area cannot be exactly the defined due to the small amount of historical data available.

It is therefore assumed that helicopter accidents on helidecks of offshore platforms occur evenly across the entire helideck. Figure 8 presents a regular octagon helideck, applied as the target structure in this study, and divided into areas so the histogram and PDF can be obtained.

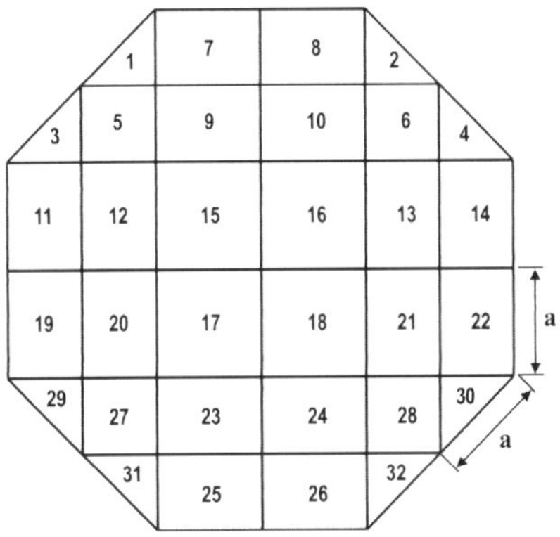

Figure 8: Divided, numbered sections for selection of the leak position on a regular octagon helideck

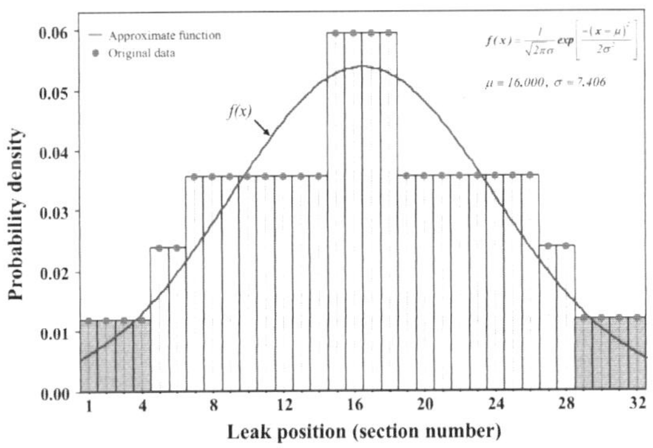

Figure 9: Probability density distribution of leak position (section number)

Table 7: Area and probability of sections

No.	Area (a^2)	Probability (%)	No.	Area (a^2)	Probability (%)
1	0.250	1.294	17	1.000	5.178
2	0.250	1.294	18	1.000	5.178
3	0.250	1.294	19	0.707	3.661
4	0.250	1.294	20	0.707	3.661
5	0.500	2.589	21	0.707	3.661
6	0.500	2.589	22	0.707	3.661
7	0.707	3.661	23	0.707	3.661
8	0.707	3.661	24	0.707	3.661
9	0.707	3.661	25	0.707	3.661
10	0.707	3.661	26	0.707	3.661
11	0.707	3.661	27	0.500	2.661
12	0.707	3.661	28	0.500	2.589
					589
13	0.707	3.661	29	0.250	1.294
14	0.707	3.661	30	0.250	1.294
15	1.000	5.178	31	0.250	1.294
16	1.000	5.178	32	0.250	1.294
Total				19.314	100.000

Table 7 and Figure 9 show the probability of each section, a histogram and the normal distribution function of the leak positions.

3.4 SELECTION OF FIRE SCENARIOS

Various sampling methods, such as Monte Carlo Simulations (MSC) [11], standard random sampling [11], stratified sampling [12], Latin Hypercube Sampling (LHS) [13], etc. are used for selecting the scenarios. Of these sampling techniques, LHS can efficiently reduce the number of scenarios and increase the accuracy of results.

Figures 10 and 11 present the scheme of the LHS technique and the method for defining the representative value.

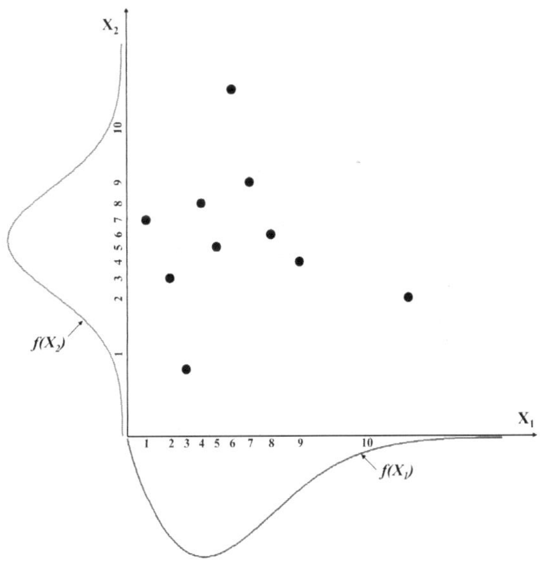

Figure 10: Scheme of the Latin Hypercube sampling technique (Ye, 1998)

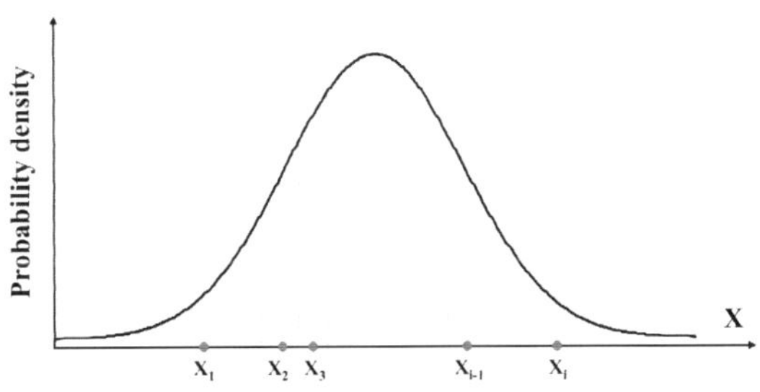

Figure 11: Method for selecting the representative value of the random variable for each parameter

Table 8: Selected fire scenarios on helidecks

Scenario No.	Wind direction (°)	Wind speed (m/s)	Leak amount (kg)	Leak position (section No.)
1	237.975	6.219	2129.880	18
2	255.749	2.636	2084.500	25
3	211.951	4.630	2146.910	3
4	187.294	2.977	2046.570	10
5	118.887	3.083	2065.200	12
6	218.334	2.243	2162.340	23
7	182.203	3.288	1556.610	7
8	140.139	5.263	2071.540	20
9	184.744	2.513	2219.840	14
10	155.847	1.738	1842.790	24
11	123.955	6.056	2077.960	21
12	106.260	0.758	2205.040	10
13	172.008	7.929	2183.510	1
14	189.860	4.832	2021.980	4
15	206.029	3.388	1638.110	14
16	59.249	2.382	2190.190	27
17	97.668	5.904	2091.170	21
18	221.756	4.149	2227.710	19
19	215.074	5.151	1601.130	30
20	158.678	4.339	1587.180	17
21	113.104	3.773	2120.540	8
22	161.440	1.926	1506.490	23
23	174.574	6.399	2028.150	22
24	276.660	6.829	2052.730	17
25	200.436	3.582	2034.290	22
26	195.070	5.380	2287.220	13
27	225.376	2.868	2235.960	20
28	85.563	5.502	2040.420	19
29	169.419	8.991	2169.620	26
30	179.666	4.532	2138.640	12
31	143.556	3.961	2275.010	6
32	197.730	3.867	2196.740	12
33	192.450	4.730	1573.180	13
34	166.800	5.042	1923.560	8
35	264.405	6.599	1666.840	18
36	233.411	1.517	2009.510	9
37	243.060	3.678	2244.690	16
38	208.941	7.101	2098.010	11
39	164.144	4.244	2212.310	15
40	177.124	2.754	1617.560	11
41	146.810	3.187	2254.000	18
42	149.930	1.234	1599.450	15
43	128.505	3.485	2058.940	5
44	203.198	7.443	2110.470	13
45	248.867	4.935	2003.170	19
46	302.975	5.762	2154.790	16
47	152.936	5.628	2264.040	9
48	229.241	4.055	1535.800	16
49	136.523	2.092	2015.770	15
50	132.665	4.435	2176.660	28

In the study, using the Latin Hypercube Sampling (LHS) technique, 50 credible fire scenarios are selected with PDFs (wind direction, wind speed, leak amount and leak position), as shown in Table 8. The centre of each section is defined as a leak position.

4. FIRE CFD SIMULATIONS

4.1 MODELLING OF FIRE CFD SIMULATIONS

To obtain the fire loads, the Kameleon FireEX [14] CFD tool is used. KFX generates control volumes (grids) automatically or by user, in association with the leak amount and/or the leak hole size. Figure 12 shows the target structure, which is the helideck, the extent of analysis and the applied grids in KFX. The total number of control volumes is 1,800,000.

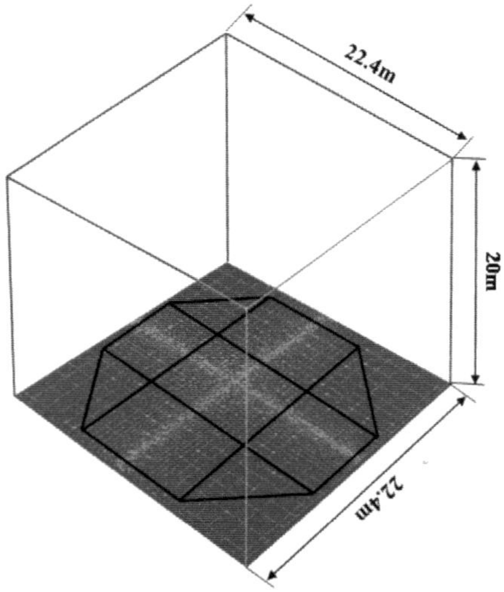

Figure 12: Target structure, extent of analysis and applied grids in KFX CFD simulations

As mentioned in Section 3.2, it is assumed that one type of helicopter fuel (JP-8) is used. Table 9 shows the gas compositions of the JP-8 fuel type.

Table 9: Gas compositions of the JP-8 fuel type

Component	Mole fraction (%)
Dodecane ($C_{12}H_{26}$)	31.58
Tetradecane ($C_{14}H_{30}$)	23.64
Decane ($C_{10}H_{22}$)	22.53
Hexadecane ($C_{16}H_{34}$)	17.12
Isooctane (C_8H_{18})	5.13
Total	100.00

4.2 OBTAINING THE FIRE LOADS

provides the fire loads at 1000 monitoring points. KFXis study, 96 points are used to obtain the fire loads. In th

fire loads, such as temperature, heat flux, flame The th, etc. differ according to the height from the heat lengce. Three elevations (0.05, 0.50 and 1.00m from the sour heat source) are therefore selected to investigate the fire loads.

Figure 13 illustrates the location of monitoring points at three elevations (Figure 13(a)). A point is located at the geometric centre of each section.

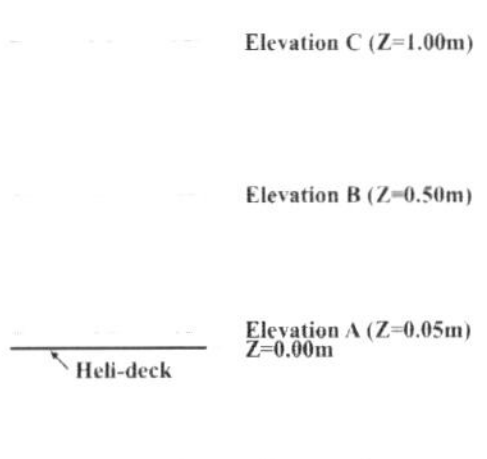

(a) Elevation view

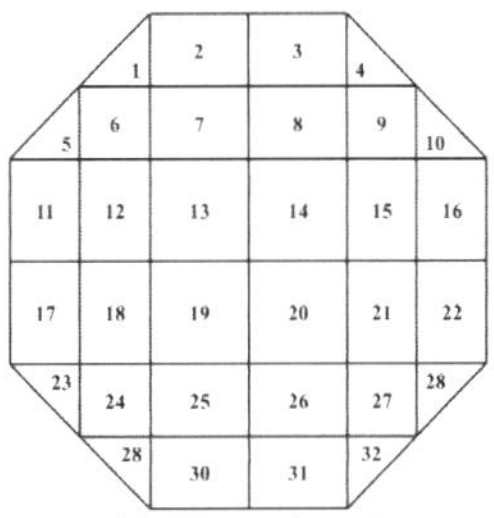

(b) Elevation A

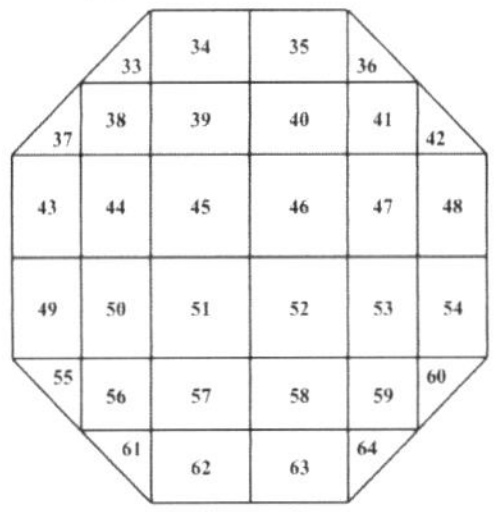

(c) Elevation B

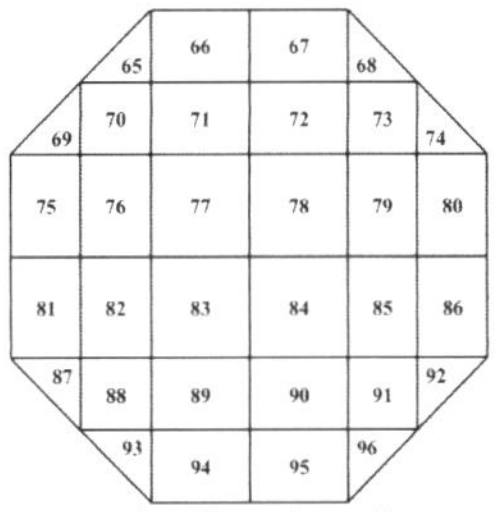

(d) Elevation C

Figure 13: Location of numbered monitoring points to investigate fire loads

4.3 RESULTS OF FIRE CFD SIMULATIONS

Figure 14 shows examples of the simulation results. Figure 14(a) presents the temperature of scenario 15 at three elevations. The temperature at the elevation nearest to the heat source is higher than the others.

Scenario 39, as presented in Figure 14(b), has a pool fire at section 15 (shown in Figure 8 and Table 8). In the first 30 seconds, the temperature at point 71 is very high compared with the other points, but then reduces and is similar to that at monitoring point 77. This is due to the effect of wind at both points.

Figure 15 illustrates the effect of the variables wind direction, wind speed, leak amount and leak position on the maximum temperature of the entire monitoring region. It shows that the wind speed has a remarkable effect on the fire loads, compared with other variables. With a wind speed of over 6 m/s, the temperature is similar to room temperature.

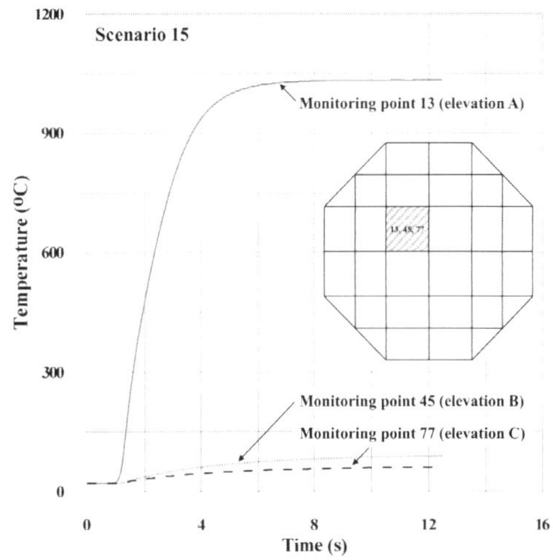

(a) Scenario 15 at monitoring points 13, 45 and 77

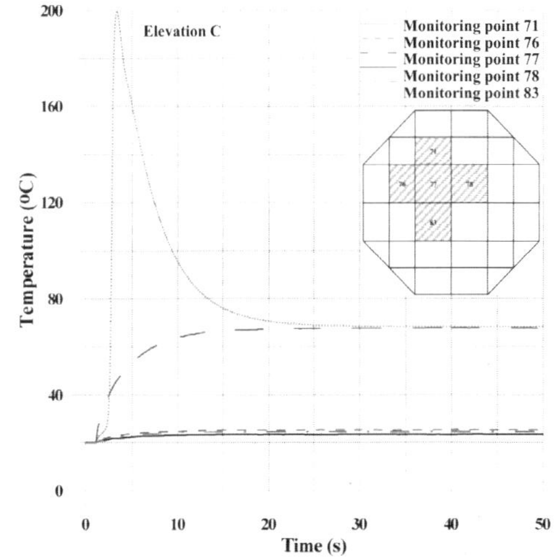

(b) Scenario 39 at elevation C

Figure 14: Examples of results of fire simulations

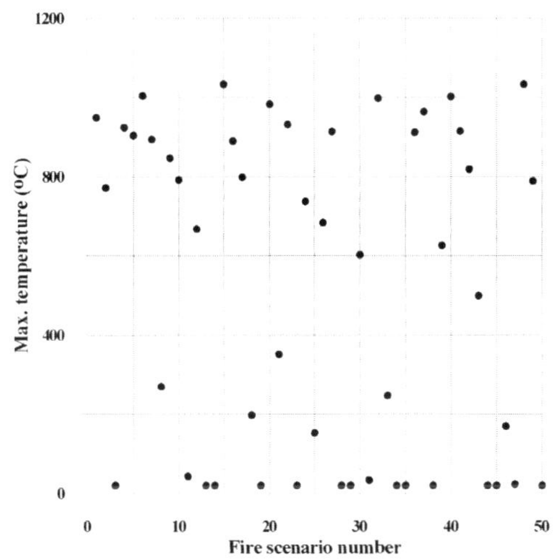

(a) Fire scenario number

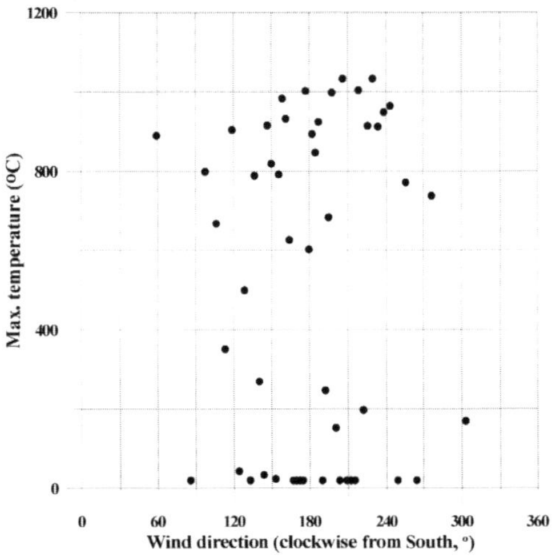

(b) Wind direction (°)

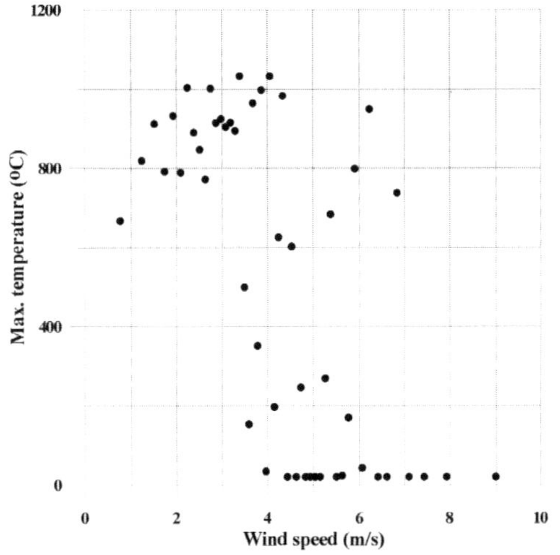

(c) Wind speed (m/s)

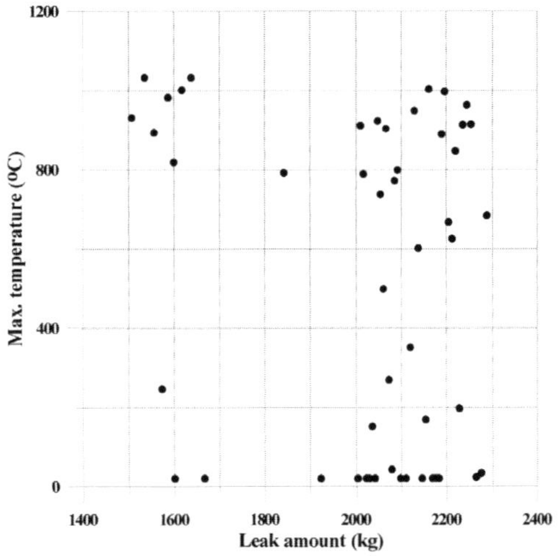

(d) Leak amount (kg)

Figure 15: Effect of variables for fire on maximum temperatures

5. CALCULATION OF FIRE FREQUENCY

To generate a fire exceedance curve, both fire load and fire frequency are required. The fire frequency can be calculated by Eq. (1).

$$F_{Fire} = F_{Release} \times P_{Ignition} \quad (1)$$

where F_{Fire}=fire frequency, $F_{Release}$=leak frequency and $P_{Ignition}$=ignition probability.

5.1 LEAK (ACCIDENTAL) FREQUENCY

OGP [1] provides historical data of heli-accidents on offshore platforms. Table 10 presents the frequency of heli-accidents according to areas. In the present study, frequencies of phase in take-off and landing are considered, as a leak from a heli-accident can occur during either. A fuel release is assumed for all accidents.

For comparison, frequencies in the North Sea and in the Gulf of Mexico (GOM) are applied. Therefore, the resulting frequencies of 8.6E-9/yr (4.3E-7/yr·50, North Sea) and 5.4E-8/yr (2.7E-6/yr·50, GOM) are equally used for leak (accidental) frequency in each fire scenario.

Table 10: Offshore helicopter transport flight accident frequency for the risk estimation model [1]

Region	Flight phase	Frequency (/yr)	Unit
North sea	In-flight	8.5E-6	per flight hour
	Take-off & landing	4.3E-7	per flight stage
Gulf of Mexico (GOM)	In-flight	8.5E-6	per flight hour
	Take-off & landing	2.7E-6	per flight stage
Rest of world	In-flight	8.5E-6	per flight hour
	Take-off & landing	2.7E-6	per flight stage

5.2 IGNITION PROBABILITY

The probability of ignition for fires from helicopter accidents has not been sufficiently investigated, but in this study it must be taken into account, so the 'offshore FPSO liquid model' suggested by OGP [15] is adopted. This model can be applied to the release of flammable liquids that do not have any significant flash fraction (10% or less) within offshore process modules or decks on FPSOs [15]. Figure 16 presents the ignition probability according to the leak rate.

n accident occurs, the fuel spills out of the tank in a If a short time, so the total leak amount is considered as veryrelease rate, shown in Figure 16, to calculate the the tion probability. igni

From Figure 16, 0.028 of initiation probability is used for all 50 fire scenarios.

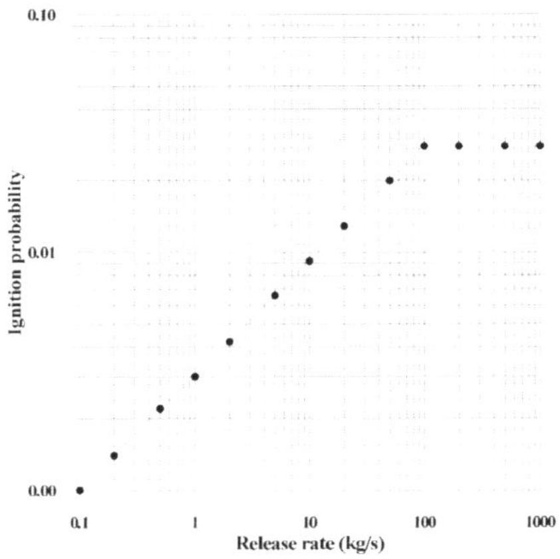

Figure 16: Ignition probability relevant to release rate [15]

5.3 FIRE FREQUENCY

As mentioned in Eq. (1), fire frequency is calculated by leak frequency and ignition probability. Table 11 shows the leak frequency, ignition probability and fire frequency of heli-accidents in the North Sea and the GOM.

Table 11: Leak frequency, ignition probability and fire frequency

Leak frequency (/yr)		Ignition probability	Fire frequency (/yr)	
North Sea	GOM		North Sea	GOM
8.6E-9	5.4E-08	2.8E-2	2.4E-10	1.5E-9

6. DESIGN ACCIDENTAL LOADS FOR FIRES ON HELIDECKS OF OFFSHORE INSTALLATIONS

The fire load exeedance curve can be generated from the fire frequency and fire loads, and design accidental loads are obtained from this curve. The maximum temperature exceedance curve is adopted to define the design accidental load.

Figures 17-19 present the maximum temperature exceedance curves at elevations A, B and C, which are derived from the study.

As shown in Figure 14(a), the temperature exceedance curve at elevation A is the highest compared with that of the air (B and C). This means that the structure, which is the helideck, is exposed to the highest temperature in the accidents. The structure should have sufficient strength to withstand the fire loads.

Helicopter accidents in the Gulf of Mexico are more frequent than in the North Sea. Consequently, the exceedance level in the GOM is almost 10 times that of the North Sea. This suggests that a design for an offshore helideck in a specific ocean area cannot be applied to one in a different area, and that offshore helidecks installed in the GOM need to be stronger for safety.

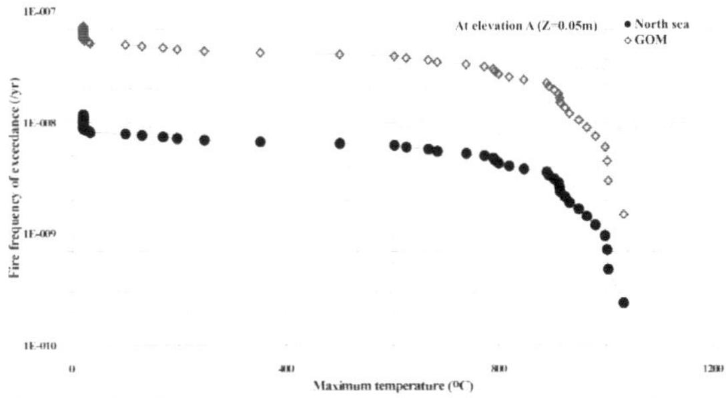

Figure 17: Probability exceedance curves of maximum temperature at elevation A

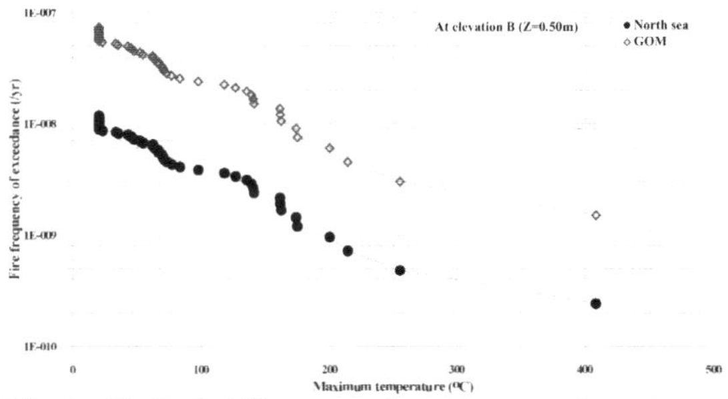

Figure 18: Probability exceedance curves of maximum temperature at elevation B

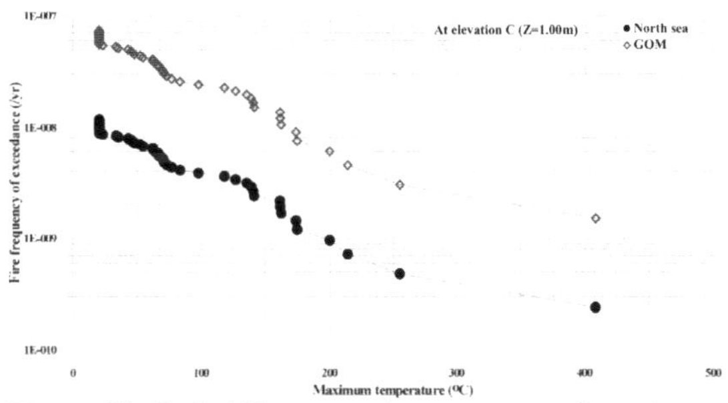

Figure 19: Probability exceedance curves of maximum temperature at elevation C

Helicopter crashes on offshore platforms are rare, so the exceedance level is very much lower than for other accidents, such as explosions, fires, collisions, groundings, etc. Different design accidental levels should therefore be adopted for the structural designs of helidecks regarding fires.

7. CONCLUSIONS

Helicopter crashes on helidecks are rare events, but there is still a risk of helidecks being exposed to fire loads from pool fires that result from heli-accidents. Defining design fire loads is therefore important for safety, and structural design should take into account design loads obtained through risk assessment.

Defining design fire loads for the safe design of helidecks on offshore platforms is not easy, as there is insufficient historical data of heli-accidents.

The aims of this study are to suggest a procedure for defining the design fire accidental loads for aluminium helideck safety, and to carry out a procedure with an applied example.

In the study, it is inevitable that assumptions were needed in the selection of scenarios, and 50 credible accidental fire (pool fire) scenarios were selected, with several assumptions, using the Latin Hypercube sampling technique. The characteristics of fire loads and temperature exceedance curves were investigated.

The conclusions of this study are as follows:

- The temperature at elevation A, the nearest elevation to the heat source, is higher than at the other elevations.
- Helidecks of offshore platform located in the Gulf of Mexico require stronger designs than those located in the North Sea.
- Less than 10^{-8}/yr (at the GOM) and/or 10^{-9}/yr (at the North Sea) of risk accepted load levels should be considered when determining fire accidental loads for aluminium safety helidecks on offshore platforms, in the rare event of a heli-accident.

8. REFERENCES

1. OGP, 'Risk Assessment Data Directory: Aviation Transport Accident Statistics', Report No.434-11.1, *International Association of Oil & Gas Producers*, London, UK, 2010.

2. API, 'Recommended Practice for Planning, Designing, and Constructing Heliports for Fixed Offshore Platforms', *American Petroleum Institute*, Washington D.C., USA, 2006.

3. DNV, 'Offshore Standard: Helicopter Decks', Report No. DNV-OS-E401, *Det Norske Veritas*, Oslo, Norway, 2001

4. SPOUGE, J., 'A Guild to Quantitative Risk Assessment for Offshore Installations', *Det Norske Veritas Technica*, Oslo, Norway, 1999.

5. VINNEM, J.E., 'Offshore Risk Assessment – Principles Modelling and Application of QRA Studies', *Springer*, Stavanger, Norway, 2007.

6. PAIK, J.K. and CZUJKO, J., 'Explosion and Fire Engineering of FPSOs (Phase I): Feasibility Study with Literature Review', Report No. EFEF-01, *The Korea Ship and Offshore Research Institute*, Pusan National University, Busan, Korea, 2009.

7. PAIK, J.K. and CZUJKO, J., Explosion and Fire Engineering of FPSOs (Phase II): Definition of Design Explosion and Fire Loads', Report No. EFEF-03, *The Korea Ship and Offshore Research Institute*, Pusan National University, Busan, Korea, 2010

8. PAIK, J.K. and CZUJKO, J., 'Assessment of Hydrocarbon Explosion and Fire Risks in Offshore Installations: Recent Advances and Future Trends', *The IES Journal Part A: Civil & Structural Engineering*, Vol. 4(3), pp. 167-179, 2011

9. PAIK, J.K. and CZUJKO, J., 'Engineering and Design Disciplines Associated with Hydrocarbon Explosion and Fire Risks in Offshore Oil and Gas Facilities', Transactions of the *Society of Naval Architects and Marine Engineers*, Vol. 120, pp. 1-39, Washington DC, USA, 2012

10. PAIK, J.K., CZUJKO, J., KIM, J.H., PARK, S.I., ISLAM, S. and LEE, D.H., 'A New Procedure for the Nonlinear Structural Response Analysis of Offshore Installations in Fires', *Transactions of the Society of Naval Architects and Marine Engineers*, Vol.121, 2013.

11. RUBINSTEIN, R.Y., 'Simulation and the Monte Carlo Method', *John Wiley*, Chichester, UK, 1981.

12. CZUJKO, J., 'Design of Offshore Facilities to Resist Gas Explosion Hazard: Engineering Handbook', *CorrOcean ASA*, Oslo, Norway, 2001

13. YE, K.Q., 'Orthogonal Column Latin Hypercube and their Application in Computer Experiments', *Journal of the American Statistical Association*, Vol.93(444), pp.121-128, 1998

14. KFX, 'User's Manual for Kameleon FireEx', *Computational Industry Technologies AS*, Stavanger, Norway, 2014

15. OGP, 'Risk Assessment Data Directory: Ignition Probabilities', Report No.434-6.1, *International Association of Oil & Gas Producers*, London, UK, 2010.

9. AUTHORS BIOGRAPHY

Mr. Sang Jin Kim and **Jeong Beom Park** are graduate students of the Department of Naval Architecture & Ocean Engineering at Pusan National University, under the academic adviser Jeom Kee Paik. They are carrying out research on fire and explosion risk assessments and the management of offshore structures.

Dr. Bong Ju Kim is an associate professor of the Korea Ship and Offshore Research Institute (The Lloyd's Register Foundation Research Centre of Excellence) at Pusan National University in Korea. His research interests are in the area of nonlinear structural mechanics and design associated with limit states and risk-based approaches. He also leads the Korean government funded project to build large-scale test facilities for hydrocarbon explosions and fires. He has been a member of the technical committee III.1 Ultimate Strength of International Ship and Offshore Structures Congress.

Professor Jeom Kee Paik is a Professor of the Department of Naval Architecture and Ocean Engineering and serves as president of the Korea Ship and Offshore Research Institute (KOSORI) and as director of the Lloyd's Register Foundation Research Centre of Excellence at Pusan National University (PNU) in Korea. In 2015, he also joined University College London's Department of Mechanical Engineering as a professor and faculty member. He is a Fellow and Vice President of SNAME, USA and a Fellow and Council Member of RINA, UK. He was awarded William Froude Medal from RINA in 2015, the David W. Taylor Medal from SNAME in 2013, and a Doctor Honoris Causa from the University of Liege in Belgium in 2012. His research interests are safety engineering and design of ships and offshore structures in extreme and accidental conditions associated with risk assessment and management.

Dr. Jung Kwan Seo is an associate professor of the Korea Ship and Offshore Research Institute (The Lloyd's Register Foundation Research Centre of Excellence) at Pusan National University in Korea. He serves as director of Subsea Research Centre of the KOSORI. He has been a member of the specialist committee V.8 Risers and Pipelines of International Ship and Offshore Structures Congress (ISSC).

Mr. Won Heaop Shin is a senior research engineer of Research Institute Team N.K co., LTD.

Dr. Joo Shin Park holds the position of Manager in the Research and Product Development Department of Samsung Heavy Industries in Geoje, Korea. He is currently responsible for managing ship and offshore R&D activities. His latest research areas focus on offshore topside engineering.

WINTERIZATION DESIGN OF ALUMINUM HELIDECK IN ARTIC ENVIRONMENT

SY Bae, GH Kang, Electric & Electronic Research Division, Korea Marin Equipment Research Institute, Korea
WH Shin, Research Institute Team, N.K. Co., Ltd., Korea
JS Park, Central Research Institute, Samsung Heavy Industries Co., Ltd., Korea
JK Seo, The Korea Ship and Offshore Research Institute (The Lloyd's Register Foundation Research Centre of Excellence), Pusan National University, Korea

SUMMARY

In recent years, the demand for ships and offshore platforms that can navigate and operate through the Arctic Ocean has been rapidly increasing due to global warming and large reservoirs of oil and natural gas in the area. Winterization design is one of the key issues to consider in the robust structural safety design and building of ships that operate in the Arctic and Sub-Arctic regions. However, international regulations for winterization design in Arctic condition regulated that only those ships and offshore platforms with a Polar Class designation and/or an alternative standard. In order to cope with the rising demand for operating in the Arctic region, existing and new Arctic vessels with a Polar Class designation are lacking to cover for adequate winterization design with HSE philosophy. Existing ships and offshore platform was not designed based on reliable data based on numerical and experiment studies. There are only designed as a performance and functional purposes. Hence, current industrial practices can be difficult to cover a wide range design purposes due to geometric/material changes, and cost saving purposes etc. with a serious risk for worker and operator. It is very important to obtain of reliable data and provide of design guidance of the anti-icing structures by taking the effects of low temperature into consideration. Therefore, the main objective of this paper reconsiders anti-icing design of aluminum helideck using the heating cable. To evaluate of reliable data and recommend of anti-icing design method, various types of analysis and methods can be applied in general. In the present study, finite element method carried out the thermal analysis with cold chamber testing for performance and capacity of heating cables. According to the results and methods can be used to inform and offer as a standardized regulation and design guidance for winterization design of equipment of ships and offshore platforms, thereby providing a more systematic, comprehensive guidance for winterization design than previously available.

1. INTRODUCTION

In the 21st century a major increase in demand is changing the global energy market. In order to meet the demand, the international oil and gas industry is exploring the globe in search of new energy resources. Oil and gas exploration companies are being challenged to provide highly technical and environmentally sensitive solutions. The Arctic region contains a large part of world's undiscovered oil and gas resources. However, exploration and production in the Arctic, where companies and humans are faced with extreme temperatures down to -60°C, is a difficult undertaking.

Recently, the Arctic region has become an attractive region. Owing to global warming, the extent of Arctic ice has decreased enough to allow more offshore activities and shipping in Arctic regions. Approximately 13% of undiscovered natural fossil resources in the Arctic region are becoming practicable. Moreover, there are some advantages in using the Northern Sea Routes (NSR) for shipping, i.e. shorter sailing distance, less fuel spend, reduction of carbon dioxide emission and high earning.

On the other hand, the IMO (International Maritime Organization) Guidelines (IMO, 2011) for ships operating in polar waters recommended that only those ships with a Polar Class designation, which were assigned based on the IACS Unified Requirements for Polar Class Ships (UR-I) (IACS, 2011), or a comparable alternative standard of ice-strengthening appropriate to the anticipated ice conditions, should operate in polar ice-covered waters. In the high north Arctic and Polar Regions the extremely low ambient temperatures and high winds create snow and ice conditions that require unique operational considerations.

Recent years, the demand for ships and offshore platforms that can navigate and operate through the Arctic Ocean has been rapidly increasing due to global warming and large reservoirs of oil and natural gas in the area. Winterization design is one of the key issues to consider in the robust structural safety design and building of ships that operate in the Arctic and Sub-Arctic regions. However, international regulations for winterization design in Arctic condition regulated that only those ships and offshore platforms with a Polar Class designation and/or an alternative standard. In order to cope with the rising demand for operating in the Arctic region, existing and new Arctic vessels with a Polar Class designation are lacking to cover for adequate winterization design with HSE philosophy.

To prevent icing of ballast system of ships operating in polar are developing heating system and circulating system with bubble (Jeong, 2011). Catalyst method has been gone for environment requirement, currently; the heating line system tends to design of winterization method with HSE philosophy.

Existing ships and offshore platform was not designed based on reliable data based on numerical and experiment studies. There are only designed as a performance and functional purposes. Hence, current industrial practices can be difficult to cover a wide range design purposes due to geometric/material changes, and cost saving purposes etc. with a serious risk for worker and operator. It is very important to obtain of reliable data and provide of design guidance of the anti-icing structures by taking the effects of low temperature into consideration.

Figure 1: Helideck in Artic region

Anti-icing or de-icing of helicopter decks can also be required. Different deck designs and materials of construction can require to evaluate each design to ensure adequate heat and reliable performance.

Therefore, the main objective of this paper reconsiders anti-icing design of aluminium helideck using the heating cable. To evaluate of reliable data and recommend of anti-icing design method, various types of analysis and methods can be applied in general. In the present study, finite element method carried out the thermal analysis with cold chamber testing for performance and capacity of heating cables. According to the results and methods can be used to inform and offer as a standardized regulation and design guidance for winterization design of equipment of ships and offshore platforms, thereby providing a more systematic, comprehensive guidance for winterization design than previously available.

2. INTERNATIONAL STANDARD

For the correct use of this standard, the winterization level (Basic, Cold or Polar) shall be specified by the end user (DnV, 2013). According to DnV standard, the Winterization Temperature (t_w) value shall be specified by the end user (in ℃) and is subject to agreement with the Society in each case. Indicative Winterization Temperature (t_w) Basic and Cold and Polar level are -15℃, -30℃ and -45℃. It can be categorized one between one and two for helideck. It should be de-icing 4~6 hour before landing. However, it should be considered in Artic condition due to icing depth (5~15 cm). Anti-icing and de-icing arrangements and equipment in DNV code, defined the heating power capacity at least 300W/m^2 for open deck areas, helicopter decks.

Figure 2 illustrates minimum input value (W/m^2) of heating cables due to temperature and wind. It can be known that the suitable input capacity tends to equal without wind while temperature is lower. Also, even temperature as -5℃, the input capacity is equalled value up to 15m/s. However, it may be having uncertainty in the actual environmental condition. A wind effect should be clarified with wind speed and same temperatures from heating cables. Therefore, the main objective of this paper reconsiders anti-icing design of aluminium helideck using the heating cable. To evaluate of reliable data and recommend of anti-icing design method, various types of analysis and methods can be applied in general. In the present study, finite element method carried out the thermal analysis with cold chamber testing for performance and capacity of heating cables.

Heat Loss from Flat Surface in w/m^2:

Wind (m/s)	Ambient (deg C) -5	-10	-15	-20
no wind	300.0	300.0	300.0	300.0
5	300.0	300.0	334.0	432.2
10	300.0	378.1	535.7	693.2
15	300.0	506.3	717.3	928.2
20	365.2	626.1	886.9	1147.8

(shaded) recommending minimum heat load of 300 w/m^2 in accordance with DNV

Figure 2: Example of DNV rules

3. TESTING FOR COLD CAMBER

Fig. 3 show a test system of heating cable. Varying heating (total 12 cables) have been as a candidate for suitable the current AL helideck winterization. Cases considered for parameterization are as shown in Fig. 4.

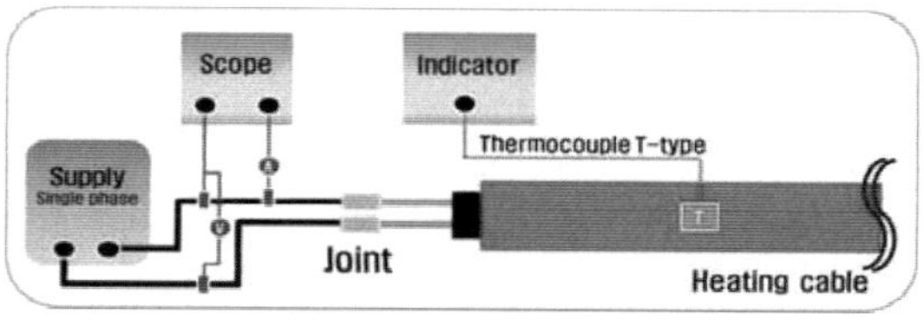

Figure 3: Heating test of heating cable

	3B	5B	8B	10B	100	150	200	4X	8X	10X	15X	20X
온도[℃] Startup: 15.5[℃]	31.1	36.4	40.7	39.8	53.3	72.9	77.7	54.2	70.1	69.0	87.1	93.9
⊿온도[℃]	15.1	20.4	24.7	23.8	38.1	57.7	62.5	38.7	54.6	53.5	71.6	78.4
입력[W]	25.76	31.51	39.79	49.91	83.5	83.0	181.1	55.4	60.7	171.1	189.7	211.3
길이[m]	2.64	2.66	2.41	3.34	3.10	2.65	3.77	2.73	2.05	4.85	3.94	3.77
단위길이 당 입력[W/m]	9.75	11.84	16.51	14.94	26.9	31.3	48.0	20.3	29.6	35.2	48.1	50.9
⊿[℃]/[W/m]	1.54	1.72	1.49	1.59	1.41	1.84	1.35	1.90	1.84	1.51	1.48	1.54

Figure 4: Results of heating test for 12 kinds of heating cable

The results of test show in Fig. 4 and Fig. 5. Various models show starting temperatures (15.5℃), variation of temperatures(∆), electronically input value (W), length of cables (m), input value per unit length (W/m), variation temperature with input value per unit length(Δ/W/m).

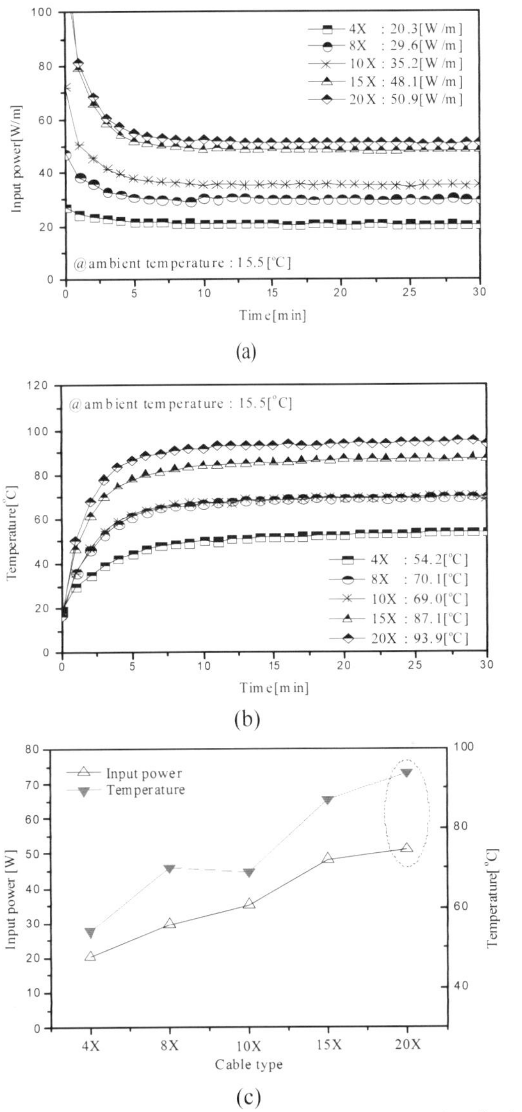

Figure 5: (a) Input power per unit length (b) The final convergence temperature (c) Comparison of heating cable type

Highly Accelerated Life Testing (HALT) techniques are important in uncovering many of the weak links inherent to the design and fabrication processes of a new product. HALT is a method aimed at discovering and then improving weak links in a product during the design phase. HALT is a discovery test as opposed to a compliance test; that is, the goal is to find problems, doing everything necessary to find and then remove the weaknesses.

In HALT, every stimulus of potential value (temperature, all-axis vibration, humidity, UV, radiation, etc.) can be used under accelerated test conditions during the development phase of a product to find the weak links in the design and fabrication processes. Accelerated stresses in combination (e.g. high-temperature ramp rates and all-axis vibration levels together) are necessary to compress or minimize the time to failure.

A highly accelerated life-test (HALT) method was employed to temperature stress test and thermal transitions test as shown in Figs. 6 and 7.

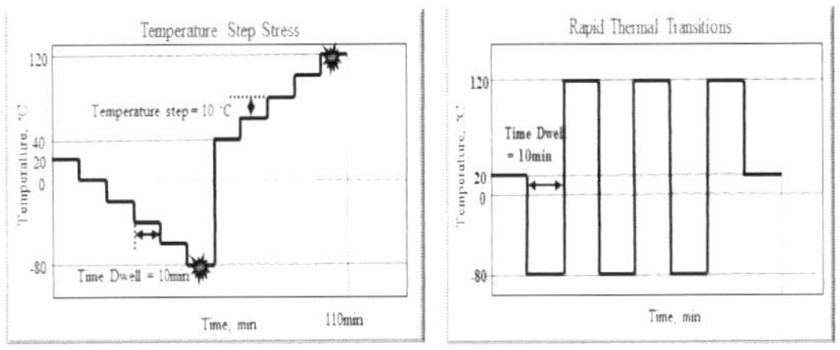

Figure 6: (a) Temperature stress test (b) Rapid thermal transitions

System Specifications

Temperature Range	+200°C to -100°C (+392°F to -148°F).
Thermal Ramp Rate	Minimum of 200°C per minute for temperature range of +85°C to -65°C.
Vibration Range	Up to 60 GRMS on table. This GRMS specification is calculated based on a cut-off frequency of 5000 Hz. This same calculation method is the basis of the GRMS number displayed by the CTI II touch-screen interface.
Cooling	Vacuum insulated liquid nitrogen system including proportional control valve and redundant safety. Direct nitrogen atomization in plenum.
Vibration Table	48" x 48" surface with 3/8-16" mounting inserts on 4" centers.
Heaters	Open element nichrome type for low thermal mass and quick response.
Exciters	Sixteen pneumatically actuated hammers, each with adjustable ball valves, which generate six degrees of freedom, non-coherent, broad band excitation.
Vibration Table Maximum Payload	600 lbs.
Doors	Two in front & two in back, open to 95° for easy access.
Door Interlocks	Inhibit thermal and vibration systems when any door is ajar.
Windows	14" x 24" multi-pane viewing on each door.
Access Port	Four 6" x 8" ports with hinged phenolic cover & foam port plugs.
Controller	PLC based controls with Chart's CTI II color touch-screen interface.
Channels	Temp: Plenum Temp, Product Temp, (2) auxiliaries (standard). Vibration: Table Control, (3) auxiliaries (standard).
Software Programs (optional)	Chart SCADA Package allows operating the chamber and logging data from a PC.
Interior Workspace Dimensions	50" w x 54" d x 32" h.
Exterior Dimensions	99" w x 71" d x 107" h.
Weight	4,500 lbs.
Power Requirements	460 VAC, 3 Phase, 140A Service.
Main Power Disconnect Switch	Rated at 50,000 AIC at 480 VAC, with lock-out capability.
Compressed Air	3/4" supply, 120 psig, 140 SCFM, 8.3 barg, 3.9 m^3/minute.
Exhaust	Two 6"/15.2 cm port.
LN_2 Connection	1" Female Chart Dissimilar Metal Bayonet.

Figure 7: Specification of Highly accelerated life test chamber

4. WINTERIZATION DESIGN BASED ON THERMAL ANALYSIS

4.1 THERMAL ANALYSIS

The finite element method (FEM) has become an essential solution technique in many areas of engineering and physics. In this thermal analysis carried out the test results for reliable data. Same as the simulation is not considered wind effect, and it measured a distributed temperature of deck frame structure due to heating input value for heating lines. The maximum input temperature of heating line selected in the experimental results (20X). The maximum temperature and input capacity are 94 ℃ and 50.9 (W/m)

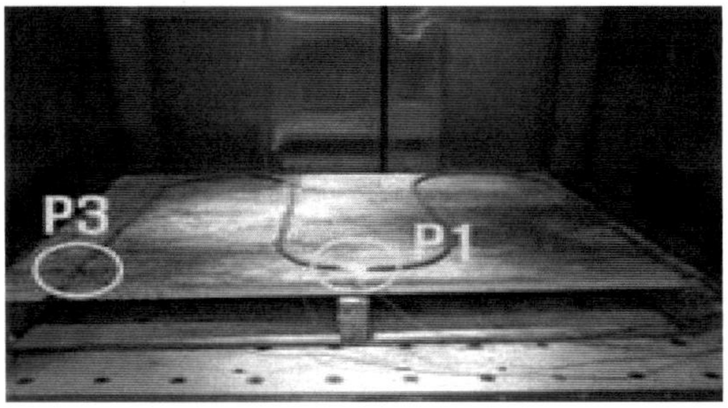

Figure 8: Temperature of heating cable area

The temperatures of heating cables measured as a attached with wooden specimens at -15℃ as shown in Fig. 8. The cold chamber has been set at -15℃, but it may be oscillated around 0.015%. P1 and P3 indicated curved part and ended point of heating cable. It is normalized temperature as average temperature at -15℃. The normalized temperature used the input value for thermal analysis.

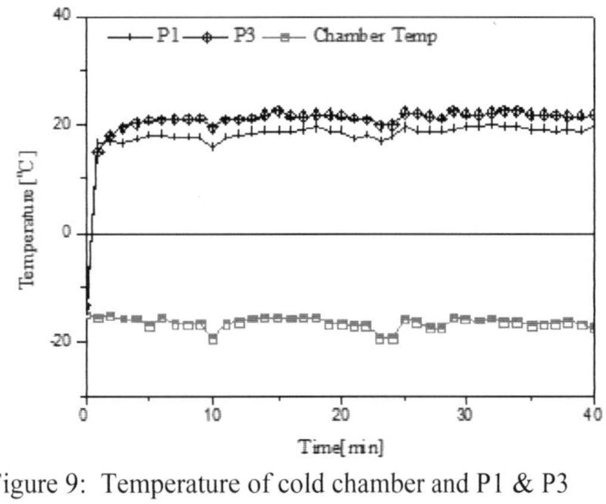

Figure 9: Temperature of cold chamber and P1 & P3

As shown in Fig. 9, the average temperature which is mean value both P1 and P2, is 20.6 °C. Measured temperature is 94 °C in above zero and 20.6 temperature at -15 °C.

A three dimensional finite element model of the helideck package is solved numerically and simulated using ANSYS Workbench to fulfil the purpose as shown in Fig. 10. Maximum T_j on the die is simulated for various parameters as mentioned in Fig. 3. Figure Fig. 10 (a) shows the three dimensional finite element model.

In the thermal analyses, eight nodes thermal elements are used. The FE modelling was considered the connected part which occur temperature reducing around 2 . Therefore, the helideck modelled two connected pancake FE structure. Five lines heating cable installed at one deck frame, total 10 lines installed for FE modelling. ANSYS Workbench is used to carry steady state thermal analysis of the mentioned the pancake using eight nodes elements to mesh the helideck model. The material properties for the analysis were evaluated at the average tested average temperature at -15℃.

Material properties are assigned to the AL helideck components in terms of AL isotropic thermal conductivity values as given in 20W/m^2·℃. It is assumed that the change in temperature of a particular component does not have any effect on its thermal conductivity.

The model is discretized into 2,132,617 elements with 4,151,169 nodes as shown in Fig. 10(c).

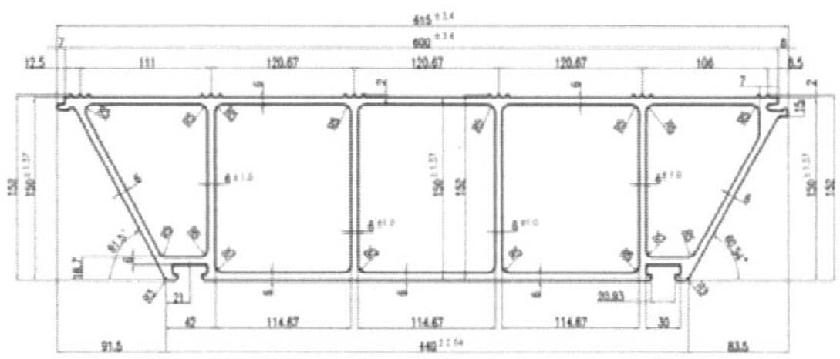

Figure 10(a): Geometry of helideck

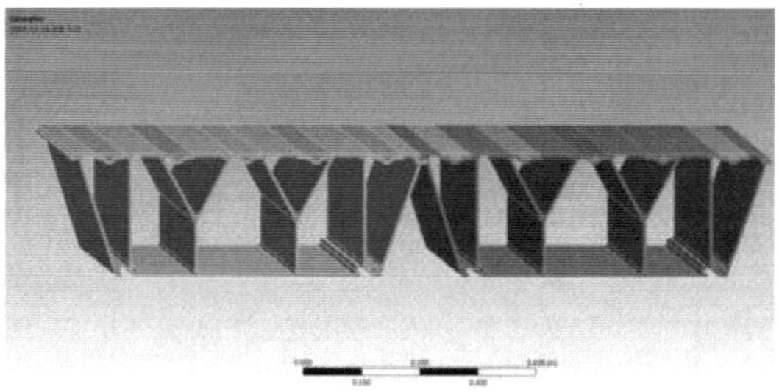

Figure 10(b): 3D modelling of helideck

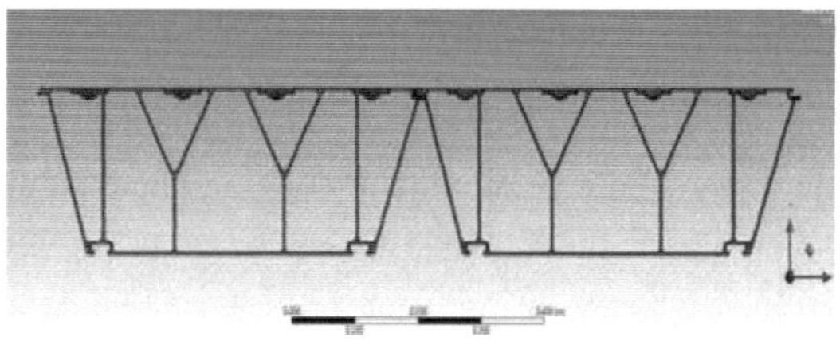

Figure 10(c): FE model of helideck

4.2 RESULTS OF THE THERMAL ANALYSIS

The calculated temperature distribution in the helideck is shown in Fig. 11. Fig.11 shows that a distributed temperature of whole pancakes, surface temperature at middle of pancakes, and distributed temperature of central part among to length.

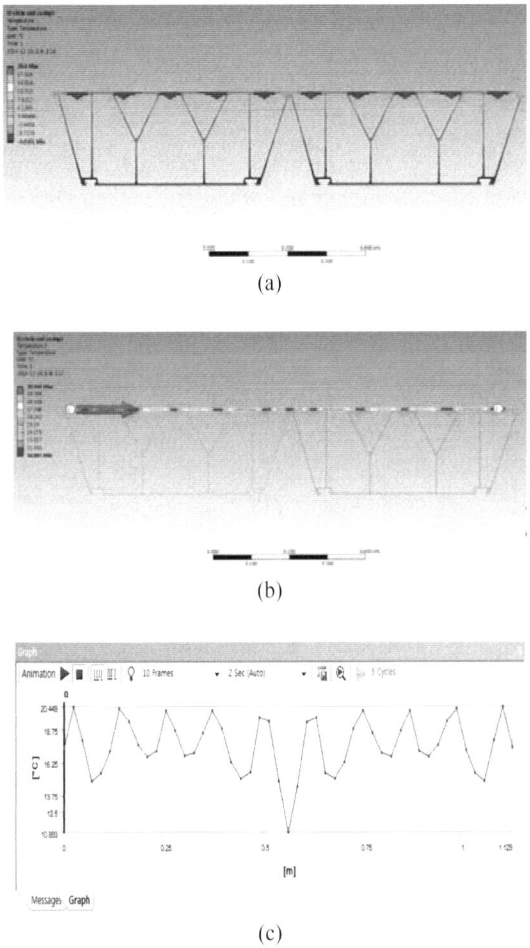

Figure 11: (a) Overall temperature distribution (b) The temperature distribution of central deck (c) Central deck temperature distribution graph

As expected results, the maximum temperature is near to heating cable, minimum temperature is the connected part which is around 10.9℃. It can be accepted winterization design requirement (more than 2℃ degree)

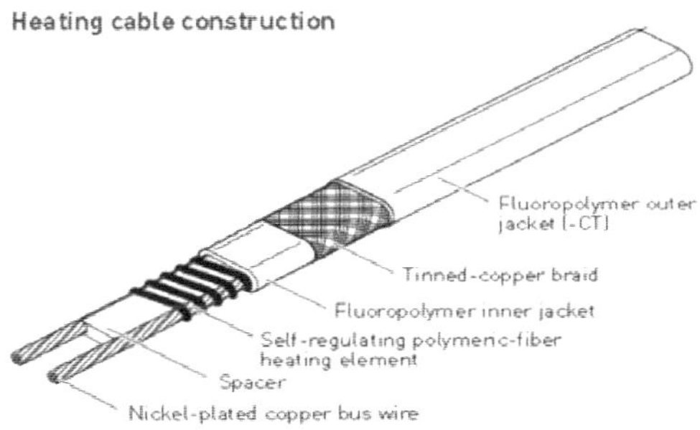

Figure 12: Heating cable construction

The suitable heating cable can be suggested, which is based on experimental test and numerical results, as shown in Fig. 12. And detail of cables parameters and electrical design data sheet shown in Fig. 13.

Table 1(a): Electrical design data sheet

Cable parameters		
Type of cable	Self-regulating cable	
Model /company	20XTV2-CT/raychem	
Available ambient temperature [1]	-15	[°C]
Startup temperature [2]	5	[°C]
Needed input power@-15°C	81	[W/m]
Total length of cable	1,382	[m]
Sizing(W • H)	11.7 • 7.6	[mm]
Weight	345.5	[kg]
Max. unjoint cable length	data sheet	
Power source parameters		
Source type	AC single phase	
Input voltage	200~277	[V]
Total input power@220[V]	111.9	[kW]
Frequency	50/60	[Hz]
Deck parameters [3]		
Width of 1 block	2.4	[m]
Height of 1 block	2.4	[m]
Number of block	36	
Number of piece per 1 block	16	
material of deck frame	Aluminium	
Total length of deck(W • H)	14.4 • 14.4	[m]
Installation conditions		
Length of straight line	14.4	[m]
Number of straight line	96	
Capa. of circuit braker per line	30	[A]
Number of control panel [4]	6	
Number of power unit [5]	6	
Ampere of each panel	85	[A]

Table 1(b): Heating cable spec

Etc.,	
name	20XTV2-CT
Certifications	IECEx BAS 06.0044X Ex e IIC T*Gb Ex tD A21 IP66 T**°C Class I, Div. 2, Groups A, B, C,D Class II, Div.2 , Groups F, G Class III CLI, ZN1, Aex e II T3[T2] 09-Iec-0005X Ex e IIC T*Gb
Max. using length:240[V] (Circuit breaker Capa. / length [m])	15 [A] / 28 [m] 20 [A] / 38 [m] 30 [A] / 57 [m] 40 [A] / 77 [m] 50 [A] / 97 [m]
Heating cable dimensions[mm]	W : 11.7, H : 7.6
Min. banding radius[mm]@20°C	12.7[mm]

It may be changeable to different materials and geometry of helideck structures due to environmental condition, different regulation and different helicopter etc. This suggestion is only given possible information of electrical design though current industrial practices.

5. CONCLUSIONS

The main objective of this paper reconsiders anti-icing design of aluminium helideck using the heating cable. To evaluate of reliable data and recommend of anti-icing design method, various types of analysis and methods can be applied in general. In the present study, finite element method carried out the thermal analysis with cold chamber testing for performance and capacity of heating cables.

The conclusions of this study are as follows:

- These results of experimental and numerical based on DnV guidance give reliable data for winterization design.
- According to DnV, it is determined minimum heating capacity of 300W/m2 without wind effect at -15. But it may be recommended new guidelines should be considered with various environmental temperature effects.
- It is suggested that electrical capacity with suitable heating cables. But it may be changeable to different materials and geometry of helideck.

According to the results and methods can be used to inform and offer as a standardized regulation and design guidance for winterization design of equipment of ships and offshore platforms, thereby providing a more systematic, comprehensive guidance for winterization design than previously available.

6. REFERENCES

1. ANSYS analysis manual (library); ANSYS is a trade mark of ANSYS Inc., Houston, PA, USA

2. Fridtjof Nansen Institute (FNI), 2007. The IMO Guidelines for Ships the structural designs of helidecks regarding fires.

3. IMO, 2011. MARPOL Annex VI. International Maritime Organization, London, UK.

4. IACS, 2011. Requirements concerning POLAR CLASS. International Association of Classification Societies, London,UK..

5. HO-KEUN KANG, KI-PYOUNG KIM, DAE-HEON KIM, 2010, Requirements for Mechanical and Environmental Conditions on Winterization of Vessels Intended for Cold Climate Operation, The Korean Society of Marine Engineering

6. HO-KEUN KANG, KI-PYOUNG KIM, DAE-HEON KIM, T.B. Ha, 2011, The Rules and Winterization Technology on Ships Intended for Cold Climate Operation, The Korean Society of Marine Engineering Vol(35), pp.721-731.

7. SEONG-YEOB JEONG, CHUN-JU LEE, SEONG-RAK CHO, 2011, A Study on Anti-Icing Technique for Ballast Water of Icebreaking Vessels Operating in Ice-Covered Water, Journal of the Society of Naval Architects of Korea , Vol.48, No.1, pp.93-97.

7. ACKNOWLEDGEMENTS

This work was supported by Industry Convergence Core Technology Development Projects (Grant no.: 10041090) funded by Ministry of Trade, industry & Energy.

8. AUTHORS BIOGRAPHY

Ms. So Young Bae is a senior researcher of Korea Marine Equipment Research Institute. She is responsible ocean IT convergence between electrical and ship and offshore R &D activities.

Dr Gyu Hong Kang is a general manager of Electric & Electronics Research Division at Korea Marine Equipment Research Institute.

Mr Won Hyeop Shin is a senior research engineer of Research Institute Team N.K co., LTD.

Dr Joo Shin Park holds the position of Manager in the Research and Product Development Department of Samsung Heavy Industries in Geoje, Korea. He is currently responsible for managing ship and offshore R&D activities. His latest research areas focus on offshore topside engineering.

Dr Jung Kwan Seo is an associate professor of the Korea Ship and Offshore Research Institute (The Lloyd's Register Foundation Research Centre of Excellence) at Pusan National University in Korea. He serves as director of Subsea Research Centre of the KOSORI. He has been a member of the specialist committee V.8 Risers and Pipelines of International Ship and Offshore Structures Congress (ISSC).

ALUMINIUM HELIDECK DESIGN FOR EUROCODE9 WITH DEFORMATION BASED DESIGN

JS PARK, YS Ha and KB Jang, Department of Noise, Vibration & Structure Research, Central Research Institute, Samsung Heavy Industries Co., Ltd., Korea

SUMMARY

Aluminium alloys are gaining increasing usage in the construction industry, offering high strength-to-weight ratios, good durability and ease of fabrication. The aluminium helideck structure should satisfy requirements in accordance with some offshore regulations and rule notations such as the Australian/New Zealand Standard and EUROCODE 9. The width-to-thickness ratio and the yield stress are recognized as the governing design parameters in the design of cross sections in these specifications. The aluminium helideck structures are composed of several structural units such as aluminium part and steel part. The aluminium part is composed of pancake, girder, safety net, and so on.

In the present paper, the aluminium helideck design with relevant EUROCODE 9 is done based on the strength calculation. It could be possible to reduce calculation time and thus provide reasonable solution in view of practical design. Static and nonlinear collapse behaviour of developed structure is investigated in this study.

The main purposes are to provide a reasonable solution that can improve the product quality by checking both strength and deformation criteria. The effect of deflection during the fabrication stage is considered in the structural design based on newly proposed EUROCODE 9. Lastly a comparison between the present results and the results by FE-Analysis is presented.

1. INTRODUCTION

Recently, the numbers of large offshore structures and platforms are rapidly increasing demand from offshore oil majors. The helideck structure should be installed in these offshore structures for the purpose of helicopter landing and take-off. It seems important to underline that nowadays the offshore applications can be considered the main future trend for aluminium alloys. In fact, they offer many benefits to offshore industry in the form of cost savings, ease of fabrication and proven performance in hostile environments. Stair towers, Mezzanine flooring, access platforms, walkways, gangways, bridges, towers and cable ladder systems can all be constructed in pre-fabricated units for simple assembly offshore or at the fabrication yard (Mei-Ni, 2014).

Helidecks have been made by using aluminium alloy since the early seventies, so they have now a fully tried experience. Moreover, they are designed to be modular and bolted connections, allowing quick erection, easy shipping and handling. In addition, they offer weight reduction of up to 70% over steel, meeting the highest safety standards and proving up to 12% cost saving.
In structural applications, 6000 series alloys are commonly used because of their favourable combination of properties. Among the 6000 series alloys, 6082 alloy is a relatively new aluminium alloy, one popular in Europe and experiencing much in used in America as well (Kissell, 2002). 6082 alloy provides a superior combination of properties such as high strength after heat treatment, satisfactory corrosion resistance, good machining properties, and good weld-ability (Mazzolani, 1995). Compared with the classic and widely-used 6061 alloy, 6082 alloy has higher strength (0-8% higher for characteristic values of tensile ultimate strength, depending on product form and alloy temper, provided by Eurocode 9(hear after EC9), better general corrosion resistance, and is approximately equivalent in term of other properties such as density, extrude-ability and anodizing response(Kissell, 2002).

During engineering stage for aluminium structure, it should be designed based on international standards such as ISO codes, Eurocodes or national standards, e.g. BS5950, NORSOK N-004 or AISC.

The various offshore structures with typical octagonal helideck are shown in Figure 1.

In the structural engineering stage of the aluminium helideck, it is found that some of design factors in the EC9 are not clearly determined. Therefore, engineering decisions for some un-cleared design factors as well as methodology are newly developed.

The main objectives of the study are as follows;

- To develop a reasonable procedure to estimate the structural safety of aluminium structure
- To propose the design based on the deformation for aluminium helideck under impact load

Figure 1: Helideck Installation and configuration

1.1 LITERATURE SURVEY

At the beginning, a brief review is made on previous research works related to structural engineering for aluminium members.

Mei-Ni Su et al. (2014) performed two series of simply supported bending tests on aluminium alloys square and rectangular hollow sections. The test program comprised of 14 three-point bending tests and 15 four-point bending tests. The combined data set from the tests and the numerical simulations were used to assess the accuracy of three international design specifications; the Aluminium Design Manual, the Australian/New Zealand Standard and Eurocode 9 and the CSM(Continuous Strength Method) for predicting the moment capacity of simply supported aluminium alloy beams. Through a deformation based approach that incorporated strain hardening, the continuous strength method was shown to offer improved predictions of capacity, up to 30% beyond those achieved in current specifications.

Yujin Wang et al. (2015) performed a series of tests on the stability of heat-treated aluminium alloy 6082-T6 circular tube columns to check the reliability of buckling strength predictions using current design rule. These comparisons shows that American Aluminium design manual predictions are too conservative for small slenderness ratios, the Australian/New Zealand Standard predictions are unsafe for large slenderness ratios, the EC9 predictions are conservative and Rasmussen-Rondal formulation provides the closest and generally conservative strength predictions.

Naiquan Ye and Torgeir Moan (2007) are investigated static and fatigue behaviour of three typed of aluminium box-stiffener/web connections. The main purposes are to provide a connection solution that can reduce the fabrication costs by changing the cutting spaces on the web frame and correspondingly the weld process and meanwhile sufficient fatigue strength can be achieved. Finite element analysis shows the influence of local geometry and weld parameters on the stress gradient near the fatigue cracking area. The influence of the weld parameters on the structural stress concentration factors is also studied.

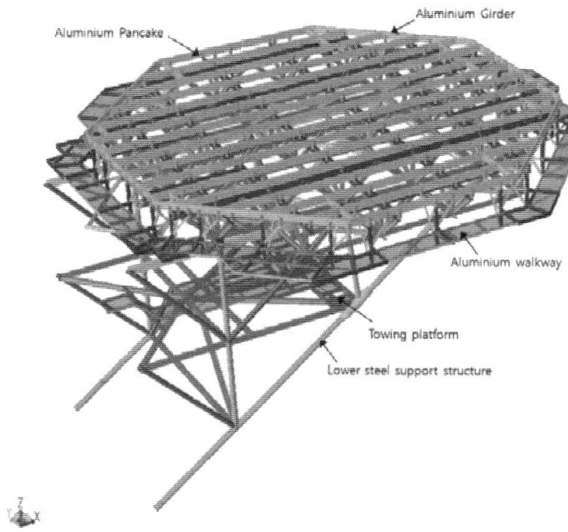

(a)Overall view of helideck structure

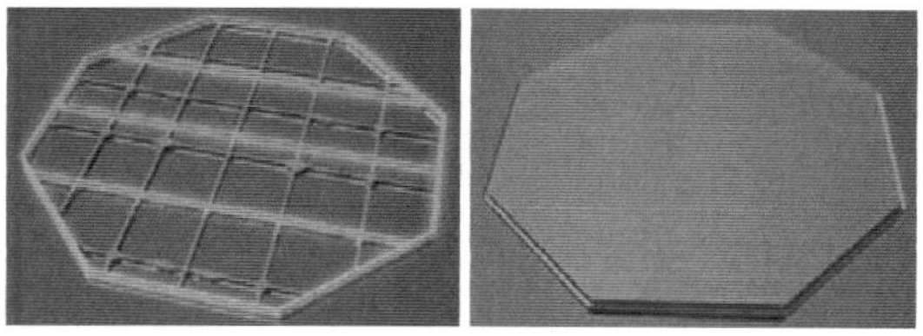

(b)Girder (c) Pancake

Figure 2: Helideck construction

The aluminium helideck structure consists of pancake, girder and steel supporting members as shown in Figure 2. The structural engineer should consider all possible basic loads and load combinations. The governing load is impact induced when helicopter landing takes place and the helideck structure has to sustain & cope with imposed loads on the deck from personnel, freight, refuelling and other temporary equipment, as well as environmental loads from wind, snow and ice, rotor downwash, etc. and its own self weight (UK CAA CAP437, 2012).

The codes and standards applicable for the structural design of the aluminium helideck will be determined by where the helideck is to be operated and the national jurisdiction governing the installation.

This paper intends to prove the structural strength for the aluminium helideck structure. International standards such as EC9 should be specified for rule based design with collaborated FE-analysis.

2. ALUMINIUM HELIDECK DESIGN

2.1 MODEL AND KEY INPUTS

The target helicopter is Sikorsky S-92 with MTOW (Maximum Take-Off Weight) of 12.6ton and rotor diameter value of 20.88m as indicated Table 1.

Table 1: Helicopter specification

Type of helicopter	Sikorsky S-92
Maximum Take-Off Weight (MTOW)	12,565 kg
Crew	2 (pilot, co-pilot)
Length	17.10 m
Width	5.26 m
Height	4.71 m
Distance between wheels	3.18 m
Applied contact area	300 x 300 mm = 90,000 mm^2

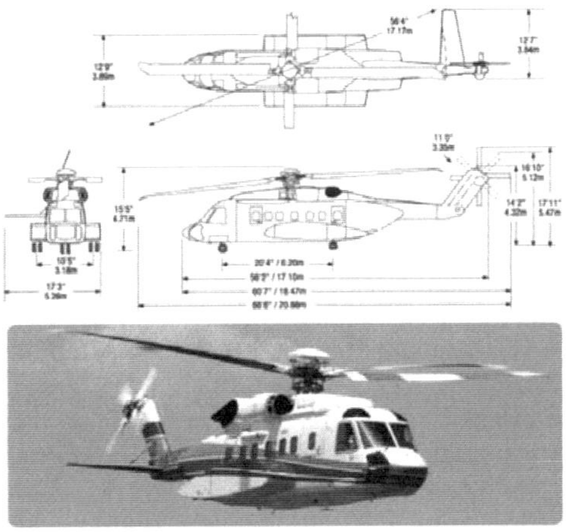

Figure 3: Design information of Sikorsky S-92

The governing load is impact load induced by helicopter landing and take-off action. In the accidental condition, an impact load of 2.5 times MTOW should be considered in overall deck (UK CAA CAP437, 2012). Also, helideck structure should be designed considering lateral impact load of half of MTOW.

Table 2: Impact loads induced by helicopter landing

Vertical impact load per a wheel (Emergency)	200.8 kN
Vertical impact load per a wheel (Operating)	123.6 kN
Lateral impact load	61.8 kN

2.2 ALUMINIUM PANCAKE MODEL

The developed aluminium pancakes have the profile of breadth equal to 615mm and height equal to 150 mm as shown Figure 4.

The aluminium pancake with Y-shaped cores is developed considering efficiency by extrusion as well as structural strength.

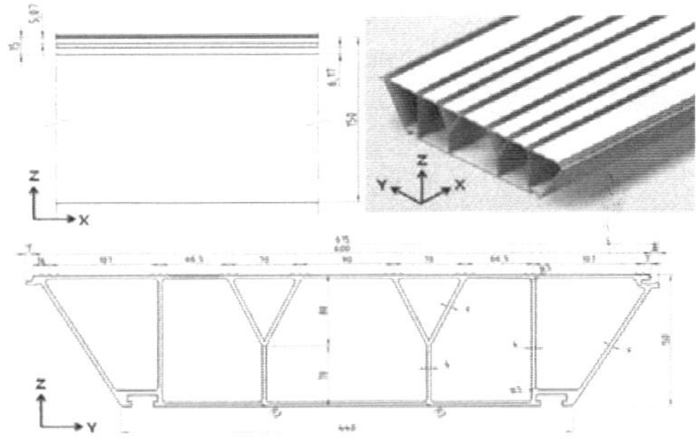

Figure 4: Aluminium pancake profile with Y-cores

The aluminium H-beam profile is defined as shown below in which the shape and dimension of the cross-section are mentioned.

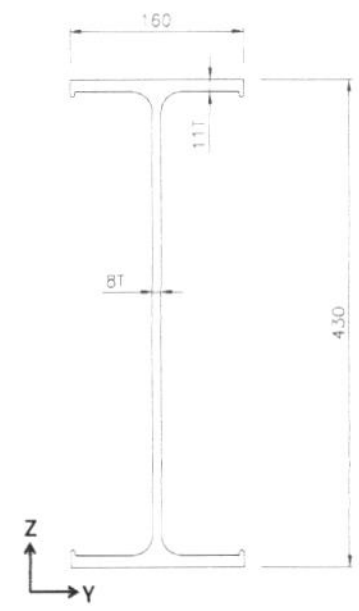

Figure 5: Dimension of aluminium H-beam profile

Section properties for the aluminium H-beam profile have been taken as follows.

Table 3: Section properties of aluminium H-beam profile

Area of cross-section	A	7,134 mm^2
Inertia moment about Y axis	I_y	213,882,693 mm^4
Inertia moment about Z axis	I_z	7,985,029 mm^4

The evaluations with regard to the structural safety and stability of the aluminium pancake are carried out by BS EN 19999-1-1:2007 to comply with the design requirement. It is found that some design factors are not clearly defined in codes and standards. For evaluating by code and defining insufficient design factors, 3-dimensional finite element analysis is performed with FEA program "ANSYS Mechanical APDL" developed by ANSYS software Inc. U.S.A.

Figure 6 shows the flow chart for engineering process of the aluminium helideck structure by EC9.

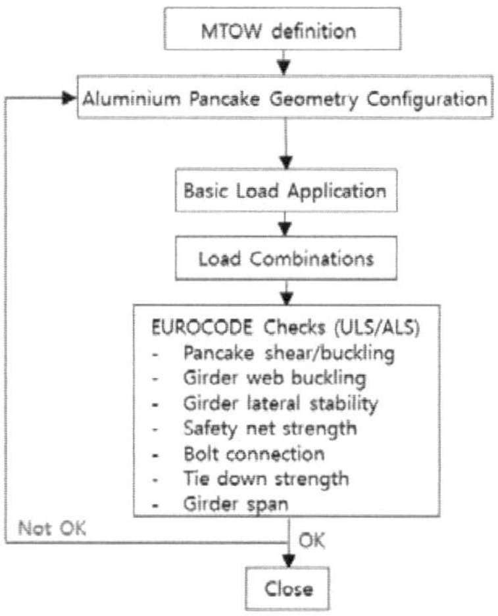

Figure 6: Flow chart for design process by classical EC9

Among the three kinds of method suggested by classical EC9, the first one is purely conventional approach, since is based on a concentrated plasticity idealization, which hardly corresponds to the actual structural behaviour at collapse.

The second one requires for the structure to be calculated two times, with a concentrated plasticity approach, as well as with a FEM numerical simulation. For this reason, it would result in a higher computation cost, which is not suitable for practical analysis of aluminium alloy structures, without disregarding the actual mechanical features of the materials.

Furthermore, because of its inherent simplicity, it can be profitably used as a design method for structures at the ultimate limit state. From practical point of view, it is quite similar to the classical plastic hinge method applied for steel structures. EC9 provides some indications about the use of such method for the plastic analysis of structures by considering the actual mechanical properties of the different aluminium alloys.

In the stage of structural engineering such as pancake shear check and girder span check, it essentially needs the number of pancakes which is not clearly defined in EC9. In order to solve these un-defined problems, the effective pancake areas are defined by using FE-analysis. The dashed red line is effective pancake area against impact load induced by helicopter take-off action in Figure 7. It is very important key factor during design stage.

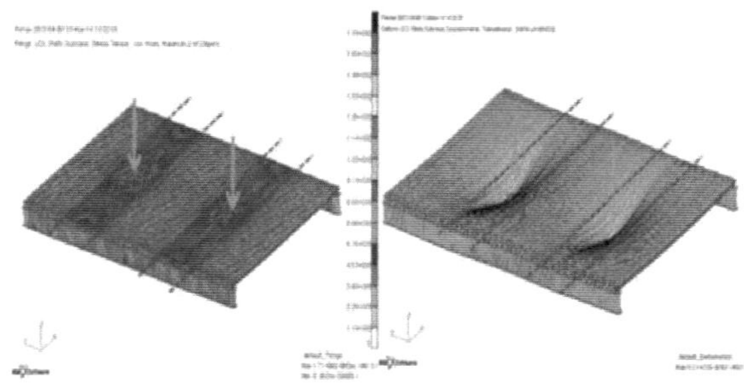

Figure 7: Defined effective pancake area based on FE analysis

In order to check the safety for aluminium helideck structure, EC9 proposed to check the strength capacity by both ULS (Ultimate Limit State) and ALS (Accidental Limit State) by using load and material factor as relevant to condition.

3. FINITE ELEMENT ANALYSIS

3.1 FE INPUTS AND METHODOLOGY

Simulation of the actual structural behaviour of the aluminium pancake requires several considerations. As the thin plate elements of structures are subjected to local and global buckling, while the plate and vertical web are subjected to in-plane and out of plane buckling effects, the chosen element must be capable of modelling these buckling phenomena and the associated behaviour. It must be capable of modelling the structural behaviour both in linear and nonlinear regions involving large displacements, elasto-plastic deformations and associated plasticity effects. A material constitutive curve using the Ramberg-Osgood law, based on the minimum requirements of the material (AL 6082 T6) is applied for the aluminium pancake.

3.2 CONSTITUTIVE LAW

A linear material model is satisfactory when only small quantities of the material are exposed to the yield stress or greater. The type of constitutive law is adopted to model the behaviour of steel; bilinear isotropic hardening. The *Bilinear Isotropic Hardening* options uses von Mises yield criteria (which includes most ductile metals) coupled with an isotropic work hardening assumption. This option is often preferred for large strain analyses. In practical designs, some engineers want to know the factor of safety of the structure for a nonlinear buckling load. Post-buckling analysis is required in such situations. The numerical simulations are carried out using the general commercial finite element code ANSYS (ANSYS.inc, 2014). The ANSYS commercial program based on the finite element method issues a sparse direct solver for all nonlinear problems. The sparse solver can address negative indefinite systems (common in nonlinear analysis due to stress stiffness and constitutive behaviour) and LaGrange multipliers from a variety of sources (such as multipoint constraints, mixed u-P elements and contact elements). The sparse solver is applicable to real, complex, symmetric and non-symmetric systems. Non-symmetric systems are critically important for contact models with significant friction. The sparse solver is a choice for all forms of nonlinear analysis (ANSYS.inc, 2014). The static limit-load analysis including all kinds of non-linearity is done in ANSYS. The shell-181element of ANSYS is used to model the shells. This element is extremely suitable for analyzing thin to moderately thick shell structures. The element is a four-node Reissener-Mindln shell element with six degrees of freedom at each node ($ux, uy, uz, \theta_x, \theta_y$ and θ_z) and is well suited for large strain nonlinear applications. Both full and reduced integration schemes are supported. Shell-181 element uses a penalty method to relate the independent rotational degrees of freedom about the normal (to the shell surface) with the in-plane components of displacements. In the present study considered both material and geometric nonlinearity. In general, the conventional Newton-Raphson method fails because of the singularity of the stiffness matrix and a diverging solution. This is arc-length method is applied in conjunction with the modified Newton-Raphson method in both standard and modified forms. It is a path following technique where both load and displacement are independent parameters. This method avoid similar situation and it is suitable for nonlinear static equilibrium solutions of unstable problems. In case of application of the are-length method can be followed by the tracing of a complex path in the load-displacement response into the buckling/post buckling regimes. For problems with sharp turns in the load-displacement behaviour or path dependent materials, it is necessary to limit the arc-length radius (arc-length load step size) using the initial arc-length radius. During the solution, the arc-length method will vary with the arc-length radius at each arc-length sub step according to the degree of nonlinearities that is involved.

To activate the arc-length method has to decide minimum and maximum multipliers for the arc-length radius. The reference arc-length radius is calculated from the load or displacement increment of the first iteration of the first sub step. This increment is determined by the following equation (2.1):

$$R_{arc} = \frac{\Delta\sigma_{total}(or\ \Delta\delta_{total})}{N_{step}} \qquad (2.1)$$

where R_{arc} is ref.arc-length radius

N_{step} is number of sub-step

$\Delta\sigma_{total}$ (or $\Delta\delta_{total}$) is total load, total displacement

The factors ARC_{min} and ARC_{max} are used to define the limits of the arc-length radius by using the following equation (2.2) and (2.3):

$$L_{min} = R_{arc} \times ARC_{min} \qquad (2.2)$$

$$L_{max} = R_{arc} \times ARC_{max} \qquad (2.3)$$

where R_{arc} is ref.arc-length radius,

L_{min} is miniumum limit of arc-length radius

L_{max} is maximum limit of arc-length radius

ARC_{min} is minimum of amplitude factor

ARC_{max} is maximum of amplitude factor

In each subsequent sub step, a newly updated arc-length radius is first calculated based on the arc-length radius of the previous sub-step and the nonlinear behavior. And then, the newly calculated arc-length radius is further modified so that it falls between the range of the upper limit and lower limit. If the calculation of nonlinear does not converge even when using the lower limit of the arc-length radius, the solution will be terminated.

3.3 BOUNDARY AND LOAD CONDITIONS

For the present nonlinear FEA, the helideck pancake model was applied in the span length (x) direction to accurately account for the effects of the rotational restraints along the edges (y), as described in Figure 8. This FE-model is suitable for the geometric properties and structural behaviour of the panel considered here because its panel deflection behaviour is symmetric between girders. In this study, we used one extreme edge condition, namely, simply supported, along the longitudinal edges (or at the bottom girders).

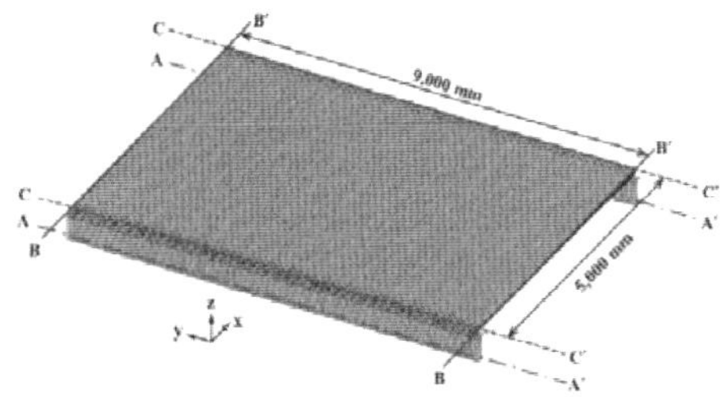

Figure 8: FE model of the aluminium pancake with H-beam girders

The vertical landing force applied was 200.8 kN and the area of the landing force acting on the pancake was 90,000 mm². Five loading cases are considered to estimate the critical influence about the vertical direction of the aluminium pancake with regard to the landing force, as shown in Figure 9.

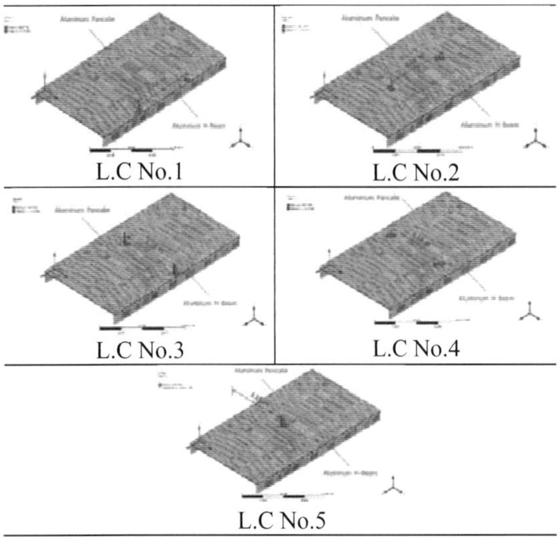

Figure 9: Load cases for FE analysis

3.4 RESULTS

The structural safety of the aluminium pancakes as well as aluminium girders is analysed using the static analysis with regard to the landing force. The safety check results based on FEA are summarised in Table 4. The allowable stress is calculated as allowable increase factor (1.33) × mesh size effect factor (1.4) × yield stress (260 MPa). From the basis of strength results, the maximum unity ratio of the structure is calculated to 0.89 where as allowable criteria is 1.0. Therefore, newly designed aluminium pancake has sufficient strength to withstand the maximum landing force of 200.8 kN from a global yield strength point of view. For the critical point of view, it was necessary to verify the structural behaviour using nonlinear structural analysis. It is clear that, as expected, the first yield point and buckling point occurred at Y-shaped webs. When the landing points are located in the middle of the pancake, yield points are equally distributed along two Y-shaped webs in the pancake. Also, it is observed that the yielding at the Y-shaped web occurred before the pancake buckling in both cases as shown in Figure 10.

The safety of the pancake is analyzed by the static analysis depending on the landing positions. From the FE results, the maximum stress of the pancake is calculated as 431.3MPa. The allowable stress of 484.1MPa for the pancake member & aluminium girder is still enough comparing with the maximum stress of 431.3MPa.

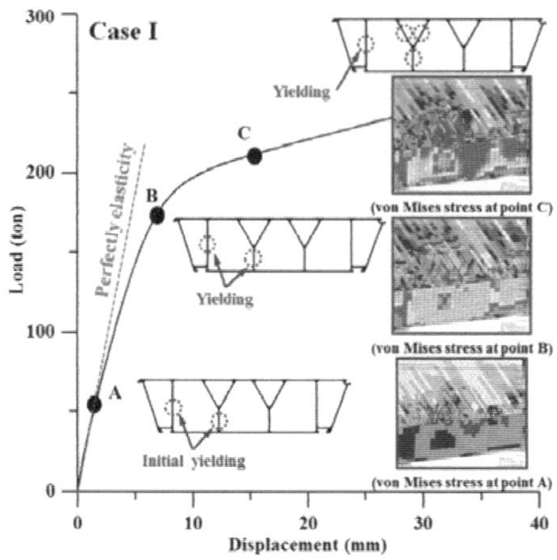

Figure 10: Progressive collapse behaviour of aluminium helideck structure

Table 4: Results of static FE-analysis

LC No.	Maximum stress (MPa)	Allowable stress (MPa)	Safety Check
1	431.3	484.1	O.K
2	307.4	484.1	O.K
3	299.7	484.1	O.K
4	331.1	484.1	O.K
5	293.8	484.1	O.K

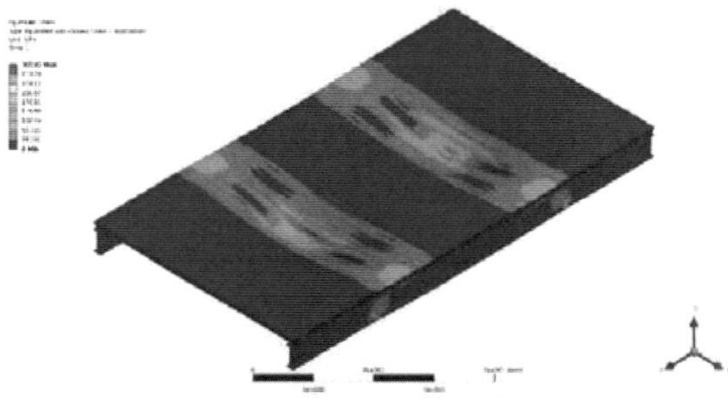

(a) Loading condition No.2

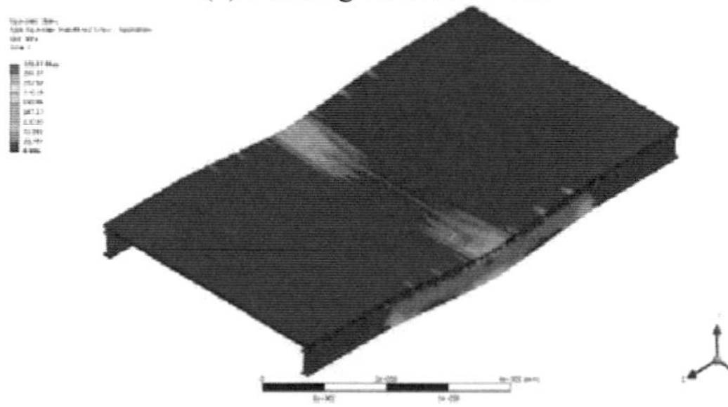

(b) Loading condition No.4

Figure 11: Stress plot and deformed shape

In case of nonlinear buckling analysis, maximum vertical force applied on the upper plate of pancake is considered as follows. These vertical forces are applied to unit load for the linear buckling analysis and ultimate load of 500kN for the nonlinear convergence buckling analysis. Where, the applied area of force is 90,000mm²

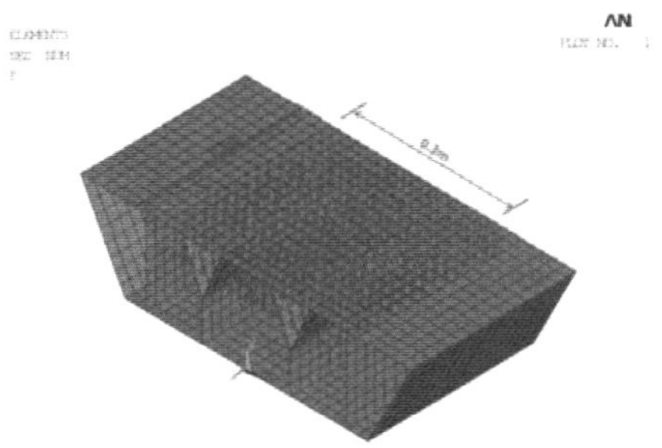

Figure 12: Loading diagram under vertical load

The elasto-plastic large deflection analysis is performed to clarify the fundamentals in progressive collapse behaviour including occurrence of buckling/yielding of thinner aluminium pancake under impact load. The load and displacement relationships are summarised in Figsure 13.

In advanced aluminium structural design, load-carrying capacity calculations of member should be considered post-weld initial imperfections as parameters of influence. In this case, only initial tolerance value of maximum 0.5mm per thickness during extrusion process because of without welding construction is considered. The initial buckling mode obtained by eigen-value analysis by using FEM is used as initial deflection mode.

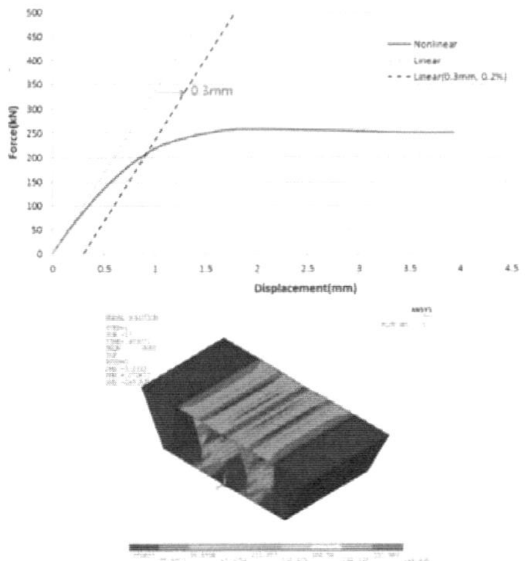

Figure 13: A relationships of between force and displacement curve under impact load

Firstly, progressive collapse behaviour of a thin deck plate subjected to compressive load is explained in Figure 13. The initial yielding takes place at the 100kN point and then, load-carrying capacity rapidly decreases due to the progressive yielding distribution at the center plate.

4. A PROCEDURE FOR DEFORMATION BASED DESIGN

4.1 DEFORMATION BASED DESIGN

The following figure shows engineering procedure with deformation based design for aluminium helideck proposed by author. It is very important to determine the reasonable deformation criteria under impact landing and then re-verify the FE-calculation.

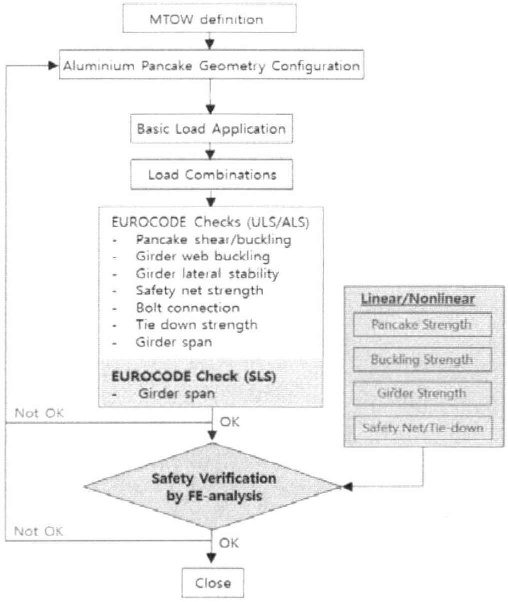

Figure 14: Flow chart for new design process

The NORSOK Standard C004 [NORSOK, 2004] is recommended in which the allowable global deflection of the cantilever length during helicopter landing and comparison results between NORSOK and new proposed criteria are indicated below Table 5.

Table 5 Comparison results between NORSOK and new proposed criteria

Rule	Allowable deflection of cantilever (mm)	Allowable deflection of simply supported condition (mm)
NORSOK C004	Length/180	-
Author	Length/150	Length/300

Table 6 Results of global deflection check according to landing positions

	Maximum deflection (mm)	Allowable deflection (mm)	Effective Length (mm)	Clamping condition
LC.1	3.7	13.5	4040.0	S.S*
LC.2	4.4	13.1	3918.0	S.S
LC.3	25.0	33.5	5030.0	CA**
LC.4	14.0	31.2	4676.0	CA
LC.5	25.0	34.4	5160.0	CA

Note ;
*S.S : Simply supported condition
**CA : Cantilever condition

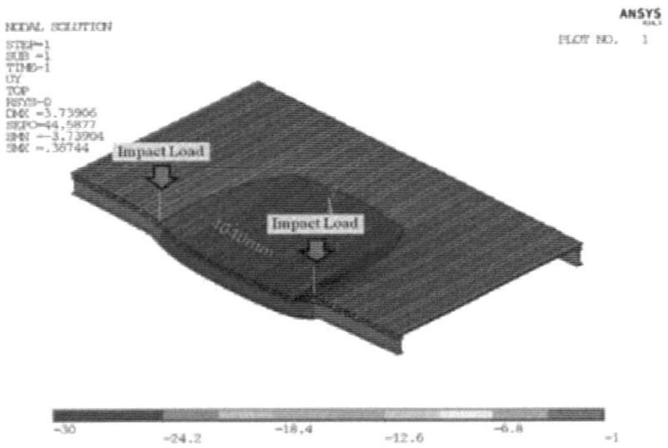

(a) Loading condition No.1

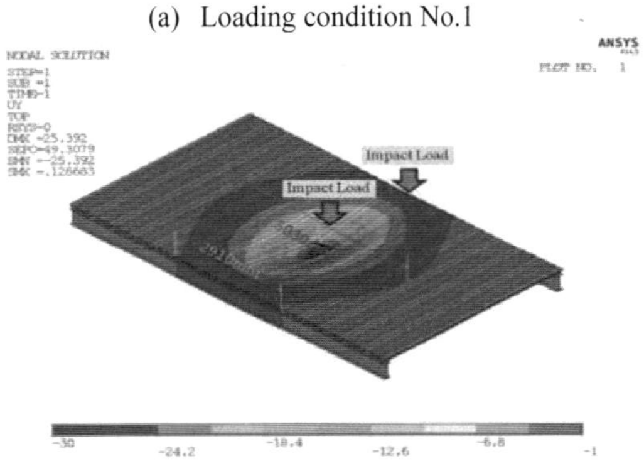

(b) Loading condition No.3

Figure 15: Global deflection shape and distribution

The global deflection shape and distribution under impact load induced by helicopter take-off action are shown in figure 15. Before determining the effective length against impact load, linear static FE-calculation by varying loading scenarios and estimation of allowable deflection are carried out.

Based on the insights noted above, the allowable deflection criteria may be empirically derived by FE-calculation as indicated in Table 5. The newly proposed criterion is possible to apply for the aluminium structural design by considering characteristics of material and structural behaviour.

5. CONCLUSIONS

The goals of this study are to suggest an engineering procedure with deformation based design in order to satisfy the helideck safety, and to perform the numerical calculation in order to verify a procedure based on the deformation criteria.

Series of nonlinear/nonlinear FEM analyses are performed to investigate into the buckling/progressive collapse behaviour and strength of the aluminium pancakes with supporting girders subjected to landing load.

The present paper focuses on the following two areas which were studied theoretically, numerically as appropriate:

- To develop a reasonable procedure to estimate the structural safety of aluminium structure
- To propose the design based on the deformation for aluminium helideck under impact load

Some collected results and conclusions developed in the present study are as follows:

(1) The Proposed deformation criteria are empirically derived from numerical results varying loading positions and accuracy of formulas for predicting allowable displacement have been confirmed by comparing the NORSOK criteria as well as calculated results with the FE-Analysis results.
(2) With the decrease of the mesh size of element, elastic stress highly increases. In order to overcome this phenomenon by the convergence test according to mesh sizes, the allowable stress should be adjusted.
(3) A deflection criterion can give a reasonable estimation of structural safety under a variety of loading conditions. The outcomes will be widely referred in the aluminium basic design to decide required span length during code check stage.
(4) The initial yielding is governing parameter in determining the capacity of thinner aluminium pancake under impact load.

6. ACKNOWLEDGEMENTS

This work is supported by the Technology Innovation Program (No.10046380, A Localization engineering development for large scale safety helideck of Al-6082-T6, width 615mm/height 200mm) funded by the Ministry of Trade, Industry & Energy (MI, Korea).

7. REFERENCES

1. MEI-NI, S., BEN., Y and LEROY., G, 'Deformation-based design of aluminium alloy beams', *Engineering Structures.*, Vol. 80, pp. 339-349, 2014.

2. KISSELL., JR, FERRY. RL, 'Aluminium alloy structures: A guide to their specifications and design', New York : John Wiley & Sons, 2002.

3. MAZZOLANI., FM, 'Aluminium alloy structures, 2nd edition, London: Champman & Hill, 1995.

4. YUJIN., W, FENG., F and SHIBIN., L, 'Experimental investigation on the stability of aluminium alloy 6082 circular tubes in axial compression', *Thin-walled Structures*. Vol. 89, pp. 54-66, 2015.

5. YUJIN., W, FENG., F and SHIBIN., L, 'Static and fatigue analysis of three types of aluminium box-stiffener/web frame connections', *International Journal of Fatigue*., Vol. 29, pp. 1426-1433, 2007.

6. CAP 437, 'Standards for Offshore Helicopter Landing Areas', *Civil Aviation Authority*, 2013.

7. BS EN1999-1-1, 'Eurocode 9: Design of aluminium structures-Part 1-1: General structural rules', 2007.

8. HSE, 'OFFSHORE HELIDECK DESIGN GUIDELINES', 2012.

9. ANSYS, 'Mechanical APDL-User's theoretical manual-Nonlinear Analysis', March, 2014.

10. NORSOK STANDARD, 'Helicopter deck on offshore installations', Rev.1, September, 2004.

8. AUTHORS BIOGRAPHY

Dr Joo Shin PARK holds the position of Manager in the Research and Product Development Department of Samsung Heavy Industries in Geoje, Korea. He is currently responsible for managing ship and offshore R&D activities. His latest research areas focus on offshore topside engineering and risk assessment against accidental cases.

Mr Yeong Su Ha holds the position of Engineer in the Research and Product Development Department of Samsung Heavy Industries in Geoje, Korea. He is currently responsible for managing ship and offshore R&D activities. His latest research areas focus on offshore structural engineering for jack-up rig.

Mr Ki Bok Jang holds the position of Principal Engineer in the Research and Product Development Department of Samsung Heavy Industries in Geoje, Korea. His research interests are in the area of nonlinear structural mechanics and design associated with limit states.

COLLAPSE MODES OF FLEXIBLE PIPES UNDER EXTERNAL PRESSURE

Yong Bai, Shuai Yuan, Peng Cheng*
Institute of Structural Engineering, College of Civil Engineering and Architecture, Zhejiang University, Hangzhou, P.R. China

SUMMARY

The application of unbounded flexible pipes increases in deep water requiring better collapse resistance of the products. The carcass is used as the innermost layer to prevent the collapse of rough bore flexible pipes under external pressure. In this paper the effects of the ratio between the bending stiffness (RBS) of the outer layers and carcass on the collapse modes of flexible pipes are investigated through numerical simulation. There are five kinds of finite element models studied by varying the RBS. The nonlinear pressure–displacement equilibrium path was traced using a Riks continuation algorithm. The numerical results agree well with the classic analytical model. The fitting formulas for the prediction to the critical pressure and corresponding collapse modes are obtained. And the sensitivity researches of related parameters.

NOMENCLATURE

b	Breadth of the cross section	mm
D	Mean diameter of a layer	mm
E	Elastic module	
F	Radial force	N
I	Inertia of moment	m^3
K	Bending stiffness	
LPF	Load proportionality factor	
P_{crr}	Critical external pressure of a ring	MPa
P_{crt}	Critical external pressure of a long tube	MPa
P_{nc}	Normalized Critical Pressure	N
R	Mean radius of a layer	mm
ϕ	Ratio	
t	Thickness of a layer	mm
δ	Displacement	mm
Subsrcipt c, s, a, o	carcass, pressure sheath, pressure armor, outer layer	
ν	Poisson's ratio	

1. INTRODUCTION

Figure 1: Structure of Unbonded Flexible Pipe
(Courtesy of GE Oil & Gas)

Unbounded flexible pipes are extensively used as various risers, flowlines and etc. in the offshore engineering due to their special structures. The typical structure of a rough bore unbounded flexible pipe shown in Figure 1 makes the pipe with low bending stiffness but relatively high radial and longitudinal stiffness. Compared with the smooth bore flexible pipes, the rough bore ones are able to resist collapse under higher external pressure thanks to the additional interlocked carcass layer. Thus, with the development of flexible pipes in ultra-deep water, much attention should be paid to the mechanical property of the carcass layer.

According to the location where the external pressure is applied, a carcass layer presents different collapse modes. When an intact flexible pipe is subjected to external pressure on the outer face of external sheath, all the layers work together to resist collapse. And then the carcass would collapse along with other layers in the manner of a single-layer tube collapse. The classic analytical model for the elastic buckling of a ring or tube under external pressure derived by Timoshenko and Gere [2] is suitable for this failure mode. The critical theoretical pressure load of a ring is

$$P_{crr} = \frac{3EI}{R^3} \tag{1}$$

And as for the long tube the critical theoretical pressure load is changed to

$$P_{crt} = \frac{3EI}{(1-\nu^2)R^3} \tag{2}$$

The above work has been the basis for most analytical models attempting to study the collapse behavior of the flexible pipe. Based on this model, Ref. [3] recommend that with full contact between layers, the elastic buckling pressure of the carcass may be determined as the sum of contributions from the carcass and the pressure armour layers as[3]:

$$P_{cr} = \sum_{i=1}^{N_p} \frac{3EI_{eq}^i}{R_c^3} \tag{3}$$

where R_c is the radius of the carcass and EI^i_{eq} is the equivalent ring bending stiffness of each layer per unit length of pipe. According to Chen et.al. [4], if the gap between the carcass and the other spiral wires in a 6" flexible pipe is greater than 2.5 mm, the EI^i_{eq} will be determined by the carcass contribution alone. API 17 B [1] suggests that the contribution from the pressure-armour layer should be verified by documented prototype tests.

However, if the external sheath happens to damage, the seawater floods into the annulus. And the external pressure is applied on the fluid barrier (also known as internal pressure sheath) directly. Then, the failure mainly depends on the collapse strength of the carcass. Considering the constraint of pressure armor outside, this failure mode of carcass is similar with the confined tube collapse. In order to learn this wet collapse profoundly, Gay Neto and Martins [5], at first, used two 3D models without the pressure armor layer to predict the limit safe external pressure. When it is achieved, a snap-through loss of stability occurs. Then, they [6] showed that the pressure layer as a constraint to the carcass loss of stability plays an important role. Similarly, Lacerda et al. [7] adopted beam elements to simulate the flexible pipe only including the interlocked armor and another considering both the interlocked armor and pressure armor. The adequacy of the numerical models is finally evaluated by experimental tests. Meanwhile, a full 3D model including the interlocked carcass, the internal polymeric layer, the pressure armor and the external polymeric layer was constructed by Gay Neto and Martins [8]. This model is adopted to deal with the wet and dry collapse of straight and curved flexible pipes. Cooke and Kenny [9] developed the full 3D model considering friction between layers and concluded that the increased friction factor can improve collapse integrity. The researches above only considered the possibility of a two symmetry planes collapse mode. Indeed, other collapse modes can also occur in the wet collapse. An investigation [10] about the shapes of the collapse modes of flexible pipes was performed. They found the heart shape buckling mode (See Figure 2) and transition buckling mode would occur depending the initial condition.

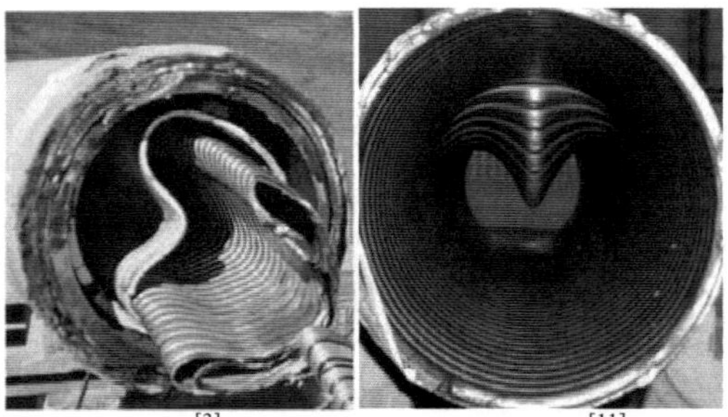

(a)[3] (b)[11]

Figure 2: Heart Shape Collapse Mode

In addition, for some dynamic flexible risers with 2 or 3 layers in the internal pressure sheath, the fluid transported in the bore of pipe can penetrate the annular space between the pressure sheath layers through the end-fitting. When the pipe bore is depressurized at a fast rate, the pressure of the fluid trapped in the pressure sheaths acting on the inner sheath could also cause the confined collapse of the carcass. Lambert et al. [11] developed a numerical model to calculate the evolution of the differential pressure during the depressurization and validated it with a large test. Thus, the carcass collapse may progress in discrete steps: a starting collapse manifested as a dent, and subsequently a total collapse at a later depressurization event in a heart shape.[3]

Among these collapse modes, Paumier et al. [12] regarded the heart collapse mode is typical of carcass collapse of rough bore pipes when the hydrostatic pressure applies on internal pressure sheath. They introduced the formula derived by Glock [13] of calculating the critical external pressure for a thin ring confined in a stiff medium without friction:

$$P_{crr} = E\left(\frac{t}{D}\right)^{\frac{11}{5}} \tag{4}$$

Also, for the long tube, the formula has changed to:

$$P_{crt} = \frac{E}{(1-\nu^2)}\left(\frac{t}{D}\right)^{\frac{11}{5}} \tag{5}$$

And Kyriakides et al. [14-16] have published a series of studies on a confined ring. They initially studied the large deflection collapse of circular rings confined in a rigid cavity under external pressure. It shows that the response beyond the limit load is unstable until the crown of the ring touches the opposite side after which it becomes stable again. And they confirmed this behavior by experiments. Then, they considered two concentric smoothly contacting rings under external pressure. The outer ring is assumed to be linearly elastic not rigid. The nonlinear response of the structure was found to be characterized by a limit load type of instability. Vasilikis and Karamanos [17] presented a 2D nonlinear FEM model of a plastic thin cylinder surrounded in an elastic medium. Several imperfections were introduced. And a plastic-hinge mechanism was developed and illustrated the post-buckling response of the cylinder in an approximate manner. All in all, the analytical research about the confined collapse is still limited in the 2D model under limited condition.

In practice, the situation outside the carcass can be varied, which would affect the confinement on the carcass to various degrees. Different confinements should be related to the different bending stiffness. Therefore, this research is started from the simplest FEM model to study the collapse modes of flexible pipes with various bending-stiffness confinements. The nonlinear pressure–displacement equilibrium path is traced using a Riks continuation algorithm. The results are compared with ones from classic analytical model. And the sensitivity researches of related parameters are also conducted.

2. FINITE ELEMENT SIMULATION

2.1 SIMPLIFICATION

In the numerical solution, some simplifications are made as follows:

1) The models simulated in this paper include the carcass, pressure sheath and pressure armor layer. The contribution to the confinement from other layers is involved in the pressure armor layer. Accordingly, the level of confinement was mainly simulated by the bending stiffness of pressure armor layer.
2) The carcass or armor is formed as a helix of folded strip to form an interlocked tube. The rectangular profile with uniform thickness was adopted instead of the real shape of tendon.
3) In all the FEM models, the load was applied using a distributed pressure on the external circumferential surface of the carcass not of the pressure sheath.
4) The manufacturing process of the metallic carcass and pressure armor layer could cause stress-induced material anisotropy. However, for this study, this effect was neglected and all the materials were considered as isotropic and elastic.

2.2 MODELING DESCRIPTION

The purpose of this modeling is to gradually study the influence of confinement on the critical external pressure and subsequent collapse modes of a flexible pipe. There were five finite element models studied including the plane strain model of two layers, 3D ring model of two layers, 3D helical-wound model of two layers, 3D ring model of three layers and 3D helical-wound model of three layers based on the simplification level. The elastic buckling of the models were simulated numerically using nonlinear finite-element tools ABAQUS version 6.14. This numerical analysis contains two steps. Firstly, a small initial imperfection is introduced. It is considered in the form of a very small downward vertical displacement applying at Node A of the carcass where $\theta = 0$ as shown in Figure 3. A small displacement value $\delta_0 = 0.075mm$ ($\frac{\delta_0}{R_c} = 0.1\%$) was selected through a short parametric study.

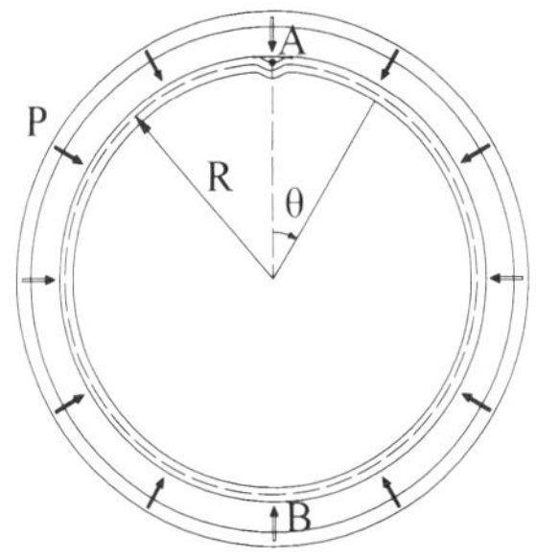

Figure 3: Initial Imperfection of Carcass

Then, this displacement is removed in the subsequent Riks analysis and the carcass at this location contains an initial imperfection. Since the Riks solution algorithm is used, the magnitude of the initial external pressure is arbitrary, but has to be chosen close enough to the expected critical pressure for avoiding numerical errors. The analysis incorporates nonlinear geometry as well as nonlinear contact algorithm. The interaction applied between layers allowed for contact, separation, and slip. The surface-to-surface formulation was used and contact pairs were defined manually considering the relative stiffness of the surfaces avoiding the penetration. The physical and equivalent geometrical parameters as shown in Table 1 are based on a 6" flexible pipe. And the following is the description of each model.

Table 1: Basic Parameters for All Models

	Carcass	Pressure Sheath	Pressure Armor
E (GPa)	200	0.6	200
ν	0.3	0.45	0.3
R (mm)	72.5	80	80/87.5*
t (mm)	5	10	10/5*
b (mm)	10	21/10**	21/10**

*The former for 2 layers model and the latter for 3 layers model
**The former for helical wound model and the latter for others

2.2(a) Model 1: 2D Plane Strain Model of Two Layers

The plane strain model is adopted firstly as it is the most close to the analytical model. In this model, the sheath and the armor layer are merged into one layer to represent the confinement outside the carcass. The related parameters are referred to Table 1. The bending stiffness of every layer is calculated according to the Ref. [3]:

$$\frac{EI_{eq}}{R^3} = \left(\frac{\pi}{4} - \frac{2}{\pi}\right)\frac{F}{\delta} \tag{6}$$

Then the ratio between the bending stiffness of outer layer and that of carcass layer in this model can be derived into:

$$\phi_{K1} = \frac{E_o I_{eqo}}{R_o^3} \Bigg/ \frac{E_c I_{eqc}}{R_c^3} = \frac{E_o\, t_o^3\, R_c^3}{E_c\, t_c^3\, R_o^3} \quad (7)$$

According to Eq.7, the ratio is related to the elastic module, thickness and radius of each layer. However, in this study, the elastic module of pressure armor is the only variable to represent different ratios. Correspondingly, 13 values of E_o are chosen to learn the various collapse modes. The sweep meshing technique was employed to discretize the geometry into rectangular elements. For E_o=0 i.e. a single carcass layer, eight-node biquadratic reduced-integration plane strain elements (CPE8R) were suitable to conduct the buckling and Riks analysis on it with fewer elements to obtain the accuracy. Four-node bilinear reduced-integration plane strain elements (CPE4R) were adopted for the modeling of the carcass and armor, since the quadratic elements were not appropriate to model the slave surface of the contact pair. A total of 800 elements were meshed on the both layers to ensure the accuracy of the numerical results. Specifically, 4 elements were applied along the thickness direction to avoid the loss of bending stiffness.

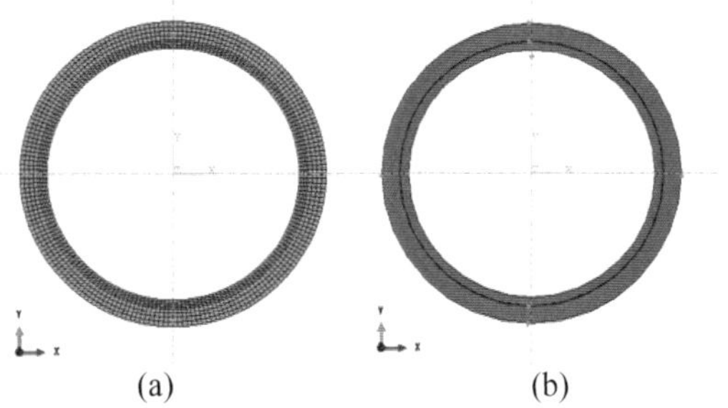

Figure 4: Mesh(a) and Boundary Conditions(b) of Model 1

Being consistent with the assumption of analytical solutions, frictionless contact was used at first. And then, a series of friction coefficient was considered to test the effect of friction on the critical pressure. The displacement boundary conditions were applied to eliminate the rigid body displacements and fit the problem simulated. In the case of E_o=0, fixed boundary conditions were applied to node sets on the outer surface of the layer at 0, 180 degree on the X direction and 90, and 270 degree on the Y direction. And in the case of $E_o = +\infty$ i.e. a rigid confinement, all nodes at the inner surface of armor layer were fully restrained in all directions to model this confinement. For the rest $0 < E_o < +\infty$, the boundary condition of armor layer was as same as the model of E_o=0. And only the X directions of Node A and B are restrained in the carcass layer. The mesh and the boundary conditions of Model 1 are shown in Figure 4.

2.2(b) Model 2a: 3D Ring Model of Two Layers

This model extends the plane strain model to the 3D space and is tested by the analytical model. Since the length of every layer is same, the Eq.7 can be adopted. There were 14 values of E_o from 0 to $+\infty$ chosen to model the collapse. The 20-node quadratic brick elements with reduced integration (C3D20R) were adopted to validate the buckling of single layer. The first order brick elements with reduced integration (C3D8R) were used on the basis of improved contact performance. Refinements were made through the thickness, circumference and longitudinal direction to ensure an optimal mesh density. In terms of the actual condition, only the models with friction coefficient of 0.3 were studied. Both the surfaces of Z=5mm and Z=-5mm adopted the same boundary conditions with the 2D model in every case. In addition, the restraint on the displacement along Z axis was applied on the surface of Z=5mm to avoid the rigid displacement of the model. Figure 5 gives the FEM mesh and the boundary conditions of this model.

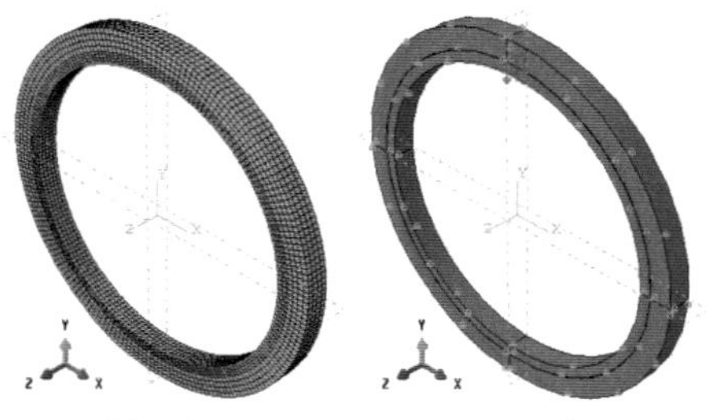

Figure 5: Mesh (a) and Boundary Conditions(b) of Model 2a

2.2(c) Model 2b: 3D Helical-wound Model of Two layers

The actual profile of carcass or pressure armor is considered in recent FEM studies [5, 6, 8, 9]. However, it is not the point in this study and rectangular profile was used instead in the carcass. Based on the Model 2a, Model 2b adopted one pitch of helical-wound carcass to replace the inner cylinder layer. To present the property of interlocking, the ratio between the pitch and breadth is small enough to be 1.1 and therefore the laying angle is close to 90°.

ϕ_{K2b} should be modified to:

$$\frac{E_o I_{eqo}}{R_o^3} \Big/ \frac{E_c I_{eqc}}{R_c^3} = \frac{E_o b_o t_o^3}{R_o^3} \Big/ \frac{E_c b_c t_c^3}{R_c^3} = \frac{E_o}{E_c} \frac{b_o}{b_c} \frac{t_o^3}{t_c^3} \frac{R_c^3}{R_o^3} \qquad (8)$$

There were 14 values of E_o from 0 to $+\infty$ selected to model the collapse. 4320 C3D8R elements were used by refining the mesh in each direction of the carcass to achieve the modeling accuracy. In order to properly account for the continuation of pitch and convergence, a coupling was established between two cutting planes by a constraint control point. And the boundary conditions were consistent with the Model 2b. The restraint on the Z axis is applied on the helical surface of the carcass layer. Figure 6 shows the coupling of carcass and boundary conditions.

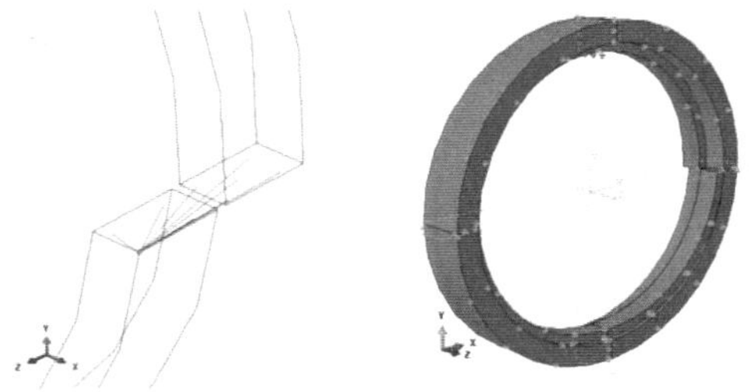

(a) (b)

Figure 6: Coupling (a) and Boundary Conditions (b) of Model 2a

2.2(d) Model 3a: 3D Ring Model of Three Layers

The pressure sheath is considered in this model to explore the effect of this layer to the critical pressure. The flexible pipe including three main structural layers was modeled starting from the 3D ring model. Considering the new layer, the ratio between bending stiffness of the outer layers and that of carcass is changed into:

$$\phi_{K3a} = \frac{E_o I_{eqo}}{R_o^3} \Big/ \frac{E_c I_{eqc}}{R_c^3} = \left(\frac{E_a t_a^3}{R_a^3} + \frac{E_s t_s^3}{R_s^3} \right) \Big/ \frac{E_c t_c^3}{R_c^3} \qquad (9)$$

E_o is still the only variable in the expression and there were 14 values of E_o from 0.2 to $+\infty$ chosen to model the confinement. The friction coefficient was changed to 0.25 according to the materials. And the other interaction method and mesh strategy kept constant with the Model 2a. Furthermore, the boundary conditions of pressure sheath also employed ones of armor layer in Model 2a. The mesh and the boundary conditions are presented in Figure 7.

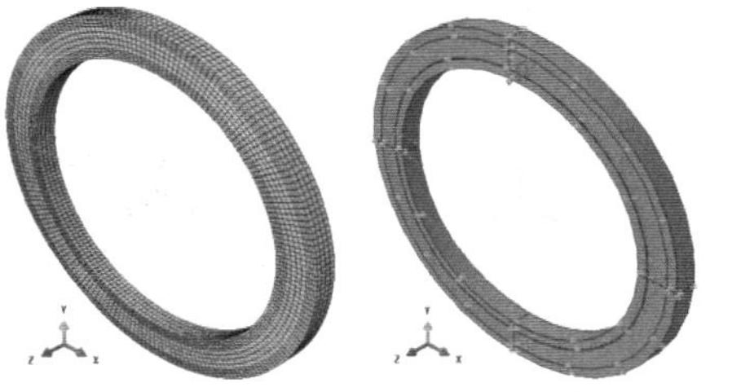

(a) (b)

Figure 7: Mesh(a) and Boundary Conditions(b) of Model 3a

2.2(e) Model 3b: 3D Helical-wound Model of Three Layers

Based on the Model 2b and Model 3a, the carcass layer and pressure armor layer were both simulated as one pitch of helical wound layers. ϕ_{K3b} in this model is:

$$\phi_{K3b} = \frac{E_o I_{eqo}}{R_o^3} \Big/ \frac{E_c I_{eqc}}{R_c^3} = \left(\frac{E_a b_a t_a^3}{R_a^3} + \frac{E_s b_s t_s^3}{R_s^3} \right) \Big/ \frac{E_c b_c t_c^3}{R_c^3} \qquad (10)$$

15 values of E_o from 0.2 to $+\infty$ are chosen to model the confinement. All the modeling methods are similar to the Model 3a. And the constraint on the Z axis is applied on the helical surface of the carcass layer and armor layer. Figure 8 shows the mesh and the boundary conditions.

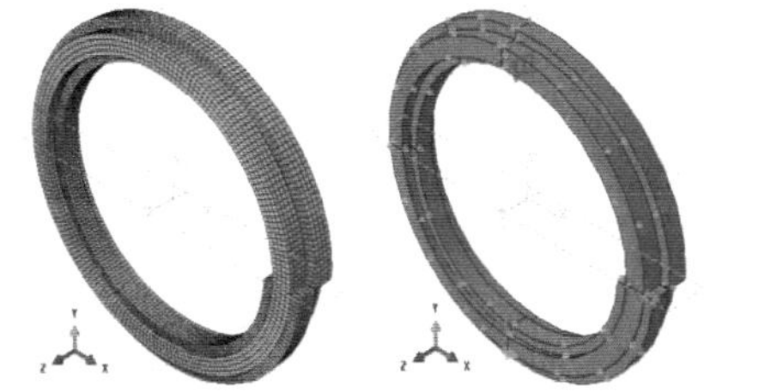

(a) (b)

Figure 8: Mesh(a) and Boundary Conditions(b) of Model 2b

3. RESULTS AND DISCUSSION

The response of numerical models with values of ϕ_K between 0 and $+\infty$ under external pressure was calculated. Since the maximum load proportionality factor was chosen as the stopping criteria of Riks and set as 1.0, the LPF at each increment can be translated directly to the external pressure by multiplying the set pressure. The numerical critical pressures were normalized by the analytical results of Glock [13] (P_{cr}/P_{GL})

when needed. The curves of normalized pressure and displacement show the equilibrium paths of buckling and collapse.

3.1 MODEL 1: 2D PLANE STRAIN MODEL OF TWO LAYERS

For $\phi_{K1}=0$ and $\phi_{K1}=+\infty$, the buckling pressures of plane strain models can be directly compared with the results of formula (2) from Timoshenko [2] and formula (5) from Glock [13], respectively , as shown in Table 2.

Table 2: Comparison Between Analytical Method and FEM of 2D Models

ϕ_{K1}	0	$+\infty$
Timoshenko(N)	18.02	
Glock(N)		133.26
FEM(N)	17.96/17.42*	137.01
Error	-0.33%/-3.3%*	2.8%

*the former for buckling analysis without imperfection and the latter for Riks analysis with initial imperfection

The comparison shows there are only slight errors between theoretical results and numerical results and thus, the finite-element methods adopted are available in these two cases which represent the limitations of the critical pressures. The errors basically originate in the assumption of the thin-wall ring. Therefore, it can be sure that the other models are available too in a large degree. Besides, the discrepancy between the buckling result and Riks one demonstrates that the imperfection reduces the buckling pressure.

The value of $\phi_{K1}=+\infty$ in Table 2 is attributed to the model without friction. Table 3 shows that the friction has obvious effect on the critical pressure. The larger the friction coefficient is, the larger the critical pressure becomes in the case of same ϕ_{K1}. Thus, it can be concluded that the analytical method of Glock is conservative to simulate the confined buckling process.

Table 3: Effect of Friction on Critical Pressure ($\phi_{K1}=+\infty$)

Friction coefficient	0.1	0.2	0.3	0.4	0.5	0.6
Critical Pressure (N)	138.27	139.65	140.11	141.70	142.15	142.57

The specific comparison of the friction effect is conducted as follows. The equilibrium curves in Figure 9 present the nonlinear relationship between the applied pressure and the downward displacement of the carcass at Node A with varied ϕ_{K1} listed on the left of the figure. The downward displacement is set as the positive one. And considering the self-contact, 70mm is set as the maximum value of the horizontal coordinates.

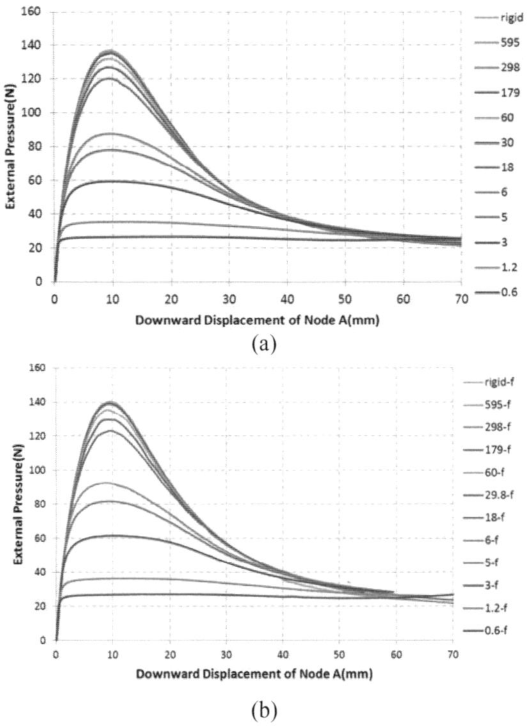

Figure 9: Displacement—Pressure curves of 2D models with varied ϕ_{K1} (a) Frictionless (b) Friction Coefficient=0.3

All the equilibrium paths in Figure 9 can be summarized as undergoing two stages. In the first stage, the equilibrium paths rapidly increase to the critical pressure. In the case of buckling, the displacements of Node A approach 10mm as the values of ϕ_{K1} increase. Based on Figure 9, the tendencies of critical pressure increasing with ϕ_{K1} are depicted in Figure 10. Since P_{nc} of ϕ_{K1} $=+\infty$ (1.028154 for frictionless) is so close to the one of $\phi_{K1}=595$ (1.024737 for frictionless), ϕ_{K1} ranges from 0 to 600.

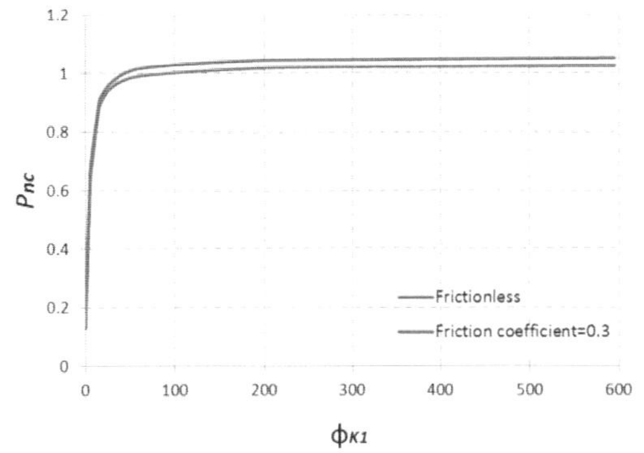

Figure 10: ϕ_{K1} — P_{nc} Curves in Model1

It has been observed that the critical pressure increases with ϕ_{K1}. Specifically, the values of P_{nc} rapidly reach to 1.0 before about 50 of ϕ_{K1} and then increase slightly. The curves prove that the friction makes the critical pressure increasing in each case which means the resistance to buckling of carcass has been improved by the friction.

Beyond the critical pressure, the carcass layers of various models exhibit different post-buckling behaviors. As shown in Figure 9, in this stage, the pressure increases slightly with the displacement increment in the case of ϕ_{K1} =0. For ratio of ϕ_{K1} = 0.6 , the pressure firstly increases at very low rate and then decreases slowly. The pressures drop at increasing rates as the rest selected values of ϕ_{K1} increase, indicating the unstable behavior. And the rate becomes very low in the final part of each path. This situation indicates varied collapse modes which can be represented by the ratio between the displacement of Node B and that of Node A ($\phi_{\delta AB}$) depicted in Figure 11. Considering the curves basically tending to be stable, 50mm is set as the maximum value of the horizontal coordinates.

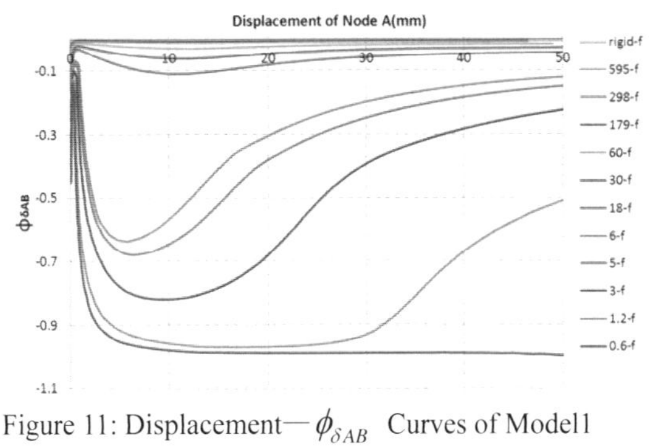

Figure 11: Displacement— $\phi_{\delta AB}$ Curves of Model1

Through Figure 11, it can be observed that at the initial condition, the ratio between displacements increases due to the decreasing level of the confinement outside. For the rigid confinement, the ratios of displacements keep in a value close to 0 which represents the collapse in a heart shape. And except this case, the other curves of confined models rapidly approach to 0 which eliminate the effect of initial displacement ratio in a degree and then decrease to different levels at different rates due to various values of ϕ_{K1}. The final parts of curves indicate the collapse modes shown in Figure 12 which represents the changing modes along with ϕ_{K1} when the displacement of Node A is 50mm. It can be probably concluded that the symmetrical shape only exists when ϕ_{K1} is lower than 1. The transition modes involve the modes with diminishing gaps around Node B.

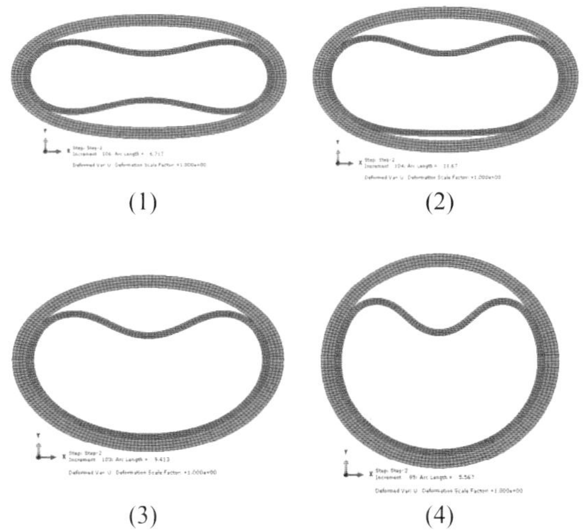

ϕ_{K1}	(1)0.6	(2)1.2	(3)3	(4)rigid
$\phi_{\delta AB}$	-0.997	-0.504	-0.223	9.03E-07

Figure 12: Collapse Modes with Different ϕ_{K1} of 2D Models

3.2 3D MODELS: MODEL 2a, MODEL 2b, MODEL 3a, MODEL 3b

For ϕ_K =0 and ϕ_K = +∞ , the critical pressures of Model 2a, 2b, 3a, 3b are compared with the results of formula (1) from Timoshenko [2] and formula (4) from Glock [13], respectively , as shown in Table 4.

Table 4: Comparison Between Analytical Method and FEM of 3D Models

E_a (GPa)	0	+∞
Timoshenko(N)	16.40	
Glock(N)		121.27
2a FEM(N)	15.96	124.65/127.88*
Error	-2.7%	2.8%/5.5%*
2b FEM(N)	16.03	124.05
Error	-2.3%	2.2%
3a FEM(N)		104.39
Error		-13.9%
3b FEM(N)		103.46
Error		-14.7%

*the former for frictionless and the latter for friction coefficient of 0.25

Same conclusions can be obtained with the plane strain models. The equilibrium curves of 4 models share the similar tendencies. Take the Model 3b as the example shown below.

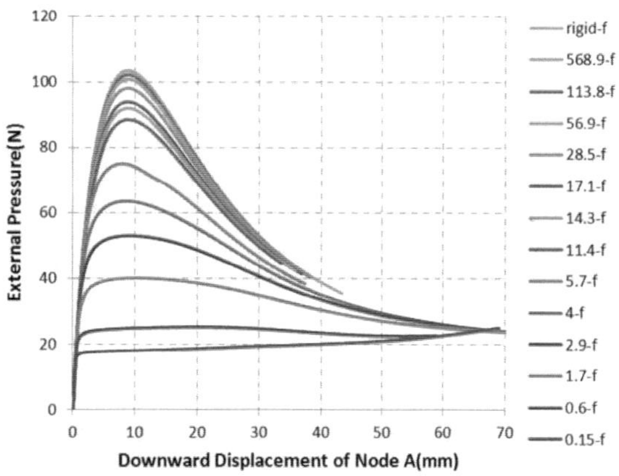

Figure 13: Displacement—Pressure curves of Model 3b with varied ϕ_{K3b}

The relationships of P_{nc} and ϕ_{K3b} in 4 kinds of 3D models are concluded in the Figure 14. Note that only the models with friction are considered. ϕ_{K3b} ranges from 0 to 100 for enough to learn the tendencies.

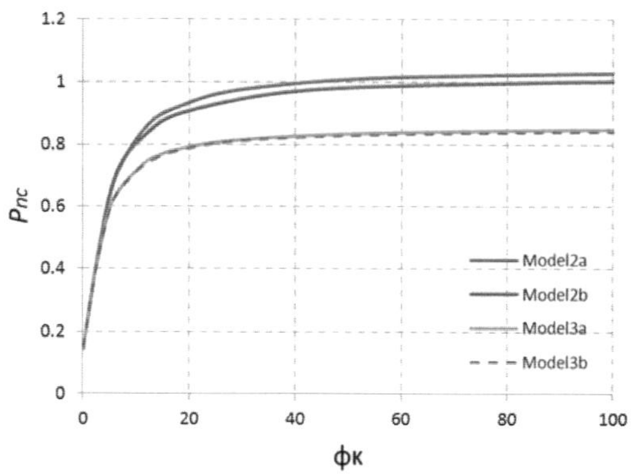

Figure 14: $\phi_K - P_{nc}$ Curves of Model 3b

The results offer that the large wound angle model can be replaced by the ring model in analysis of collapse under external pressure since the curves are close to each other. Besides, critical loads of 3-layer models are lower than ones of 2-layer models which means the property of the layer next to the carcass affects the critical pressure and the pressure sheath cannot be neglect without validation. As the Model 3b is the most close to the actual flexible pipe, the fitting formula of Model 3b is obtained to predict the critical pressures with varied ϕ_{K3b} under external pressure by the software DataFit.

$$P_{nc} = \frac{0.87 \times \phi_{K3b}}{(2.46 + \phi_{K3b})} \tag{11}$$

In this formula, the adjusted coefficient of multiple determination is 97.98% which verifies the availability of this formula. The single layer model is not discussed in this formula.

The ratio of displacements of Node B and Node A is still adopted to account for the collapse modes. Similar tendencies with Model 1 can be obtained in Figure 15. Considering the $\phi_{\delta AB}$ basically tending to be stable, 30mm of Node A displacement is set as the standard to learn the collapse modes. Four typical collapse shapes are selected in Figure 16.

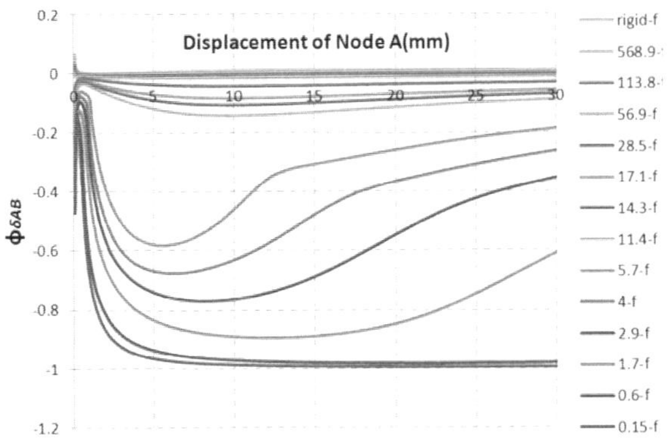

Figure 15: Displacement- $\phi_{\delta AB}$ Curves of Model 3b

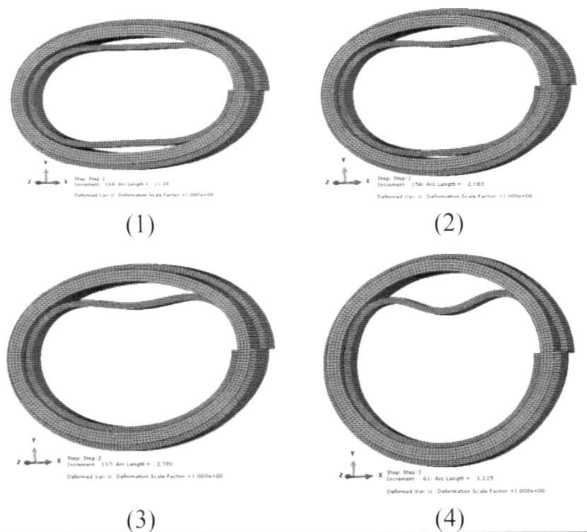

ϕ_{K3b}	(1)0.6	(2)1.7	(3)2.9	(4)rigid
$\phi_{\delta AB}$	-0.980	-0.598	-0.353	0.00871

Figure 16: Collapse Modes with Different ϕ_{K3b} of Model 3b

As a result, Figure 17 presents the relationship between ϕ_{K3b} and $\phi_{\delta AB}$. The fitting formula (12) is obtained to predict the collapse modes through ϕ_{K3b}.

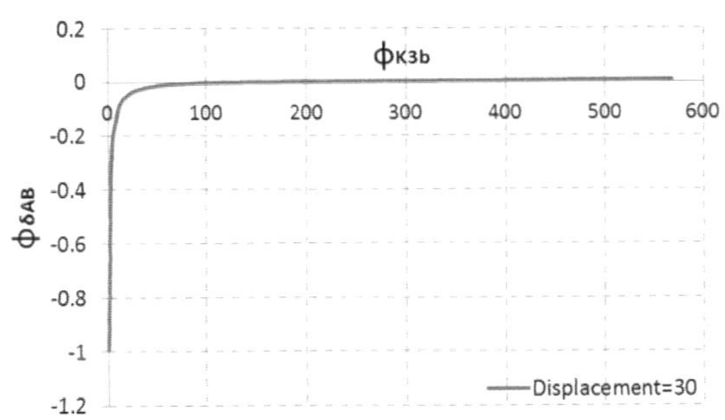

Figure 17: $\phi_{K3b} - \phi_{\delta AB}$ Curve of Model 3b

$$\phi_{\delta AB} = -1.102 \times 0.709^{\phi_{K3b}} \quad (12)$$

In Formula (12), the adjusted coefficient of multiple determination is 98.11% which verifies the availability of this formula.

For the 3D helical-wound model of three layers, except the effect of friction discussed in Model 1, the other parameters also have influence on the buckling behavior of carcass. The Poisson's ratio of each layer except the carcass and thickness of pressure sheath are examined below. ϕ_{K3b} of model is chosen as 17.1.

Table 5: Effect of Poisson's Ratio on P_{nc}

Poisson's Ratio	0.2	0.25	0.3	0.45
Pressure Sheath	0.7534	0.7558	0.7582	0.7753
Pressure Armor	0.7747	0.7752	0.7753	0.7755

Table 5 shows that the larger the Poisson's ratio becomes, the lager the critical pressure is in both layers. And the Poisson's ratio of pressure sheath has larger effect than the pressure armor layer which shows the effect of sheath is not negligible.

To further study the effect of pressure sheath, the thickness of sheath is changed in the case of ϕ_{Kac} kept constant. The results are shown in the Table 6.

Table 6: Effect of Thickness of Pressure Sheath On P_{nc}

t(mm)	5	10	15	20
P_{nc}	0.8222	0.7753	0.755	0.7353

From the Table 6, it has been observed that the critical pressure decreases with the thickness of pressure sheath increasing. It may be explained that the thicker pressure sheath can reduce the local deformation of the carcass.

4. CONCLUSIONS

This paper presents the preliminary study of the confined collapse of flexible pipe. From the research we can learn that:

(1) The critical pressure increases with ϕ_K. The values of P_{nc} rapidly reach to maximum values before about 50 of ϕ_K.

(2) The results of 2-layer models including Model 1, Model 2a, and Model 2b are close to the analytical results while the results of 3-layer models involving Model 3a and Model 3b are lower than the

analytical results. Formula (11) and Formula (12) can be as references to predict the critical pressure and collapse mode only considering the effect of ϕ_K.

(4) The friction between each layer can improve the resistance to buckling of carcass.

(5) The property of the layer next to the carcass affects the critical pressure. The critical pressure increases with the Poisson's ratio of pressure sheath or pressure armor increasing. And with ϕ_{Kac} kept constant, the critical pressure decreases with the thickness of pressure sheath increasing.

In the future work, the elastic-plastic behavior of carcass layer should be considered to compare with the result of Vasilikis[17]. It also needs to be investigated that the manufacture process could affect the collapse behavior of carcass. More parametric studies should be conducted to improve Formula (11) and Formula (12). And the corresponding theoretical analysis and tests are needed to validate the results of FEM in the case of ϕ_K between 0 and $+\infty$.

5. REFERENCES

1. API RP 17B, American Petroleum Institute, 2014.

2. TIMOSHENKO, S. P., GERE, J. M., 'Theory of Elastic Stability', McGraw-Hill, 1961.

3. FERGESTAD, D., LØTVEIT, S.A., 'Handbook on Design and Operation of Flexible Pipes', NTNU, 4Subsea and MARINTEK, 2014.

4. CHEN, B., NIELSEN, R., COLQUHOUN, R., 'Theoretical Models for Prediction of Burst and Collapse and Their Verification by Testing'. In Flexible pipe technology, Int. seminar on Research and Development 18-20 February, NTH, 1992.

5. GAY NETO, A., MARTINS, C.A., 'A Comparative Wet Collapse Buckling Study for the Carcass Layer of Flexible Pipes', Journal of Offshore Mechanics and Arctic Engineering, 2012.

6. GAY NETO, A., Martins, C.A., 'Flexible Pipes: Influence of the Pressure Armor in the Wet Collapse Resistance', Proceedings of the 30th International Conference on Offshore Mechanics and Arctic Engineering, 2014.

7. LACERDA, J., LOURENÇO, M.I., NETTO, T.A., 'Development and Experimental Calibration of Numerical Model Based on Beam Theory to Estimate the Collapse Pressure of Flexible Pipes', 34th International Conference on Ocean, Offshore and Arctic Engineering, 2015.

8. GAY NETO, A., MARTINS, C. A., MALTA, E. R., GODINHO, C. A. F., BARBOSA NETO, T.F., and LIMA, E. A., 'Wet and Dry Collapse of Straight and Curved Flexible Pipes: A 3D FEM Modeling,' 22nd International Ocean and Polar Engineering Conference, 2012.

9. COOKE, N., KENNY, S., 'Comparative Study on the Collapse Response of Flexible Pipe Using Finite Element Methods, 33rd International Conference on Ocean, Offshore and Arctic Engineering, 2014.

10. MALTA, E. R., de ARRUDA MARTINS, C., NETO, A. G., and TONI, F. G. 'An Investigation about the Shape of the Collapse Mode of Flexible Pipes', The Twenty-second International Offshore and Polar Engineering Conference. International Society of Offshore and Polar Engineers, 2012.

11. LAMBERT A, FELIX-HENRY A, GILBERT P, et al. Experimental and Numerical Study of a Multi-Layer Flexible Pipe Depressurization', 31st International Conference on Ocean, Offshore and Arctic Engineering. American Society of Mechanical Engineers, 2012.

12. PAUMIER, L., AVERBUCH, D., and FELIX-HENRY, A., 2009, 'Flexible Pipe Curved Collapse Resistance Calculation', 28th International Conference on Ocean, Offshore and Arctic Engineering, 2014.

13. GLOCK, D., Uberkritisches Verhalten eines Starr Ummantelten Kreisrohres bei Wasserdrunck von Aussen und Temperaturdehnung. (Post-critical behavior of a rigidly encased circular pipe subject to external water pressure and thermal extension). Der Stahlbau 1977.

14. KYRIAKIDES, S. and YOUN, S. K., 'On the Collapse of Circular Confined Rings under External Pressure', International Journal of Solids and Structures, 1984.

15. KYRIAKIDES, S. 'Propagating Buckles in Long Confined Cylindrical Shells'. International Journal of Solids and Structures, 1986.

16. LI, F. S and KYRIAKIDES, S. 'On the Response and Stability of Two Concentric, Contacting Rings Under External Pressure'. International Journal of Solids and Structures,

1991.

17. VASILIKIS, D., and KARAMANOS, S. A., 'Stability of Confined Thin-Walled Steel Cylinders under External Pressure', Int. J. Mech. Sci., 2009.

6. AUTHORS BIOGRAPHY

Yong Bai holds the current position of professor at Institute of Structural Engineering in Zhejiang University. His previous experience includes pipelines, subsea engineering etc.

Shuai Yuan holds the current position of Ph.D. candidate at Institute of Structural Engineering in Zhejiang University. His previous experience includes pipeline engineering.

Peng Cheng holds the current position of lecturer at Institute of Structural Engineering in Zhejiang University. His previous experience includes pipeline engineering. He is also the corresponding author of this paper.

ICSOT Korea: Safety of Offshore and Subsea Structures in Extreme and Accidental Conditions,
15-16 September 2015, Busan, Korea

PRESCREENING OF OCEAN ENVIRONMENT FOR PREDICTING EXTREME TENSION LOADS ON OFFSHORE STRUCTURES

DH Lim and **Y Kim,** Seoul National University, Korea
T Kim, Samsung Heavy Industries, Korea

SUMMARY

For estimating dynamic behaviors of an offshore structure and extreme loads on mooring lines, numerous simulation cases should be considered to deal with all the environmental and operational conditions to which the structure may be subjected during its lifetime. Since a sophisticated time-domain coupled analysis cannot handle this massive computational load, there is a demand for an efficient preliminary process to reduce the number of simulations. This paper proposes a pre-screening procedure to reduce the number of environmental conditions for the dynamic analyses of offshore structures. To make this procedure efficient, frequency-domain theories were applied for estimating platform offsets, by using quasi-static analyses to predict line tension. The results were compared with the time-domain solution obtained through a platform-mooring line coupled analysis, and the results were found to be in fair agreement. In addition, the characteristics of the environmental conditions classified as extreme cases were investigated by applying the procedure to actual environmental conditions.

1. INTRODUCTION

Most offshore structures are operated in the fixed locations for a long period. In particular, in the case of offshore production facilities, the service time may stretch to several decades depending on the well condition. Therefore, offshore structures and mooring systems must be designed to endure the various environmental conditions that they may encounter during their lifetime. For this purpose, an accurate estimation of the extreme values of design parameters such as the offset of the platforms and the tension in the mooring lines and risers is necessary. Numerical simulations are mostly preferred over experimental methods for such estimation because simulations are relatively easy and inexpensive to implement and have a reasonable degree of accuracy as proved by many researchers.

The numerical analysis of floating offshore platforms moored at sea includes certain nonlinearities, which originate mostly from the effects of coupling between the platform and the mooring lines and from those of the nonlinear viscous drag forces. The geometric nonlinearity in the mooring lines arising from large displacement caused by the slow-drift motion of the platform also makes the numerical analysis further complex. This is because wave frequency and the low frequency motions cannot be analysed separately owing to instantly varying stiffness, inertia and damping properties. Hence, in many of the recent studies, the analysis has been carried out in the time domain where a direct consideration of the nonlinearities is feasible. For example, Ormberg and Larsen[1] compared the results of coupled and uncoupled analyses in the time domain, and observed a tendency of design parameters such as platform displacements and the line tensions being underestimated when uncoupled methods are used. Emmerhoff[2] and Kim and Sclavounos[3] introduced a new numerical code for fully coupled dynamic simulations, later named as SML (SWIM-MOTION-LINES). Wichers and Devlin[4] also conducted a fully coupled time domain analysis of the dynamics of a turret-moored FPSO (Floating Production Storage Offloading). They concluded that the coupling effects are significant especially when a loop-current exists. Kim et al.[5] developed a numerical tool for fully coupled analyses in the time domain and validated the simulation results through comparison of the results with the findings of the experiment carried by Offshore Technology Research Center (OTRC).

Although fully coupled time domain analyses have been proven successful, as shown by the abovementioned studies, owing to their high computational cost, there is still demand for simpler and more practical analysis methods. In particular, in the initial stage of the design procedure, fully coupled time domain analyses may require simulations up to the order of one million in order to take into account all the environmental and operational conditions that the structure may experience; none of the present time-domain coupled analysis methods can deal with such a large number of simulations. Hence, there is a demand for a pre-screening procedure to reduce the number of simulations drastically.

In this study, a systemized pre-screening procedure to reduce efficiently the number of environmental combinations in the design analyses of offshore structures and mooring systems is proposed. Using this procedure, risk factors such as platform offset and mooring line tension can be easily estimated. Since in the present study, the focus is on realizing efficiency and expeditiousness rather than a high degree of accuracy, a quasi-static uncoupled analysis in the frequency domain was employed. A series of time-domain coupled analyses were performed with the commercial software SML as the basis for the applied methods and for the validation of the pre-screening results. Finally, actual pre-screening of numerous environmental conditions was performed, and the characteristics of environmental conditions classified as severe to the system have been investigated.

2. PRESCREENING PROCEDURE

The proposed pre-screening procedure consists of three major steps, namely, heading angle analysis, mean offset analysis, and the quasi-static motion analysis. The flowchart in Figure 1 describes the procedure, and details of each step are provided in the following subsections.

2.1 HEADING ANGLE ANALYSIS

The heading angle analysis is performed only when a weathervaning turret mooring system is applied. This step involves determining the platform's mean stable heading angle in which the resultant static yaw moment due to wave, wind, and current becomes zero. The static yaw moments attributed to wind and current are calculated using platform specific drag coefficients, or if the coefficients are not available, using OCIMF[6] experiment results. The moment attributed to wave drift load is estimated using a numerical code based on potential theory. Generally, two mean heading angles—one stable and the other, unstable—are found. The stability of a mean heading angle is judged by the sign of the derivative at the zero-crossing point of the resultant yaw moment curve versus heading angle. A zero-crossing point with negative derivative is considered stable since a slight disturbance of the heading angle at this point causes a restoring moment toward this point. There can also be more than one stable mean heading, depending on the environmental conditions. Once a stable mean heading is found, the platform is set to maintain this heading angle throughout the rest of the pre-screening procedure.

2.2 STATIC OFFSET ANALYSIS

In the static offset analysis, the platform's horizontal mean offset position, for which the environmental loads and the restoring force of the mooring system cancel each other, is estimated. The loads calculated in this step are time-invariant static loads in horizontal directions. The equation for the horizontal mean offset is given as follows:

$$\sum \bar{\mathbf{F}} = \sum\left(\bar{F}_x, \bar{F}_y\right) = \bar{\mathbf{F}}^{\text{wind}} + \bar{\mathbf{F}}^{\text{wave}} + \bar{\mathbf{F}}^{\text{current}} + \bar{\mathbf{F}}^{\text{mooring}}(\vec{x}) = \mathbf{0} \quad (1)$$

where $\bar{\mathbf{F}}$ is the horizontal static force vector, and $\vec{x}$ is the horizontal mean offset vector of the platform. For wind and current, the horizontal static forces are estimated using the platform-specific drag coefficients or OCIMF experiment results. For the static wave load, wave drift forces for individual wave frequencies are pre-calculated and then superposed using the corresponding wave spectra according to the following equation.

$$\bar{F}_x^{\text{wave}} = 2\int_0^{\infty} Q(\omega,\omega)S(\omega)d\omega \quad (2)$$

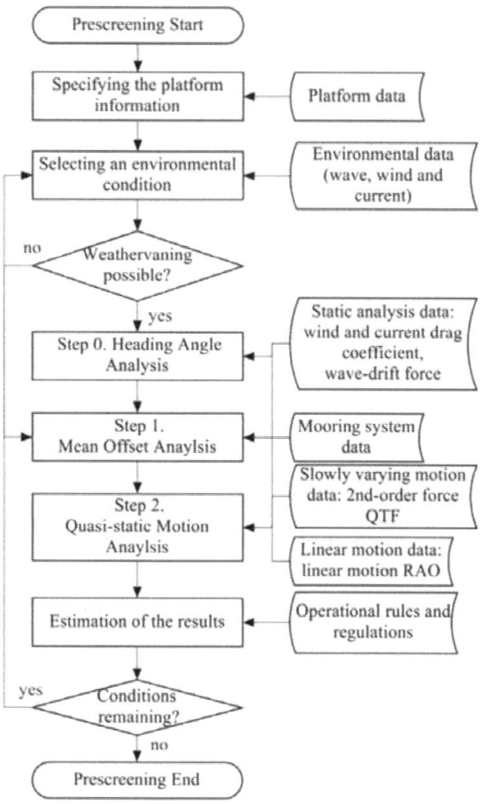

Figure 1: Flowchart of the pre-screening procedure

In equation (2), $Q(\omega,\omega)$ is the mean wave drift force per unit wave amplitude attributed to the wave frequency component ω, and $S(\omega)$ is the wave spectrum. The restoring force of the mooring system is estimated using the pre-calculated static properties of individual mooring lines.

After the mean position of the platform is determined, linearized restoring coefficients in the horizontal directions attributed to the mooring lines are estimated for the slowly varying motion analysis. Since slowly varying motion is a large-displacement problem, the restoring coefficients were calculated by applying the least-square method to several points, rather than by adopting the first derivative at the mean position.

2.3 QUASI-STATIC ANALYSIS

2.3 (a) Wave-frequency Motion

The wave-frequency motion is estimated in the traditional manner by using the motion response amplitude operator (RAO) and the wave spectrum. The variance of the wave-frequency motion is given as follows:

$$\sigma_{i,\text{WF}}^{\ 2} = \int_0^{\infty} \left|H_i(\omega)\right|^2 S(\omega)d\omega \quad (3)$$

Where, $H_i(\omega)$ is the motion RAO and $S(\omega)$ is the wave spectrum.

2.3 (b) Slowly Varying Motion

In this step, the slowly varying motion of the platform induced by second-order difference-frequency wave load and the time-varying wind load are estimated in the frequency domain. For this purpose, the low-frequency force spectrum is utilized. The difference-frequency wave load spectrum is calculated as follows:

$$S_F^{\text{wave}}(\mu) = 8\int_0^{\infty} S(\omega)S(\omega+\mu)\left|Q(\omega,\omega+\mu)\right|^2 d\omega \tag{4}$$

Where, $Q(\omega,\omega+\mu)$ is the quadratic transfer function (QTF) for the difference-frequency wave load.

The low-frequency wind load spectrum is derived as follows. Since the wind load is proportional to the square of the wind velocity, the total wind load is given as,

$$F^{\text{wind}}(t) = C^{\text{wind}}\left\{\bar{U}+\tilde{u}(t)\right\}^2 \cong C^{\text{wind}}\bar{U}^2 + 2C^{\text{wind}}\bar{U}\tilde{u}(t) \tag{5}$$

Where, C^{wind} is the coefficient of the wind drag force; $\bar{U}$ is the mean velocity of the wind; and $\tilde{u}(t)$ is the dynamic wind velocity component that is assumed to be much smaller than $\bar{U}$. Hence, if the wind velocity spectrum is denoted as $S^{\text{wind}}(\mu)$, the wind load spectrum is given as shown in equation (6) below.

$$S_F^{\text{wind}}(\mu) = \left(2C^{\text{wind}}\bar{U}\right)^2 S^{\text{wind}}(\mu) \tag{6}$$

The total low-frequency load spectrum is given by summing up equation (4) and equation (6).

$$S_F^{\text{LF}}(\mu) = S_F^{\text{wave}}(\mu) + S_F^{\text{wind}}(\mu) \tag{7}$$

Variance of the low-frequency motion of the platform is estimated as follows:

$$\sigma_{i,\text{LF}}^{\ 2} = \int_0^{\infty} \frac{S_{F_i}^{\text{LF}}(\mu)}{(c_i - m_i\mu^2)^2 + b_i^2\mu^2} d\mu \tag{8}$$

Where, i denotes the direction of the load and the motion; c_i, m_i, and b_i are the restoring coefficient, mass, and damping coefficient, respectively. Since only the horizontal components of the platform motion are considered, only the mooring system contributes to c_i.

Through these steps, the mean and standard deviation of the horizontal offset of the platform are roughly estimated. Thereafter, the maximum displacement during the prescribed duration is predicted using a statistical distribution model. In the case of the slow-drift motion of offshore structures, nonlinearities originating from many different sources may make the statistics of the problem deviate from those of simple models such as the Gaussian model. For instance, the second-order slow-drift excitation force derived from the Volterra series model is known to have an asymmetric, non-Gaussian probability density function; hence, special formulations have been adopted to calculate its probability density function (Neal[7]; Naess[8]; Kim[9]). Viscous drag forces on slender elements of the platform and the mooring lines, wave-drift damping forces, and the nonlinear restoring effect of the mooring system also complicate the statistics of slow-drift motion. Hence, an appropriate statistical distribution model must be selected to predict the maximum value, as described in the following section.

3. STATISTICAL MODEL FOR SLOW-DRIFT MOTION OF THE PLATFORM

In this study, a series of time-domain analyses were performed using the coupled analysis program SML for the platform and lines to identify a suitable probability distribution model for the slow-drift motion. The platform used for the test is a semi-submersible model shown in Figure 2. The significant wave height and the spectral peak period were set as 10 m and 12 s, respectively, throughout the test. Three simulation cases were considered depending on the degree of nonlinearity. In Case 1, the viscous drag coefficients and wave-drift damping were zero, and the mooring system was of linear uncoupled type; hence, the equation of motion was completely linear. In Case 2, the viscous drag on the platform was considered as Morison's drag, and wave-drift damping was considered. In Case 3, the mooring lines were assumed to be slender rods, and a coupled analysis was performed for the platform and lines; Morison's drag and wave-drift damping were also considered. Table 1 presents the details of the abovementioned simulation cases.

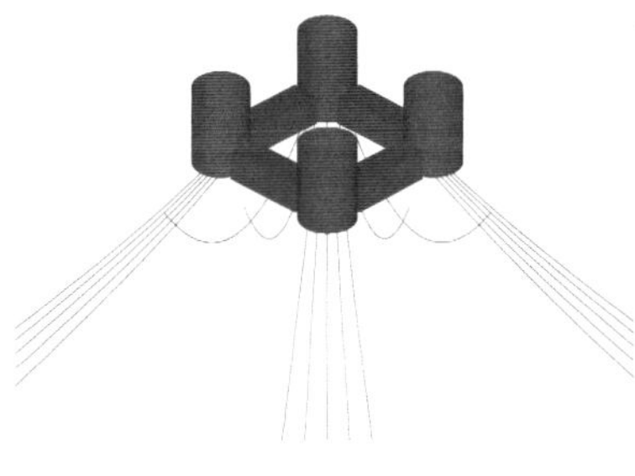

Figure 2: Semi-submersible model used for the statistical test of slow-drift motion

Table 1: Simulation cases for the time-domain slow-drift motion analysis with SML

Case #	Viscous drag	Wave-drift damping	Mooring lines
Case 1	X	X	Linear/uncoupled
Case 2	O	O	Linear/uncoupled
Case 3	O	O	Coupled

Figure 3 shows the probability distribution and the exceedance probability distribution of the peaks of the difference-frequency excitation force in the simulations.

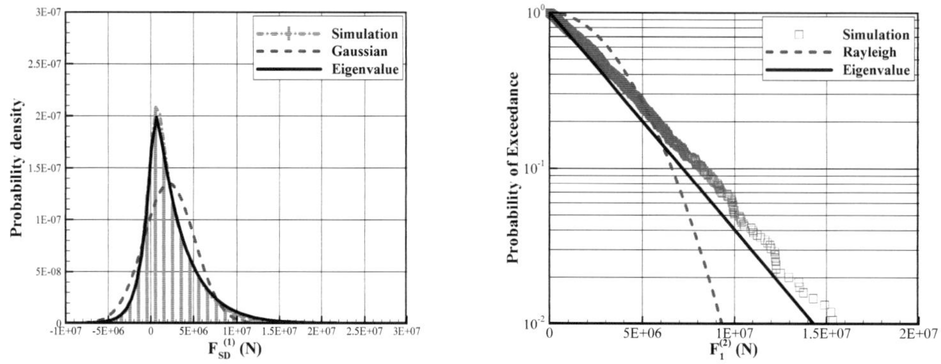

Figure 3: Probability distribution (left) and the exceedance probability distribution of the peaks (right) of the difference-frequency excitation force

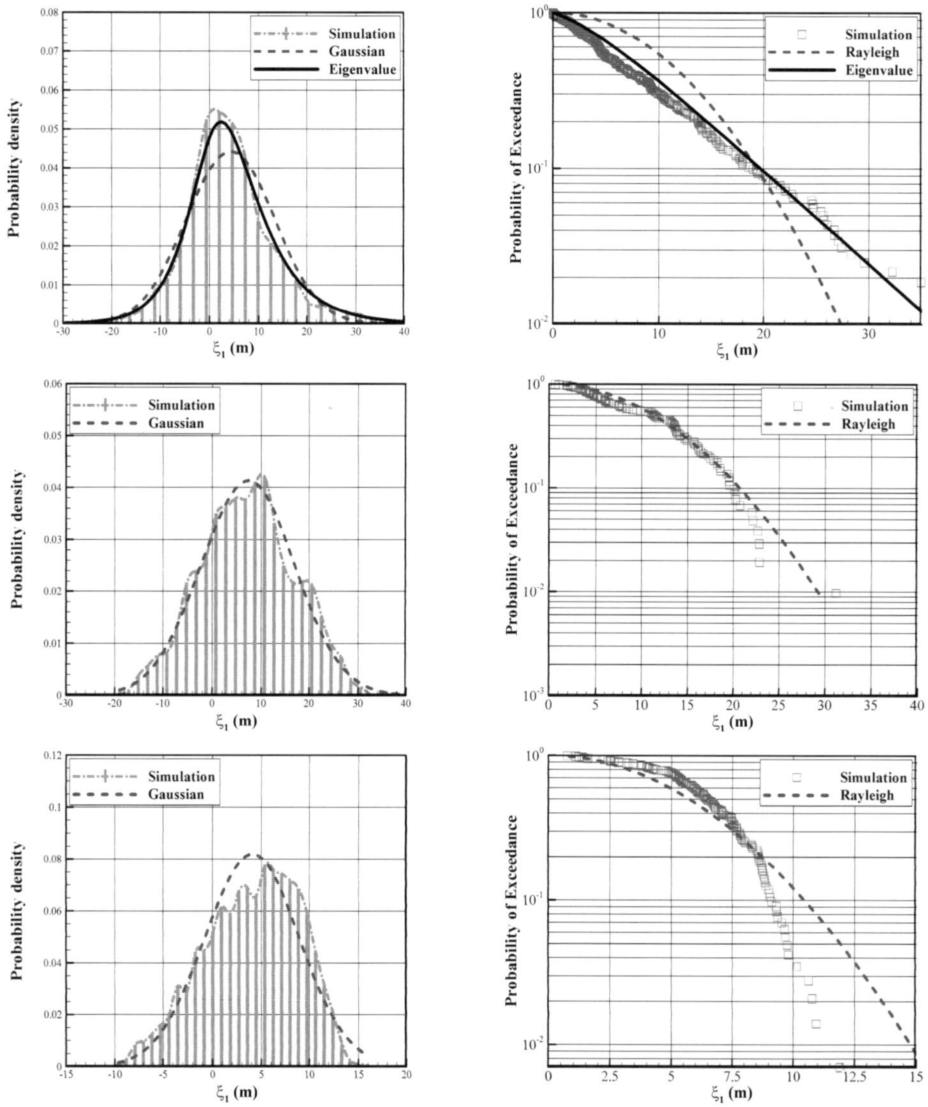

Figure 4: Probability distributions (left) and exceedance probability distributions of peaks (right) for the slow-drift surge motion for case 1 (top), case 2 (middle), and case 3 (bottom)

The black solid lines in the figure indicate the probability distribution functions computed by an eigenvalue analysis of the Hermitian kernel, and the detailed formulations can be found in literature (Neal[7]; Naess[8]; Kim[9]). From the figure, it can be seen that the probability distribution of the difference-frequency excitation force is strictly non-Gaussian and asymmetric.

Figure 4 shows the probability distributions for slow-drift motions computed for the three simulation cases. In Case 1, where nonlinearity arises only from the second-order wave excitation force, the probability distributions again deviate from the Gaussian and Rayleigh model and follow the eigenvalue analysis results. This is because the slow-drift motion can be represented as a second-order Volterra series, just like the second-order difference-frequency excitation force, as the equation of motion is completely linear. In Case 2, however, the probability distribution quite similar to the Gaussian model, and the peaks are positioned near those of the Rayleigh model. This behavior may be explained by the interaction of many different nonlinear factors, resulting in convergence to the most natural probability distribution model. In the last case in which the dynamic behaviors of the mooring lines are solved simultaneously along with the platform motion, the probability distribution becomes asymmetric such that the extreme values are smaller than those for the Rayleigh model. This is because of the nonlinear hardening caused by the mooring line restoring forces, and this phenomenon depends completely upon the characteristics of the mooring system.

These analyses lead to the conclusion that the probability distribution model suitable for the nonlinear slow-drift motion of a platform is the Gaussian and Rayleigh distributions when considering nonlinear factors other than the difference-frequency wave loads. Nonlinear hardening of the mooring line restoring coefficient may affect the probability distribution significantly, and hence, care should be taken when the restoring curve is highly nonlinear.

4. VALIDATION OF THE PRESCREENING RESULTS

For validating the accuracy of the proposed pre-screening procedure, its results were compared to the results of time-domain coupled analyses. The platform used for the comparison test is an FPSO model, as shown in Figure 5, and the particulars of the mooring system are listed in Table 2. The difference-frequency surge excitation QTF was replaced with the mean wave drift force results by Newman's approximation[10], which is a reasonable substitution for the second-order QTFs that are tedious to calculate with a high degree of accuracy. Figure 6 indicates the motion RAOs and the wave drift force for the FPSO computed in the frequency domain by a commercial software WADAM developed by DNV.

The environmental conditions considered for the comparison test are summarized in Table 3, where U_C is the current speed, and the wave spectrum is the JONSWAP spectrum ($\gamma = 3.3$). In Case 1, the spectral peak period was set to induce significant heave motion to magnify the dynamic effects. Case 2 was set up to have a small linear motion for ease of comparison with Case 1. In Case 3, the current load was added to observe the effect of current on the accuracy of pre-screening. Case 4 was characterized by large wave heights. For each of the cases, the time-domain coupled analyses were performed four times with different random phases of the irregular wave, and their results are denoted as CA1–CA4 in the following histograms. The simulation time was set to 8,192 s throughout the simulations.

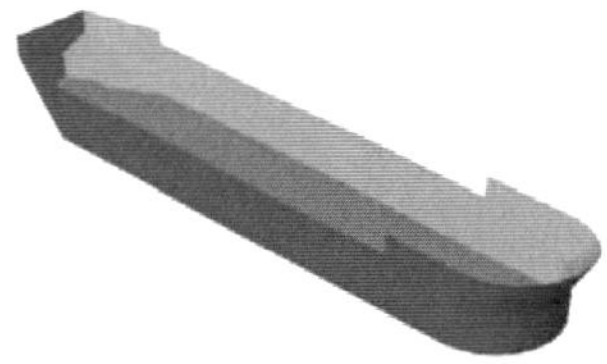

Figure 5: FPSO model used for validating the pre-screening results

Table 2: Particulars of the mooring system for the FPSO model

Designation	Unit	Value
Water depth	m	3,048
Pretension	kN	1,691
Number of lines		4×3
Degrees between the 3 lines	degree	5
Length of mooring line	m	4,267.2
Segment 1 (ground section): Chain		K4 studless
Length	m	121.9
Diameter	mm	101.6
Dry weight	N/m	2,112.5
Wet weight	N/m	1,837.5
Stiffness EA	kN	1,037,720
Segment 2 : wire		Polyester
Length	m	4,053.8
Diameter	mm	180.1
Dry weight	N/m	213
Wet weight	N/m	55.6
Stiffness EA	kN	240,192
Segment 3 (top section): Chain		K4 studless
Length	m	91.44
Diameter	mm	101.6
Dry weight	N/m	2,112.5
Wet weight	N/m	1,837.5
Stiffness EA	kN	1,037,720

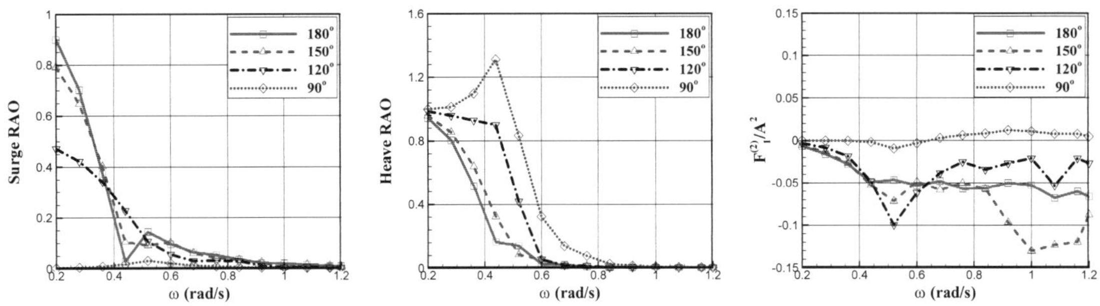

Figure 6: Surge and heave RAOs and the wave drift force for the FPSO model in head sea

Table 3: Environmental conditions for each simulation case in the time-domain analysis of the FPSO

	Wave heading (°)	H_S (m)	T_P (sec)	Spectrum	U_C (m/s)	Current heading (°)	Note
Case 1	180	10	12	JONSWAP	0	None	Significant heave motion
Case 2	180	10	8	JONSWAP	0	None	Small linear motion
Case 3	180	10	12	JONSWAP	1	180	Current existing case
Case 4	180	16	16	JONSWAP	0	None	Severe wave condition

Figure 7 shows the histograms of the results of the comparison test. The first four bars in these histograms indicate the time domain coupled analysis results denoted by CA1–CA4 as mentioned above, and the last bar indicates the results of the pre-screening procedure. For Case 1, the results of the time domain analyses and of the pre-screening appear to agree reasonably even in the presence of significant heave motions. The mean and standard deviation of the surge motion predicted by each method are within a very narrow range. The maximum values from the time domain analyses are shown to have some discrepancies with each other because of different random phases, but the maximum value among the four cases is very close to the maximum surge predicted by the pre-screening. The mooring line tensions show a similar tendency. The mean values in the time domain analyses and pre-screening are almost identical, while the maximum value predicted by the pre-screening corresponds well to the maximum among the four time domain analyses results. The results of Case 2 for small linear motions, which was included for comparison with Case 1, also display a pattern analogous to that of Case 1.

In Case 3, where a current load is considered in addition to the wave condition of Case 1, no significant difference with the previous cases was observed with regard the prediction of the surge and tension. The mean and standard deviation were again predicted well by pre-screening, and the maximum value obtained by pre-screening corresponds to the maximum among the four time domain simulations. Although this phenomenon of little effect of the current on the result may originate from the exclusion of risers in the simulations, it can be concluded that the pre-screening procedure predicts the design parameters with a reasonable degree of accuracy even when the current is considered. In Case 4, where the surge motion is largest among the four cases, again, the same trend was found, and the displacement and the tension were reasonably predicted.

Thus, the results of this validation test indicate that the pre-screening procedure has a reasonable degree of accuracy.

5. AN EXAMPLE OF PRE-SCREENING APPLICATIONS

As an example of actual application, the responses under a large number of environmental conditions were investigated for the same FPSO model shown in Figure 5 and for the mooring system described in Table 2. Wind and current drag forces were calculated using OCIMF coefficients. A diverse range of environmental conditions were selected to cover most of the states of the sea that the platform may encounter through its lifetime. Table 4 summarizes the considered environmental conditions. The total number of the environmental conditions was 90,720.

Table 4: Input environmental conditions for pre-screening (total number: 90,720)

Wind	· V_W : 10–40m/s, 10 m/s interval · Direction : 0–180°, 30°interval · Spectrum : NPD wind gust spectrum
Wave	· H_S : 6–15 m, 1 m interval · T_P : 8–16 s, 1 s interval · Direction : 180° · Spectrum : JONSWAP spectrum
Current	· V_C : 0.5–1.5 m/s, 0.5 m/s interval · Direction : 0–330°, 30° interval

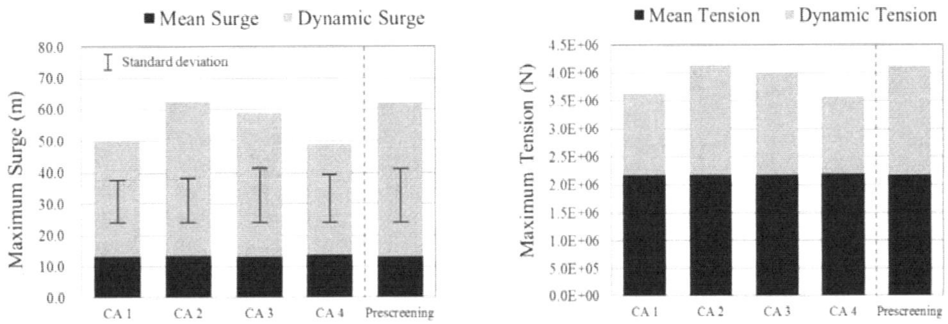

(a) Comparison test results for surge and mooring line tension in simulation case 1

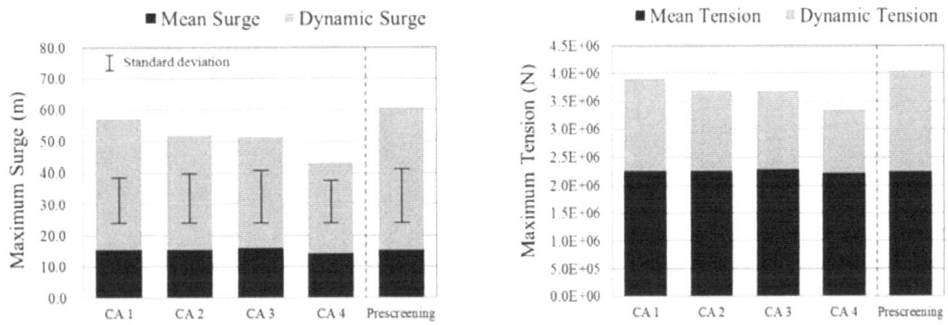

(b) Comparison test results for surge and mooring line tension in simulation case 2

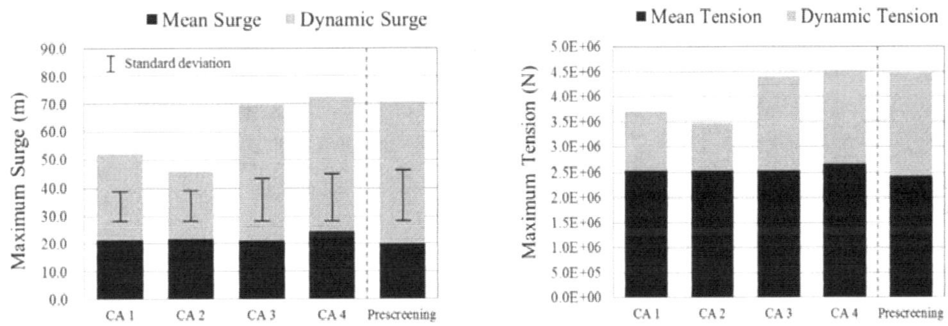

(c) Comparison test results for surge and mooring line tension in simulation case 3

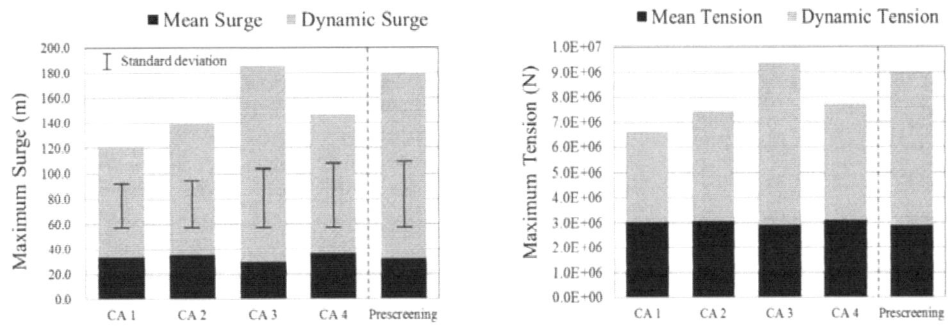

(d) Comparison test results for surge and mooring line tension in simulation case 4

Figure 7: Comparison of prescreening and time-domain simulations for surge and mooring line tension

Table 5: Environmental conditions, and mean and maximum offsets of the top 10 largest maximum offset values

U_{wind} (m/s)	α_{wind} (°)	$U_{current}$ (m/s)	$\alpha_{current}$ (°)	H_S (m)	T_P (sec)	α_{wave} (°)	**Mean heading (°)**	**Mean offset (m)**	**Maximum offset (m)**
40	90	1.5	60	15	13	180	328.9	39.9	158.5
40	90	1.5	30	15	13	180	331	41.3	157.4
40	60	1.5	60	15	13	180	328.3	38.1	156.9
40	60	1.5	30	15	13	180	329.9	39.9	156.8
40	120	1.5	60	15	13	180	331.6	41.3	156.8
40	30	1.5	60	15	13	180	328.4	36.8	155.0
40	30	1.5	30	15	13	180	330.2	38.6	154.7
40	150	1.5	60	15	13	180	334.8	41.6	153.6
30	90	1.5	60	15	13	180	331.2	38.6	153.5
40	90	1.5	60	15	12	180	331.9	40.4	153.5

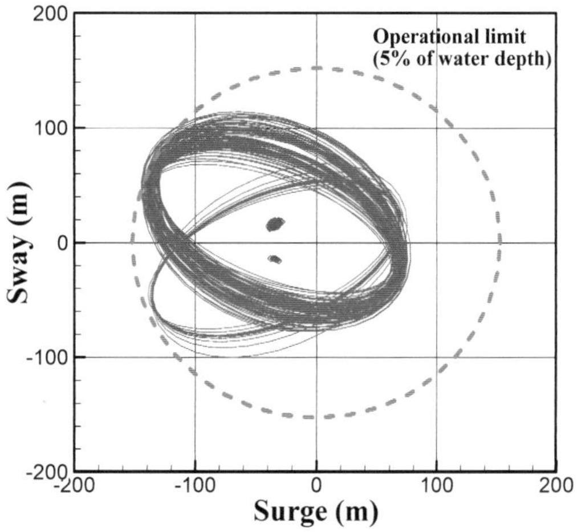

Figure 8: Static and dynamic offsets of top 100 largest offset cases from pre-screening procedure

Figure 8 shows the platform offset prediction results for top 100 largest horizontal maximum offset among the 90,720 environmental conditions. The scattered symbols represent the horizontal mean offset positions for the corresponding environmental conditions, while the solid ellipses indicate the expected boundaries of the maximum horizontal motions for each condition. The figure shows that the platform offset may be significantly underestimated if the dynamic components are not considered. For better understanding the severe environmental conditions, the top 10 cases of the largest maximum horizontal offset conditions are arranged in Table 5. Here α denotes the incident angle of each environmental load. The data in Table 5 reveals a noticeable feature of most of the environmental conditions: the directions of the environmental loads are strictly non-collinear. Owing to this feature, the angle of attack becomes significantly large, drastically enhancing the environmental loads in the transverse direction of the platform. In other words, the weathervaning of the turret mooring system fails in most of the environmental conditions where severe offsets occur. That is, the weathervaning of the platform and mean heading angles are considerably related to the severity of environmental conditions.

In terms of the computational efficiency, approximately 10 min of simulation time was required for the pre-screening of 90,720 environmental conditions using a PC with a 2.80 GHz CPU. In comparison, the time domain analysis with SML takes about 10 min for one simulation; thus, as a rough estimate, approximately 600 days are required to simulate the same number of cases with same hardware.

6. CONCLUSIONS

In this study, a pre-screening procedure for the design of offshore structure and mooring system is proposed and validated. The achievements and conclusions of this study are as follows:

- A pre-screening procedure to reduce the massive number of environmental conditions was suggested; this procedure can be applied in the early stage of design of offshore structures and mooring lines. It comprises heading angle analysis, static analysis, and dynamic analysis, all of which are focused on the efficiency and speed of calculations.

- The statistical behavior of the slow-drift motion of offshore platforms was investigated through a series of time-domain simulations. It was found that the slow-drift motion follows the Gaussian and Rayleigh distribution models as the nonlinearity of the problem increases. Nonlinear hardening of the restoring forces from mooring system is also a key factor affecting the statistical behavior.

- The accuracy of the pre-screening procedure was validated by comparing its results with the results of direct dynamic simulation in the time domain. Pre-screening was found to have a reasonable degree of accuracy.

- A large number of environmental conditions were processed by the suggested pre-screening procedure. It was shown that the dynamic component of the horizontal offset is dominant. More specifically, by observing cases of large horizontal offsets, it was shown that in most of the cases, the angles of attack for environmental loads were large; this indicates the failure of weathervaning.

- The present pre-screening procedure is practically applicable in the early design stage. By applying this procedure, the ocean environmental conditions that cause large horizontal offset and mooring tension can be predicted.

7. ACKNOWLEDGEMENTS

This work was supported by the Technology Innovation Program, Project No. 10042556, Core Technology Development for Deepwater O&G Production System FEED Engineering & Floating Systems, funded By the Ministry of Trade, Industry & Energy(MTIE), Korea. Their support is acknowledged.

8. REFERENCES

1. H. ORMBERG and K. LARSEN, 'Coupled Analysis of Floater Motion and Mooring Dynamics for a Turret-moored Ship', *Applied Ocean Research*, 1998.

2. O.J. EMMERHOFF, 'The Slow-Drift Motions of Offshore Structures', *Ph.D. Thesis, MIT*, 1994.

3. S. KIM and P.D. SCLAVONOUS, 'Fully Coupled Response Simulations of Theme Offshore Structures in Water Depths of Up to 10,000 Feet', *International Offshore and Polar Engineering Conference*, 2001.

4. J.E.W. WICHERS and P.V. DEVLIN, 'Effect of coupling of mooring lines and risers on the design values for a turret moored FPSO in deep water of the Gulf of Mexico', *International Offshore and Polar Engineering Conference*, 2001.

5. M.H. KIM, B.J. KOO, R.M. MERCIER and E.G. WARD, 'Vessel/mooring/riser coupled dynamic analysis of a turret-moored FPSO compared with OTRC experiment', *Ocean Engineering 32*, 2005.

6. OCIMF, 'Prediction of Wind and Current Loads on VLCCs, 2nd ed', *Witherby & Co.Ltd*, 1994.

7. E. NEAL, 'Second-order Hydrodynamic Forces Due to Stochastic Excitation', *Tenth Naval Hydrodynamics Symposium*, 1974.

8. A. NAESS, 'The statistical distribution of second-order slowly-varying forces and motions', *Applied Ocean Research*, 1986.

9. M.H. KIM, 'The Complete Second-order Diffraction and Radiation Solutions for a Vertically Axisymmetric Body', *Ph.D. Thesis, MIT*, 1988.

10. J.M. NEWMAN, 'Second order, slowly varying forces on vessels in irregular waves', *Int. Symp. Dynamics of Marine Vehicles and Structures in Waves*, 1974.

9. AUTHORS BIOGRAPHY

Dong-Hyun Lim is a Ph.D. candidate at the Marine Hydrodynamics Laboratory in Seoul National University. He obtained a Bachelor's degree in naval architecture and ocean engineering at Seoul National University.

Yonghwan Kim currently holds the position of a professor in the Department of Naval Architecture and Ocean Engineering at Seoul National University. He obtained his Bachelor's and Master's degrees in naval architecture and ocean engineering at Seoul National University, and his Ph.D. in ocean engineering at MIT.

Taeyoung Kim currently holds the position of a research engineer at Samsung Heavy Industries. He obtained his Bachelor's degree and Ph.D. in naval architecture and ocean engineering at Seoul National University.

ICSOT Korea: Safety of Offshore and Subsea Structures in Extreme and Accidental Conditions, 15-16 September 2015, Busan, Korea

SENSITIVITY ANALYSIS OF WTIV LEG DESIGN CONSIDERING SOIL CHARACTERISTICS FOR KOREAN WEST-SOUTH OFFSHORE WIND ZONE

T M Cho, J S PARK, Y S Ha, J H Jeon and **K B Jang**, Samsung Heavy Industries Co., Ltd, Korea

SUMMARY

In this study, sensitivity analysis of WTIV (Wind Turbine Installation Vessel) leg for the Korean west-south offshore wind zone is performed considering soil characteristics. Firstly, environmental conditions and seabed characteristics of the Korean west-south offshore wind zone is collected and investigated. Based on these data, design specifications are established and the overall basic design is performed. The sensitivity analysis of WTIV leg design for the Korean west-south offshore wind zone due to changes of soil characteristics is performed. It is observed that the WTIV leg design is very sensitive to soil characteristics. The structural integrity of the WTIV leg is verified through the code check and the adequate safety margin is observed. The results of this study can be expected as practical and useful data for the design of the WTIV for the Korean west-south offshore wind zone.

1. INTRODUCTION

In order to prepare the global warming and shortages of fossil fuels, the Korean government planned to develop offshore wind zones located at the west-south part of the Korean Peninsula. The Korean west-south offshore wind zone, shown in Figure 1, is regarded as the most suitable area for the generation of wind energy in Korea. The development schedule is divided into 3 steps such as verification, pilot, and expansion phase as shown in Table 1. During step 1(verification phase), environmental conditions, seabed characteristics, estimation of spudcan penetration force, and design of leg and jacking system for the WTIV (Wind Turbine Installation Vessel) are to be developed.

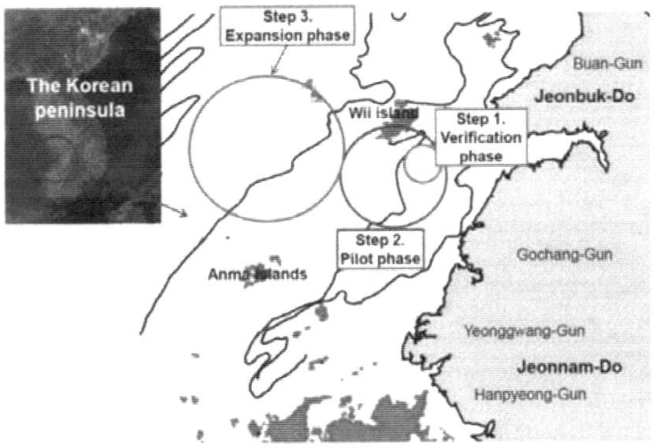

Figure 1: Korean west-south offshore wind zone

Table 1: Development schedule of Korean west-south offshore wind zone

Item	Development phase		
	Verification	Pilot	Expansion
Capacity	100 MW	400 MW	2,000 MW
Development schedule*	2012-2015	2016-2017	2018-2020

*Development schedule can be changed accordingly

In this study, sensitivity analysis of the WTIV leg for the Korean west-south offshore wind zone is performed considering soil characteristics. The environmental conditions and seabed characteristics of the Korean west-south offshore wind zone are first collected and investigated. Then the design specifications are established, and the overall basic design is performed. The sensitivity analysis of the WTIV leg design for the Korean west-south offshore wind zone due to changes of soil characteristics is performed. The structural integrity of the WTIV leg is verified through finite element analysis and the code check. The conclusions are discussed in the last section.

2. ENVIRONMENTAL CONDITIONS

The environmental conditions of the Korean west-south offshore wind zone are investigated based on direct measurements of that area and literature surveys.

2.1 DIRECT MEASUREMENTS

The wave measurement points are shown in Figure 2 and the statistical results of the wave measurement are shown in Table 2. In Table 2, Tz, Hs, and Dir. mean the zero-up crossing period, significant wave height, and wave direction, respectively. The water depth of the Korean west-south offshore wind zone is also measured, as shown in Figure 3 and 4. As depicted in Figure 4, about 40 m is observed as the maximum water depth.

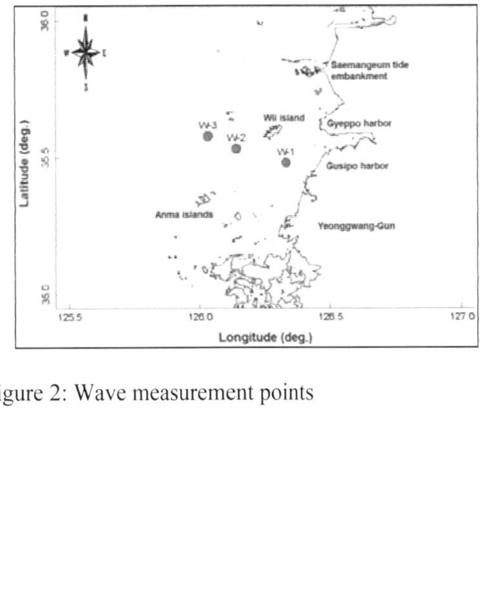

Figure 2: Wave measurement points

Table 2: Statistical results of wave measurement

Point	Tz(sec) / Hs(m)	< 4.0	4.0 ~	5.0 ~	6.0 ~	7.0 ~	8.0 ~	9.0 ~	10.0 ~	11.0 ~	12.0 ~	13.0 ~	14.0 ~	Total
		Probability of occurrence for each sea state (%)												
W-1	< 0.5	3.6	37.2	10.3	0.6	0.0	0.0	0.0	0.0	0.0	0.0	0.0	0.0	51.7
	0.5 ~	0.5	8.7	15.4	5.1	0.1	0.0	0.0	0.0	0.0	0.0	0.0	0.0	29.8
	1.0 ~	0.0	2.8	2.2	4.8	1.1	0.0	0.0	0.0	0.0	0.0	0.0	0.0	10.9
	1.5 ~	0.0	0.0	1.2	2.2	1.0	0.0	0.0	0.0	0.0	0.0	0.0	0.0	4.4
	2.0 ~	0.0	0.0	0.0	0.4	0.7	0.6	0.2	0.0	0.0	0.0	0.0	0.0	1.9
	3.0 ~	0.0	0.0	0.0	0.0	0.5	0.2	0.4	0.0	0.0	0.0	0.0	0.0	1.1
	4.0 ~	0.0	0.0	0.0	0.0	0.0	0.0	0.2	0.0	0.0	0.0	0.0	0.0	0.2
	Total	4.1	48.7	29.1	13.1	3.4	0.8	0.8	0.0	0.0	0.0	0.0	0.0	100
W-2	< 0.5	0.0	3.9	33.4	13.1	2.8	0.7	0.0	0.0	0.0	0.0	0.0	0.0	53.9
	0.5 ~	0.0	0.6	13.4	8.7	0.9	0.0	0.0	0.0	0.0	0.0	0.0	0.0	23.6
	1.0 ~	0.0	0.1	2.3	7.1	2.0	0.0	0.0	0.0	0.0	0.0	0.0	0.0	11.5
	1.5 ~	0.0	0.0	0.5	2.8	3.7	0.0	0.0	0.0	0.0	0.0	0.0	0.0	7.0
	2.0 ~	0.0	0.0	0.0	0.2	0.9	0.6	0.0	0.0	0.0	0.0	0.0	0.0	1.7
	3.0 ~	0.0	0.0	0.0	0.0	0.1	1.0	0.4	0.0	0.0	0.0	0.0	0.0	1.5
	4.0 ~	0.0	0.0	0.0	0.0	0.0	0.1	0.7	0.0	0.0	0.0	0.0	0.0	0.8
	Total	0.0	4.6	49.6	31.9	10.4	2.4	1.1	0.0	0.0	0.0	0.0	0.0	100
W-3	< 0.5	21.2	17.9	8.0	4.3	1.8	0.6	0.6	0.0	0.1	0.0	0.0	0.0	54.5
	0.5 ~	4.0	7.1	9.1	7.3	0.9	0.2	0.1	0.0	0.0	0.0	0.0	0.0	28.7
	1.0 ~	0.6	1.2	2.3	3.9	0.7	0.0	0.0	0.0	0.0	0.0	0.0	0.0	8.7
	1.5 ~	0.0	0.2	0.6	1.8	0.9	0.1	0.0	0.0	0.0	0.0	0.0	0.0	3.6
	2.0 ~	0.0	0.0	0.1	1.0	1.5	0.4	0.1	0.0	0.0	0.0	0.0	0.0	3.1
	3.0 ~	0.0	0.0	0.0	0.0	0.0	0.3	0.4	0.0	0.0	0.0	0.0	0.0	0.7
	4.0 ~	0.0	0.0	0.0	0.0	0.1	0.1	0.5	0.0	0.0	0.0	0.0	0.0	0.7
	Total	25.8	26.4	20.1	18.3	5.9	1.7	1.7	0.0	0.1	0.0	0.0	0.0	100

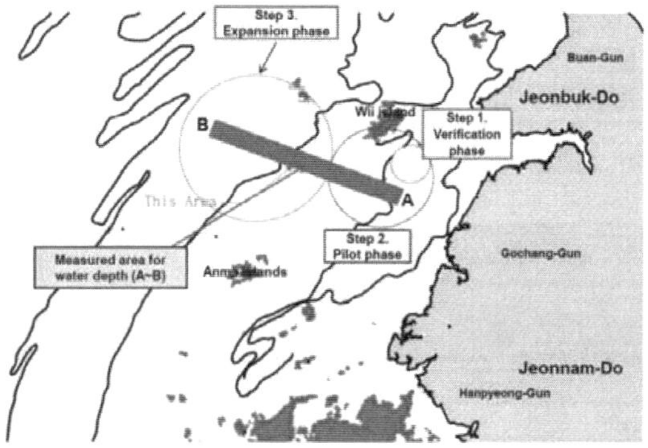

Figure 3: Wave region for water depth

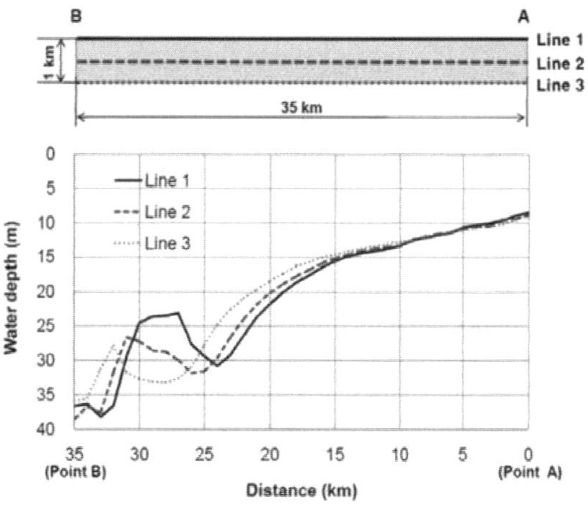

Figure 4: Results of water depth

2.2 LITERATURE SURVEYS

The environmental conditions of the Korean west-south offshore wind zone are also investigated through literature surveys. Tables 3 and 4 show the cumulative probability of occurrence for sea states and design conditions based on a 50-year return period, respectively. These data are obtained from the Korea Institute of Ocean Science & Technology (KIOST) report [1]. The design conditions for each return period for the Korean west-south offshore wind zone, which were studied by Jeong et al. [2], are summarized in Table 5. The design conditions for the wind velocity are determined to be 41 m/sec based on the literature surveys [3].

Table 3: Cumulative probability of occurrence for sea state [1]

Sea state		Cumulative probability of occurrence (%)
Hs (m)	Tz (sec)	
2.0~2.5	8~9	95
1.5~2.0	7~8	90
1.5~2.0	5~6	85
1.0~1.5	6~7	80
1.0~1.5	5~6	75
1.0~1.5	~4	70

Table 4: Design conditions [1]

Sea state	Design conditions for each wave direction (50-year return period)			
	SW	W	WNW	NW
Hs (m)	6.06	5.56	6.91	7.00
Tz (sec)	11.92	10.67	11.86	11.76

Table 5: Design conditions for each return period [2]

Return period (year)	Wave direction	Hs (m)	Tz (sec)
1	NW	3.64	10.24
10	NW	6.18	12.51
20	NW	6.50	12.76
50	NW	6.86	13.02
100	NW	7.11	13.20

3. OVERALL BASIC DESIGN OF WTIV

The environmental conditions have been investigated through direct measurements and literature surveys. Based on these data, the design specifications and overall basic design of the WTIV for the Korean west-south offshore wind zone are determined.

In the basic design, a 4-leg type of WTIV is considered. The total leg length is determined to be 105 m, as shown in Table 6. The air gap is determined to be 9.0 m based on the environmental conditions. In Figure 4, 40 m is observed as the maximum water depth. However, 55 m of the water depth is selected for the design of the total leg length, with a consideration of some design margin. Figure 5 shows the composition of the total leg length described in Table 6.

Table 7 shows the design specifications of the transit and DP (dynamic positioning), reloading, crane operation, and survival conditions. Among all the conditions, the survival condition is generally the most severe case for the majority of the structures. The maximum wave height and wind velocity are determined to be 13.76 m and 41 m/sec for the survival condition, respectively.

Table 6: Total leg length

Parameter	Data
Leg reserve	2.5 m
Jack case height	12.5 m
WTIV depth	11.0 m
Air gap	9.0 m
Max. water depth	55.0 m
Penetration depth	15.0 m
Total	Approximately 105.0 m

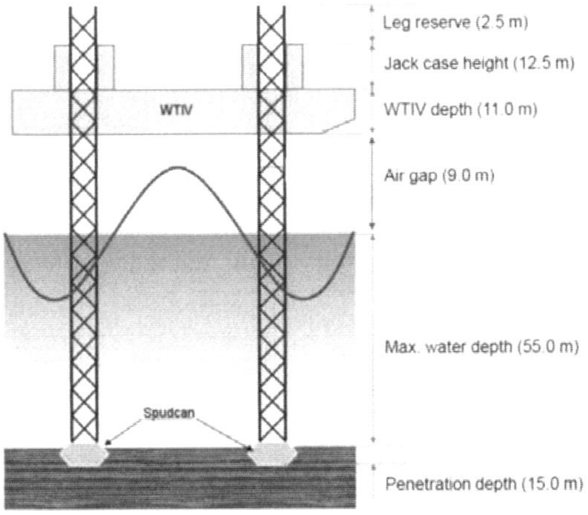

Figure 5: Composition of total leg length

Table 7: Design specification of WTIV

Parameter		Unit	Transit & DP condition	Preloading condition	Crane operation	Survival condition
Cargo weight		Ton	3,700	3,700	3,700	2,600
Consumable		Ton	2,300	2,300	2,300	2,300
Maximum payload		Ton	6,000	6,000	6,000	4,900
Lightweight		Ton	12,500	12,500	12,500	12,500
Maximum elevated weight		Ton	20,350	20,350	20,350	19,140
Water depth		m	55.0	55.0	55.0	55.0
Air gap		m	-	9.0	9.0	9.0
Wave height	H_{max}	m	3.70	3.70	4.70	13.76
	Hs	m	2.00	2.00	2.50	7.40
Wave period	Min.	sec	4.86	4.86	5.44	9.36
	Max.	sec	6.25	6.25	6.99	12.02
Current speed	Surface	m/sec	1.025	1.025	1.025	1.150
	Bottom	m/sec	0.50	0.50	0.50	0.85
Wind speed		m/sec	20.0	20.0	13.0	41.0

4. SEABED CHARACTERISTICS

4.1 POSSIBILITY OF PUNCH-THROUGH

The seabed characteristics of the Korean west-south offshore wind zone are investigated through direct measurements and literature surveys. Figure 6 shows the measurement points for the seabed characteristics. In Fig. 6, 8 points (NO.1-NO.8) have been measured by the Korea Electric Power Research Institute [4]. Other points (OW.1-OW.8 and OWW.1-OWW.8) have been measured by the authors through the national project acknowledged at the end of this paper.

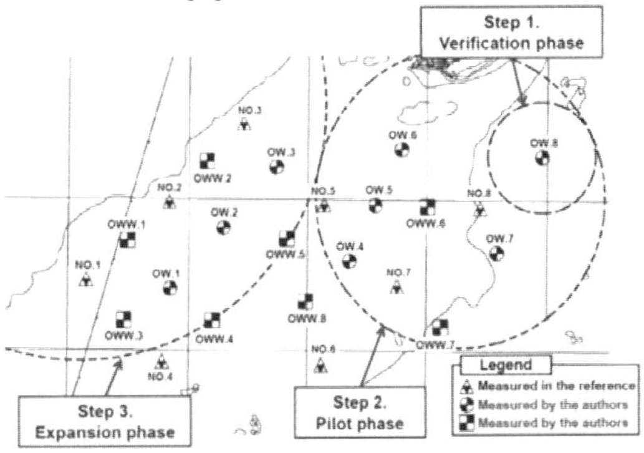

Figure 6: Measurement points for seabed characteristics

Figure 7 shows the seabed characteristics of the measurement points. In the standard penetration test, the N value, shown in Figure 7, is the number of hammer strikes required to drive the split spoon sampler into the soil to a certain depth. The N value is generally corrected based on the experimental conditions such as the hammer and sampler efficiency. In Figure 7, the measured points NO.1-NO.4, NO.6, OW.3, OW.4, and OW.8 show stable characteristics, whereas there is the possibility of a punch-through at the measured points NO.5, NO.7, NO.8, OW.1, OW.2, OW.5-OW.7, and OWW.1-OWW.8. Table 8 shows the soil properties at the measured points NO.1 and NO.8 [4]. In Table 8, φ, c_u, and γ' represent the angle of internal friction for the sand, submerged unit weight, and undrained shear strength, respectively. These data are required to estimate the spudcan penetration force.

Table 8: Soil properties of Korean west-south offshore wind zone [4]

Position	Layer	Soil	Depth (m) Top of unit	Depth (m) Base of unit	φ (°)	c_u (kPa)	γ' (kN/m^2) (assumed)
NO.1	1	Sand	0.0	32.0	35	-	10
	2	Clay	32.0	42.0	-	20	10
	3	Sand	42.0	63.0	30	-	10
	4	Weathered soil	63.0	67.0	25	-	10
NO.8	1	Sand	0.0	6.0	41	-	10
	2	Clay	6.0	27.0	-	13	10
	3	Sand	27.0	39.0	36	-	10
	4	Weathered rock	39.0	41.0	43	-	10

In the design of the jack-up or WTIV, the estimation of the spudcan penetration force is an important step. So far, many studies have been performed on the design of the spudcan and the spudcan penetration force [5~6]. In this study, the spudcan penetration force for the Korean west-south offshore wind zone is estimated based on the SNAME rule [7]. The SNAME rule is widely accepted in offshore engineering, and the conservative results are obtained through the SNAME rule.

The spudcan penetration force has already been studied by Cho et al [5]. The spudcan is assumed to be a flat, circular, cross-sectional area of 100 m^2 with height of 2 m. At the measured point NO.1, the seabed can resist about 14,000 tons without the punch-through phenomenon occurring. However, at the measured point NO.8, the seabed can resist only 1,500 tons and then the punch-through phenomenon occurs. The estimated spudcan penetration force can be useful and practical data for the design of the spudcan.

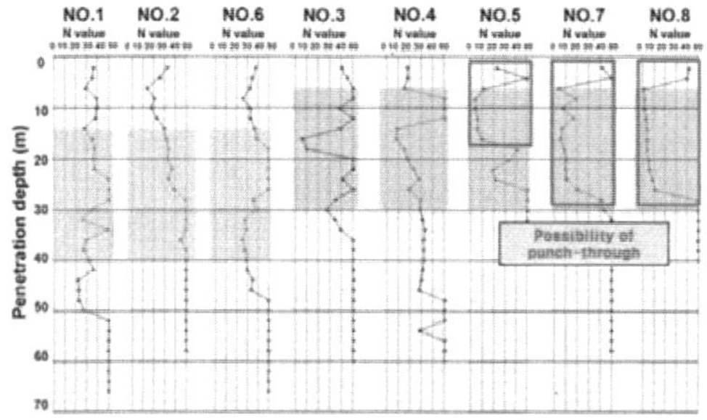

(a) NO.1~NO.8 [4]

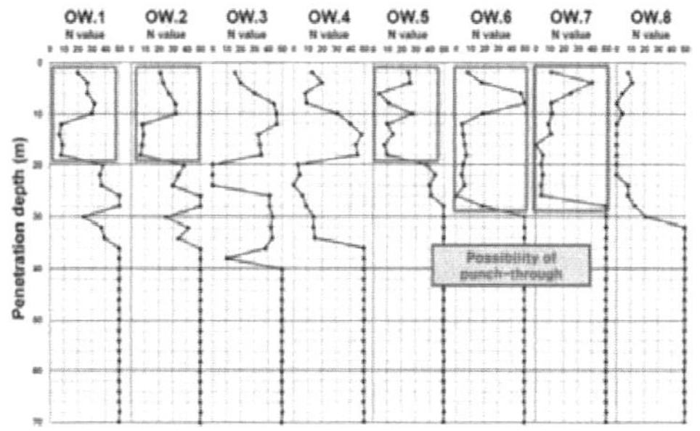

(b) OW.1~OW.8

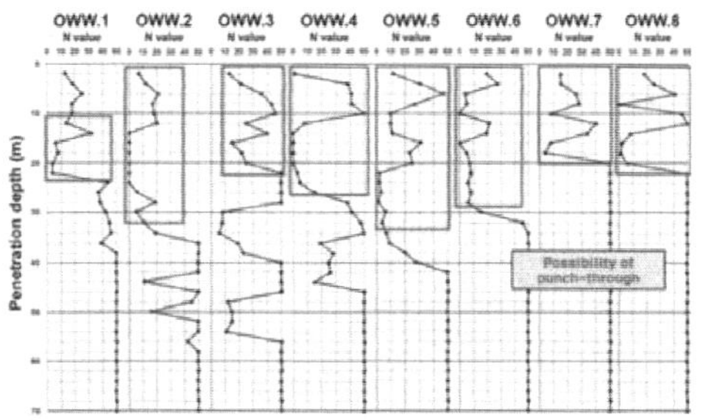

(c) OWW.1~OWW.8

Figure 7: N value vs penetration depth curve

4.2 SOIL STIFFNESS OF KOREAN WEST-SOUTH OFFSHORE WIND ZONE

Figure 9 shows effect of spudcan-soil fixity [8]. Figure 10 shows bending moment diagram for pinned and fixed conditions. If soil fixity effect is considered, the bending moment at the lower guide can be reduced. Therefore lighter scantlings are possible in the upper portion of leg and hull-leg interface region. Even though pinned constraint is regarded as the more conservative method, partially-fixed constraint can be also useful for the strength estimation of the WTIV leg.

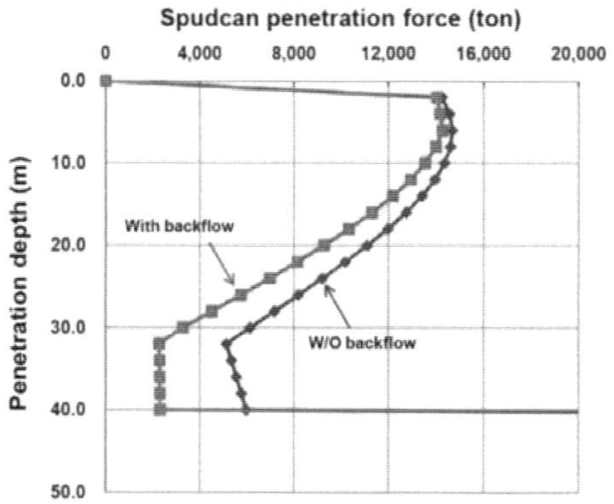

(a) NO.1 point [5]

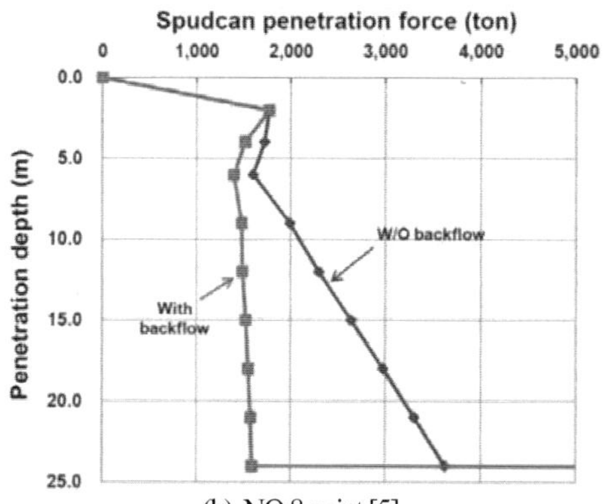

(b) NO.8 point [5]

Figure 8: Spudcan penetration force

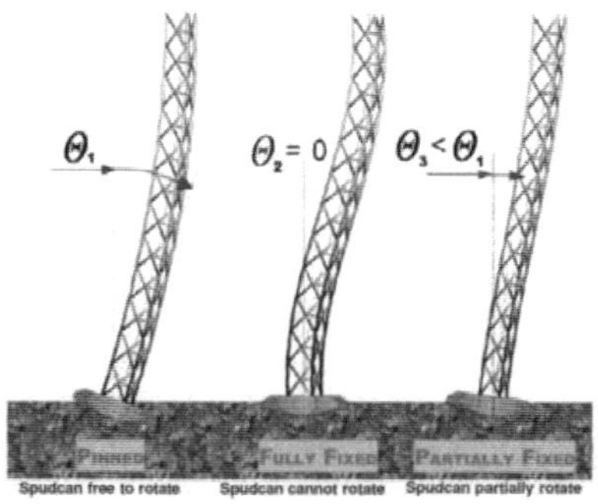

Figure 9: Effect of spudcan-soil fixity [8]

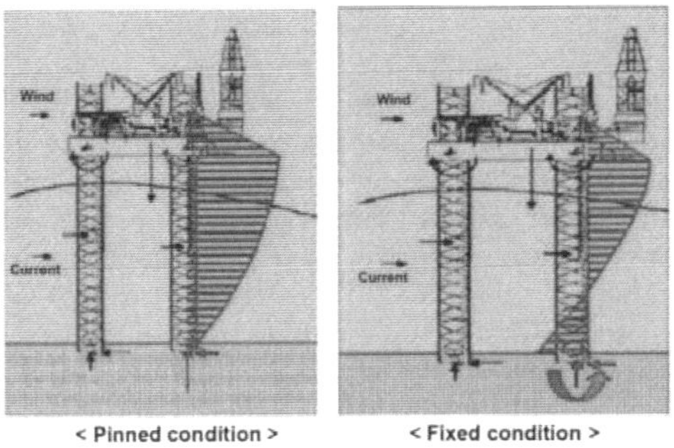

Figure 10: Bending moment for pinned and fixed conditions

In order to perform sensitivity analysis of the WTIV leg for the Korean west-south offshore wind zone, the soil stiffness needs to be obtained. In this study DNV CN 30.4 [9] is used. Most of the parameters are obtained through the actual soil investigation for the Korean west-south offshore wind zone and some parameters are obtained from the references [7, 9]. In DNV CN 30.4 [9], the initial stiffness can be calculated through Equations (1)~(3). These parameters can be possible to represent as actual overturning behavior against environmental loads such as wind, current and wind.

- Vertical (Z) :

$$K_Z = \frac{4GR}{1-\nu}\left(1+\frac{1}{2}\frac{D}{R}\right) \tag{1}$$

- Horizontal (X)

$$K_X = \frac{8GR}{2-\nu}\left(1+\frac{2}{3}\frac{D}{R}\right) \tag{2}$$

- Rocking (θ)

$$K_\theta = \frac{8GR}{3(1-\nu)}\left(1+2\frac{D}{R}\right) \tag{3}$$

Where,
R : radius of foundation in contact with soil
G : initial shear modulus of soil for infinitesimal strains
ν : Poisson's ration
D : embedment of the maximum diameter section

5. SENSITIVITY ANALYSIS OF WTIV LEG DESIGN CONSIDERING SOIL STIFFNESS

5.1 PREPARATION OF STRENGTH ANALSYS FOR WTIV LEG

In order to perform the global in-place analysis of the WTIV leg, all kinds of engineering data, such as the main dimension of the WTIV leg, analysis parameters, material properties, allowable criteria, and loading conditions (static and dynamic components), are required.

The parameters for the strength analysis of the WTIV leg are summarized in Table 9. The material properties and allowable stresses are summarized in Tables 10 and 11, respectively. In Table 11, the code checks are calculated based on the ARI RP 2A-WSD 21st criterion for the tubular members such as the brace and the AISC 13th criterion for the non-tubular members such as the chord and rack.

Table 9: Parameters for strength analysis of WTIV leg

Parameter	Data
Hull main dimension (L×B×D)	124.8 m×40 m×11 m
Lightweight (MT)	12,000
Payload (MT)	6,000
Crane load	Not operating
LCG, TCG, VCG (m)*	73.79, 0.0, 20.0
Design leg length (m)	105
Water depth (m)	55
Air gap (m)	9.0
Leg chord distance (m)	8.0
Leg bay height (m)	4.2
Wave height (m)	13.7
Wave period (s)	12.0
Surface current (m/s)	1.029 (1/7 power law)
Wind speed (m/s)	41.0
Wave reduction factor	0.86
Current blockage factor	1.0
Spudcan penetration (m)	3.0-15.0
Marine growth	No marine growth
CD/CM for rack and chord**	1.0, 1.8/2.0
CD/CM for braces	0.65/2.0
Wave theory	Stream function 5th order

*LCG (longitudinal center of gravity), TCG (transverse center of gravity), VCG (vertical center of gravity)
**CD (drag coefficient), CM (inertia coefficient)

Table 10: Material property of WTIV leg

Part	Material grade	Minimum yield strength (MPa)
Rack plate	Dillimax 690 MOD	690.0
Chord pipe	ASTM A-514H MOD	690.0
Chord stiffener pipe	ASTM A-514H MOD	690.0
Braces	EN S355	355.0

Table 11: Allowable stress

Stress type	Operating condition	Survival condition
Axial and bending stress (MPa)	552.0	600.0
Shear stress (MPa)	363.1	379.1
Equivalent stress (MPa)	627.2	657.1

The leg property data are indicated in Table 12. Figure 11 shows actual 3D view of one bay of the WTIV leg structure. In case of the strength analysis model of the leg structure, the equivalent section properties as rectangular shape are used in the SACS software.

Table 12: Leg property data

Type of leg	3-chorded truss type
Chord distance	Approximately 8.0 m
Type of chords	Split pipe with opposed teeth rack
Thickness of rack in chord	177 mm thick and 995 mm wide
Bracing type	Fully-crossed X-type bracing
Chord steel min. yielding stress	690 MPa
Penetration depth	Min. 3 m/Max. 15 m

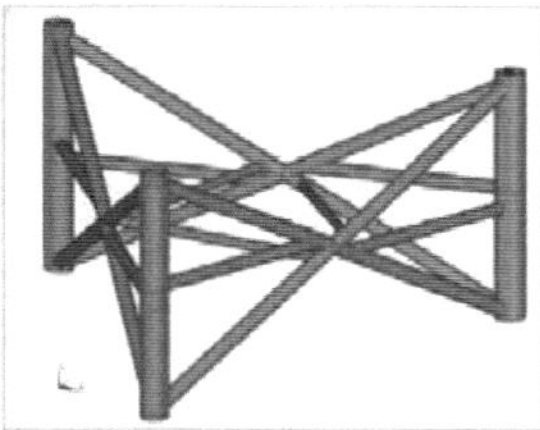

Figure 11: Actual one bay model in leg structure

In the strength analysis, the following methodologies are applied:

- All the members in the leg structure are modelled as beam elements. The hull is idealized as a grillage, and equivalent properties of the hull are assigned.
- The members forming the leg-to-hull connection are modelled as axial compressive elements with respective initial gaps between the rack and gear distance.
- The spudcans are modelled as rigid beams connecting the leg chords to the pinned support that is at half the depth of the spudcan penetration.
- The gravity loads, such as the self-weight of the legs, lightship weight, and payload, are applied to the structure. The lightship weight and payload are applied as uniform loads, and global moments are applied to correct the center of gravity (COG) of the applied loads.
- The basic environmental loads are applied to the structure in various combinations with the gravity loads.
- The structure is analysed under the applied loads, with the P-Δ large deflection effect being taken into account.
- 7% damping is assumed on the basis of the SNAME T&RB 5-5A guidelines [7].
- The structure is also checked for the available factor of safety against overturning on the basis of the DNV-RP-C104 recommended practice [10].

The flow chart of the global in-place analysis of the WTIV leg is shown in Figure 12. During the first step of the calculation, the static results, such as of both the base shear and Overturning Moment (OTM) at the end of the spudcan, are estimated. Then the natural frequency based on the SDOF (Single Degree Of Freedom) model is estimated during the second step of the calculation. The Dynamic Amplification Factor (DAF) is defined by the use of the classical method based on the previous results depending on the wave heading angles. After these procedures, a reanalysis is performed to consider the newly updated inertia loads.

The strength analysis of the WTIV leg is performed with the use of Structural Analysis Computer System (SACS) software and code checks. The SACS model of the WTIV leg for the strength analysis is shown in Figure 13. The considered loading conditions in the global in-place analysis of the WTIV leg are shown in Table 13. A total of four cases of loading directions are considered, as shown in Figure 14.

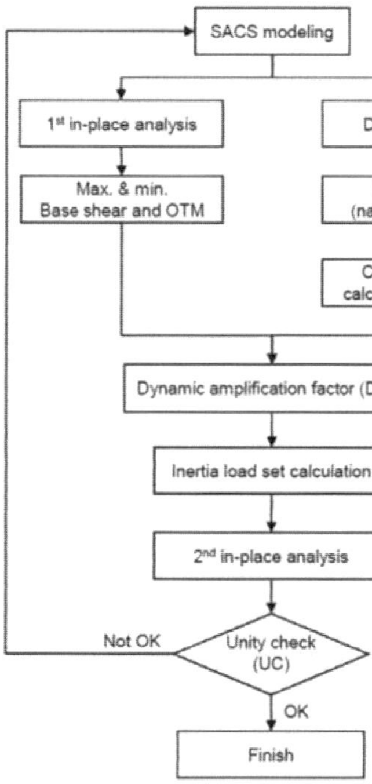

Figure 12: Flow chart of strength analysis for WTIV leg

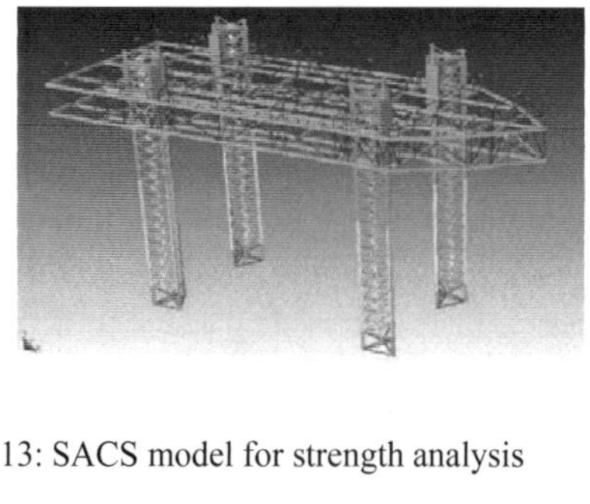

Figure 13: SACS model for strength analysis

Table 13: Loading conditions in global in-place analysis of the WTIV leg

Load case	Subcase	Loading combination					
		Self-weight	Pay load	Con-sumable	Wind load	Inertia load	Wave load
1	1	O	O	O	-	-	-
2	4 cases of loading directions	-	-	-	-	-	O(1)
3		O	O	O	O	-	-
4		O	O	O	-	O	-
5		O	O	O	-	O	O(1)
6		O	O	O	-	-	-
7		O	O	O	O	O	O(1)

LC 1: Modal analysis for the estimation of DAF.
LC 2: Estimation of max. base shear amplitude for inertia load.
LC 3 to LC 6: Overturning moment analysis and strength analysis of leg for partial combination.
LC 7: Overturning moment analysis and strength analysis of leg.
(1) Wave load: 15° interval of phase angle is applied from 0° to 360°.

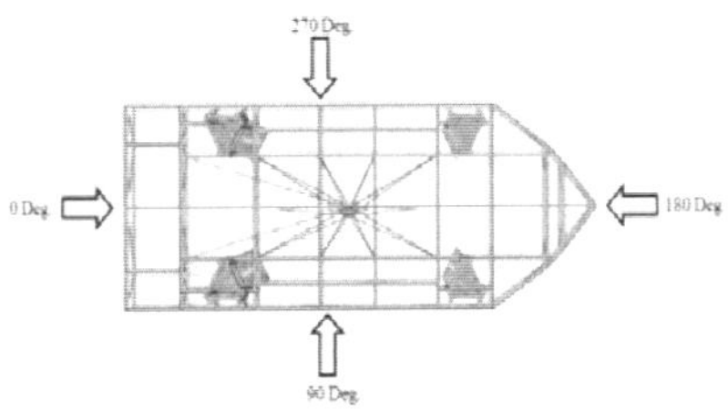

Figure 14: Four cases of loading directions

5.2 STABILITY AGAINT OVERTURNING

In the calculation of overturning stability against environmental loads, the pinned conditions are applied at the bottom of the legs as shown in Figure 15. Table 14 shows the overturning stability calculation and the factor of safety. The 1.97 of the minimum safety factor is calculated and compared the allowable criteria of 1.05. It is seen that the designed WTIV leg has enough safety margin against environmental loads [5].

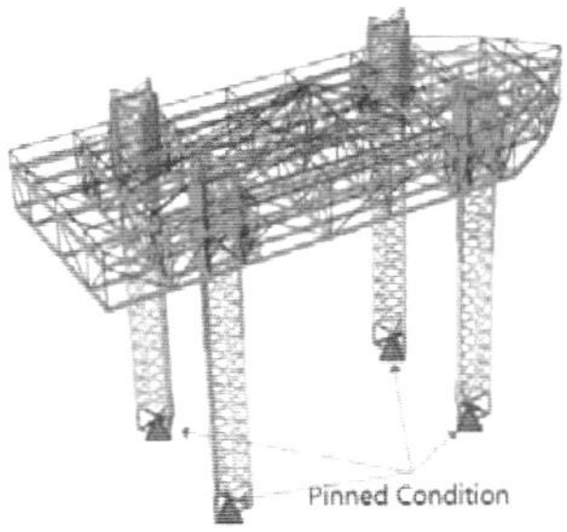

Figure 15: Analysis model with pinned constraint condition

Table 14: Overturning stability calculation [5]

Angle [deg.]	Mx [kN-m]	My [kN-m]	Overturning moment [kN-m]	dx [cm]	dy [cm]	e [m]	SMRF* (kN-m)	M_{so} [kN-m]	Stabilizing moment [kN-m]	Factor of safety	Remark (1)
0	-2.06E+04	1.34E+06	1.34E+06	69.98	1.43	0.700	2.40E+05	8.17E+06	7.93E+06	5.93	OK
30	-8.10E+05	1.28E+06	1.52E+06	67.4	60.39	0.905	2.88E+05	9.68E+06	9.39E+06	6.19	OK
60	-1.72E+06	8.22E+05	1.91E+06	50.55	116.92	1.274	3.74E+05	8.18E+06	7.81E+06	4.09	OK
90	-2.00E+06	-1.27E+05	2.01E+06	0.97	137.32	1.373	3.97E+05	4.35E+06	3.95E+06	1.97	OK
120	-1.66E+06	-7.68E+05	1.83E+06	-48.57	111.63	1.217	3.61E+05	8.26E+06	7.90E+06	4.33	OK
150	-7.66E+05	-1.27E+06	1.49E+06	-67.02	54.42	0.863	2.78E+05	9.56E+06	9.28E+06	6.25	OK
180	-1.51E+04	-1.39E+06	1.39E+06	-71.62	1.08	0.716	2.44E+05	8.33E+06	8.08E+06	5.83	OK
210	8.31E+05	-1.30E+06	1.54E+06	-66.61	-59.69	0.894	2.86E+05	9.18E+06	8.90E+06	5.78	OK
240	1.69E+06	-7.75E+05	1.86E+06	-47.18	-113.84	1.232	3.65E+05	7.59E+06	7.23E+06	3.89	OK
270	1.98E+06	-1.34E+05	1.98E+06	2.04	-135.77	1.358	3.94E+05	4.35E+06	3.96E+06	2.00	OK
300	1.68E+06	7.75E+05	1.85E+06	48.19	-114.06	1.238	3.66E+05	7.52E+06	7.15E+06	3.87	OK
330	7.43E+05	1.23E+06	1.44E+06	64.92	-55.56	0.854	2.76E+05	9.05E+06	8.77E+06	6.10	OK

*SMRF : stabilizing moment reduction factor (stabilizing moment= M_{so}-SMRF)
(1)Allowable safety factor: 1.05

5.3 GLOBAL IN-PLACE ANALYSIS OF LEG WITHOUT SOIL STIFFNESS

In case of without soil stiffness, the pinned conditions are also applied as shown in Figure 15. For all the considered loading conditions, the results of the unity check for all the members satisfy the allowable value of 1.0. The maximum combined unity check plot is shown in Figure 16 [5]. The 0.75 of the maximum value of the unity check is calculated at the connection of between lower guide and chord members.

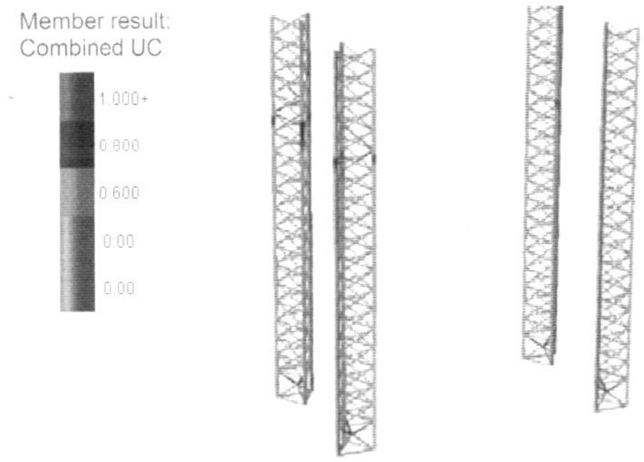

Figure 16: Maximum combined unity check (UC) plot (without considering soil stiffness) [5]

5.4 GLOBAL IN-PLACE ANALYSIS OF LEG WITH SOIL STIFFNESS

In this case, the calculated soil stiffness in Section 4.2, is applied as shown in Figure 17. The maximum combined unity check plot is shown in Figure 18. The 0.303 of the maximum value of the unity check is calculated. The maximum value of the unity check is calculated at the connection of between lower guide and chord members as well as brace member at the lower part in leg, since the diagram of bending moment is rapidly changed according to change the boundary condition as mentioned Figures 9~10.

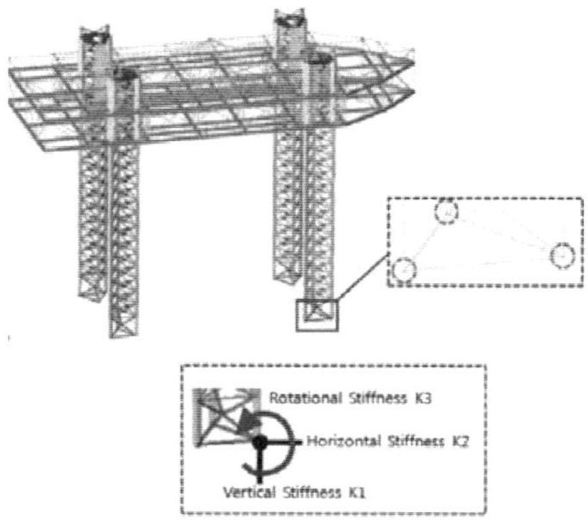

Figure 17: Analysis model with soil stiffness condition

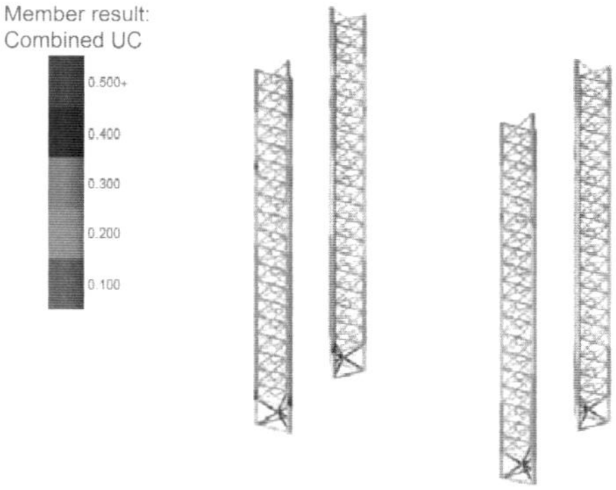

Figure 18: Maximum combined unity check (UC) plot (with considering soil stiffness)

5.5 DISCUSSION

Both the two cases (without/with considering soil stiffness), the maximum values of the unity check are greater than the allowable value. It is seen that the designed WTIV leg has enough safety margin against the overturning moment and the structural integrity of the WTIV leg is verified through the code checks. The WTIV leg of the Korean west-south offshore wind zone is conservatively design to account for the various kinds of uncertainties during the developing schedule. The code check results of the WTIV leg are dramatically reduced from 0.75 to 0.303. Through this study, it is observed that the WTIV leg design is very sensitive to soil characteristics. Even though it's hard to obtain the actual soil stiffness of the target area but it should be deserved to try it.

6. CONCLUSIONS

The sensitivity analysis of the WTIV leg of the Korean west-south offshore wind zone is performed considering soil characteristics. The environmental conditions and seabed characteristics of the Korean west-south offshore wind zone are firstly investigated and collected through both direct measurements and literature surveys. Based on these data, design specifications are established and the overall basic design is performed. Through this study, it can be seen that the code check results of the WTIV leg for the Korean west-south offshore wind zone are much improved if the soil stiffness is considered. The results of this study can be expected as practical and useful data for the development of the Korean west-south offshore wind zone.

7. ACKNOWLEDGEMENTS

This work was supported by the New & Renewable Energy of the Korea Institute of Energy Technology Evaluation and Planning (KETEP) grant (No. 20123010020090), which was funded by the Ministry of Knowledge Economy of the Government of South Korea.

8. REFERENCES

1. Korea Institute of Ocean Science and Technology, 'Deep Water Design Waves for Whole Sea Areas of the Korean Peninsula', *Report No. BSPE95100-1767-2*, 2005.

2. JEONG, S. T., KO, D. H., PARK, H. J. and LEE, I. H., 'An Analysis of the Marine Design Conditions for Development of Offshore Wind Farm in the Western Sea of Korea', *Proceedings of the Korea Wind Energy Association 2012 Spring Conference, pp. 11-15*, 2012.

3. LEE, Y. K., LEE, S. S., HAM, H. J. and BIENKIEWICZ, B., 'Estimation of Design Wind Speed for Offshore Structures', *Trans. of The Wind Engineering Institute of Korea, Vol. 13, No. 3, pp. 129-135*, 2009.

4. Korea Electric Power Research Institute, 'Geotechnical Report of Korean West-South Offshore Wind Zone', *Report No. TM.8034. P2011.0600*, 2011.

5. CHO, T. M., PARK, J. S., HA, Y. S., KIM, B. J. and JANG, K. B., 'Global in-place analysis of WTIV leg for Korean west-south offshore wind zone', *Journal of Ocean and Wind Energy, Vol. 2, No. 2, pp. 121-127*, 2015.

6. MERIFIELD, R. S., LYAMIN, A. V., SLOAN, S. W. and YU, H. S., 'Three-dimensional lower bound solutions for stability of plate anchors in clay', *Journal of Geotechnical and Geo-environmental Engineering, Vol. 129, No. 3, pp. 243-253*, 2003.

7. The Society of Naval Architects and Marine Engineers, 'SNAME Technical & Research Bulletin 5-5A : Guidelines for Site Specific Assessment of Mobile Jack-Up Units', 2002.

8. ABS Seminar, 'Fundamentals Self-Elevating MODU', 2012.

9. Det Norske Veritas, 'Classification Notes – No. 30.4 : Foundation', 1992.

10. Det Norske Veritas, 'Recommended Practice DNV-RP-C104: Self-Elevating Units', 2011.

9. AUTHORS BIOGRAPHY

T M Cho holds the current position of Senior Engineer at Samsung Heavy Industries Co., Ltd, Korea. He is responsible for structural analysis of ships and offshore structures. His experience includes reliability-based design optimization and finite element analysis.

J S Park holds the current position of Senior Engineer at Samsung Heavy Industries Co., Ltd, Korea. He is currently responsible for managing ship and offshore R&D activities. His latest research areas focus on offshore topside engineering and risk assessment against accidental cases.

Y S Ha holds the current position of Engineer at Samsung Heavy Industries Co., Ltd, Korea. He is responsible for structural analysis of ships and offshore structures. His experience includes fatigue and fracture of steel structure.

J H Jeon holds the current position of Assistant Engineer at Samsung Heavy Industries Co., Ltd, Korea. He is responsible for structural analysis of ships and offshore structures. His experience includes finite element analysis and optimization.

K B Jang holds the current position of Principal Engineer at Samsung Heavy Industries Co., Ltd, Korea. He is responsible for structural analysis of ships and offshore structures. His experience includes structural safety and optimization.

THE BETTER DESIGN OF HIPPS

Jinhyung Park, Yokogawa Electric Korea, South Korea

SUMMARY

HIPPS is the abbreviation of High Integrity Pressure Protection System. The gas out of the gas/oil well is compressed to really high pressure and the design pressure of equipment in main process for gas treatment is much lower than the gas pressure compressed by compressor. So the failure of pressure operation results in very serious consequence in main process. HIPPS is installed to protect the main process from high pressure by compressor after well head. Considering this importance of HIPPS, the technology of HIPPS needs to be diversified and separated from other safety system and be more reliable. The better design with diverse technology of HIPPS and the reason of better design of HIPPS will be described in this paper.

NOMENCLATURE

HIPPS	High Integrity Pressure Protection System
HIPS	High Integrity Protection System
HIOPS	High Integrity Overpressure Protection System
SIS	Safety Instrumented System
SCM	Subsea Control Module
ROV	Remotely Operated Vehicle
SIL	Safety Integrity Level

1. INTRODUCTION

HIPPS is the most important system to protect the pressure vessels in production area from high pressure on well head. The pressure on well head in deep-water is normally several hundred bar and even though the pressure drop on the way to production area, there is no pressure vessels to stand this high pressure. For example, the well-head pressure and temperature initially on the Kristin field which subsea HIPPS was delivered in mid-2005 were 10,700 psi (738 bar) and 324 degree F (162 degree C). [1] This article describes the criteria of HIPPS for better design of HIPPS.

2. THE DEFINITION OF HIPPS

HIPPS is a generic term applied to active Safety Instrumented Systems (SIS) which protect process plant from damage caused by Pressure Transfer from a suitably designed area to a de-rated area. [2]
ISO/TR 12489 also defines HIPPS or OPPS as "a HIPS exclusively devoted to protection against overpressure. [3] HIPPS is sometimes called HIOPS, High Integrity Overpressure Protection System.

3. THE GENERAL FEATURES OF HIPPS

The HIPPS is located at the Specification Break Point. i.e., the first location where the plant design cannot withstand the potential maximum process pressure. This typically applies to wells however it can be applied to any location where a Specification Break exists. The justification for using HIPPS in this fashion is to reduce the cost of the process plant without compromising the tolerable risk. The HIPPS may be located on land or subsea. [2] Subsea HIPPS is usually installed in Subsea Control Module (SCM). [1]

4. THE HISTORY OF SUBSEA HIPPS

Yokogawa has supplied many HIPPS with a magnetic core logic based ProSafe-SLS system (see figure 1), certified for applications up to safety integrity level (SIL) 4.

In 1996, the Kingfisher project in the UK sector of the North Sea seemed financially questionable unless the operator could apply innovative techniques to reduce the capital and operating expenditures. One technology selected was the use of HIPPS on the seabed that would allow for de-rating the flow lines. Traditionally, project designs have flow lines that can withstand the wellhead pressure and HIPPS installed on the topsides to protect the process.

In this instance, based on earlier conceptual studies, the designers realized that relocation of the HIPPS closer to the wellheads would save major capital expenditures. This was a simple concept, but the realization required technology that would combine the highest integrity with the highest reliability and availability.

The combination of these three features is often a balancing act. Increased integrity and nuisance failures reduce availability. Incorporating redundancy to improve availability and reliability requires more components that may fail. Because this was an innovative approach, the UK Health & Safety Executive set the integrity level for the first system at SIL4, a very rare demand, although it accepted later systems at SIL 3.

Suggestions at first were to build a new system with integrated electronics; however, the client quickly rejected this and referred to its experience with the magnetic core logic systems used on many North Sea platforms.

The client approached Yokogawa to develop a HIPPS prototype solution to satisfy the constraints of the subsea environment with magnetic technology, followed by a system for deploying the HIPPS.

Yokogawa standardized the system design so that future subsea HIPPS could use the same design if it were proved on Kingfisher. This basic design has been used on all subsea projects including the Kristin field in mid-2005 with some project-specific adjustments. [1]

Figure 1: ProSafe-SLS module circuit board

5. THE SYSTEM DESIGN OF HIPPS

The design constraints for all subsea HIPPS is summarised as follows:

- Diverse control functions
- Remote communications inbuilt
- Autonomous shutdown functions
- Remote testability
- Remote diagnostics
- Additional diagnostics for transmitters and actuators
- Space; must fit into existing SCM
- Weight, SCM and remotely operated vehicle (ROV) friendly
- Environmental strength
- Power, minimum increase on standard SCM
- No batteries
- Guaranteed cold start-up [1]

6. THE OPERATION BASIS OF HIPPS

Yokogawa's ProSafe-SLS HIPPS uses tiny toroidal magnetic cores (ceramic with a magnetic film). A magnetic field can magnetize each toroidal core in a clockwise or counter clockwise direction, the direction being controlled by passing current pulses through windings on the toroid.

One winding on each toroid is a clock pulse with a 1-ms period, although pulse duration is only 50 µs.

The design's combined windings on a toroid and combined toroids provide all the derived basic logic functions. A key feature is that a logic one occurs when the magnetic field is alternating, while a logic zero occurs when the field is fixed clockwise or counter clockwise. Tests have shown that it is impossible to generate an output from the system when the input is not pulsing at the required interval. No recorded incidents have been noted in which this logic system failed in an unsafe state.

For analogue inputs, the system uses a limited variability programmable trip amplifier with a library of certified applications. For HIPPS applications, this trip amplifier's configuration provides one trip output via a magnetic core, plus diagnostic outputs for over current, undercurrent, and transmitter noise levels. [1]

7. ENVIRONMENTAL ISSUES OF HIPPS

Initially, it was not certain whether a subsea HIPPS could meet the environmental requirements, specifically shock and vibration but also temperature and humidity. The original design required a 24-inch rack for the complete HIPPS logic and input-output processing, although some later systems used two racks.

The only enhancement in the subsea unit to the standard HIPPS rack is the inclusion of a locking compound on all screws. All subsea HIPPS racks are subjected to shock and vibration testing. [1]

8. POWER OF HIPPS

Each complete subsea HIPPS logic, including the powering of transmitters, solenoid valves, and the communications interface typically requires 4 amps at 24V DC. The Yokogawa ProSafe-SLS logic system has magnetic core technology with dynamic current pulses on a 1-ms clock frequency as previously stated.

The current pulses are 0.5 amps in amplitude to provide extremely high immunity, but they are only 50- µs in duration, which results in the very low overall consumption and low heat dissipation.

Some systems included additional digital input loops that are typically 12 mA so that they have minimal additional loading. [1]

9. OPERATIONAL ISSUES OF HIPPS

The Kingfisher project went on line in late 1997 and received much notice as the world's first subsea HIPPS installation. The system operated without problems until the HIPPS reported a series of transmitter faults.

Despite the subsea HIPPS attracting much attention for saving several million dollars and the fact that the project was brought on stream early and under budget, the next subsea HIPPS was installed 5 years later.

The Kingfisher project was to prove out the concept, with Yokogawa providing a solution to enhance the capability of the subsea control systems supplier. After 5 years of successful operation, more systems finally followed in the

North Sea, culminating with the unit for the Kristin field in mid-2005.

The design of the Kristin system clearly relates to the original Kingfishef system, although it has an enhanced and smaller communications element and an unchanged logic system.

To date, operators have recovered two SCM's with subsea HIPPS that have had equipment problems. In neither case were the problems with the subsea HIPPS.

Industry has shown interest in the using of subsea HIPPS in the Gulf of Mexico and in South America. The technology and the solution are now well proven (no system failures recorded on operating system to date) so that all other factors permitting, subsea HIPPS should be a first consideration rather than an alternative. [1]

10. WHY IS HIPPS USED?

The simplest way to deal with overpressure situations is allow a relief valve to vent to atmosphere or into flare. In several cases as below, such venting cannot be used:

- If flow rates through relief valve would be excessive
- If toxic gases would be release
- If there would be a significant environmental impact
- If plant has non-flaring operational objectives
- If it is desirable to minimize loss of product
- If operating company has other requirements (strategic protection of major production installations)

Figure 2. HIPPS failure consequence [2]

The alternative to a mechanical pressure relief valve is a HIPPS designed for SIL 3 or SIL 4. [2]

The gas from well head contains toxic gases and the overpressure in production area will result in relief valve open in the process before the treatment of toxic gases and the toxic gases will be directly vented into flare. That is the reason why HIPPS is indispensable between well head and gas production plant. HIPPS failure consequence is shown in figure 2.

11. THE NORMAL DESIGN OF HIPPS

The design of HIPPS varies according to SIL 3 or SIL 4, but typical design of HIPPS is as figure 3.

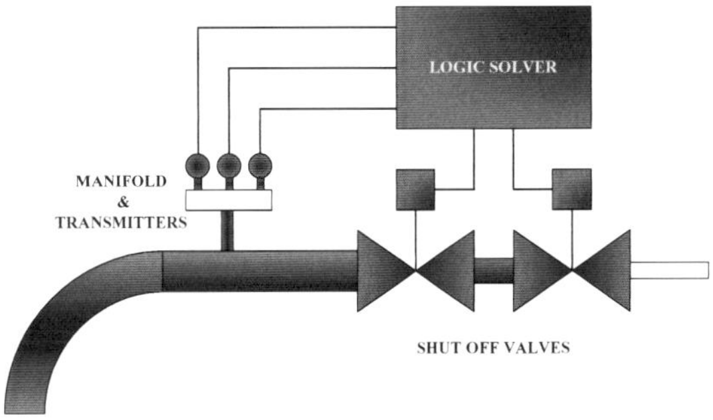

Figure 3. Typical design of HIPPS [2]

12. HIPPS IN STANDARDS

12.1 ANSI /API 521, ISO 23251

- Reference is made to code case 2211, which is included in ASME VIII div 1.
- Safety instrumented system may be accepted: Local regulation or user criteria, if not present then minimum SIL3 required para 4.2.4
- HIPS is mentioned and defined as High Integrity Protection System.
- Informative Annex E provides discussion and procedures which should be followed
- IEC 61508 / 61511 are specifically mentioned. [2][4][5]

12.2 API 14 C

- Indicates that an SDV (in addition to the SSV) with independent PSH is an acceptable alternate for the PSV
- Instrumented systems or HIPS not specifically mentioned
- But an addition in the final phase [2]

12.3 API RP 14 C

- Section 3.4, para b:
  "The safety system should provide two levels of protection to prevent or minimise the effects of an equipment failure within the process. (...) In general, the two levels should be provided by functionally different types of safety devices for wider spectrum of coverage. (...)"

- In effect, two layers of protection should be diverse. [2][6]

12.4 API RP 17O

- For subsea only:
  Integrates 14C, 61508 and 61511 for HIPPS
- Definition of HIPPS:
  Mechanical and electrical-hydraulic SIS used to protect production assets from high-pressure upsets
- Discussed Partial Stroke Testing:
  PST shall normally be treated as a functional test which covers only a fraction of the possible failures and not as self-test with diagnostic coverage.
- Indicates requirements for redundancy
- Reference is made to IEC 61508 as well
- Minerals Management Service has indicated that a HIPPS for the Gulf of Mexico should have a mandated minimum SIL of 3. [2][7]

13. CONCLUSIONS

HIPPS is the last defence of plant, so the safer design of HIPPS is really important. The summary of this paper is as below.

- The concept of Independent Layers is clear in the IEC safety standards
- HIPPS is effectively a defined Safety Function for Active Over Pressure Protection with specific architectures and performance constraints
- Future standards need to address this Safety Function.

14. ACKNOWLEDGEMENTS

I appreciate Ian Ramsay-Connell providing the papers and materials about HIPPS.

15. REFERENCES

1. Ian Ramsay-Connell, High-integrity pressure protection lowers subsea costs, OIL&GAS JOURNAL, Aug. 21, 2006

2. Ian Ramsay-Connell, 'HIPPS Challenges & Solutions', Yokogawa Middle East & Africa, 16th September 2013

3. International Association of Oil & Gas Producers (IOGP), 'High Integrity Protection Systems – Recommended Practice', REPORT 443, FEBRUARY 2015

4. ANSI/API, 'ANSI/API STANDARD 521 – Pressure-relieving and Depressuring Systems', ANSI/API, 5th edition, January 2007

5. ISO, 'ISO 23251 – Petroleum and natural gas industries-Pressure-relieving and depressuring systems', ISO, 2006

6. API, 'API RECOMMENDED PRACTICE 14C – Recommended Practice for Analysis, Design, Installation, and Testing of Basic Surface Safety Systems for Offshore Production Platforms', API, March 2007

7. API, 'API STD 17O – Recommended Practice for Subsea High Integrity Pressure Protection System (HIPPS)', API, Second Edition, 2014

16. AUTHORS BIOGRAPHY

Jin Hyung Park holds the safety expert at Yokogawa Electric Korea and holds Certified Functional Safety Expert, TUV Rheinland Functional Safety Expert. He is responsible for the functional safety consulting including HAZOP, SIL classification and SIL calculation and responsible for the trainer of TUV Rheinland Functional Safety Engineer Course and the lecturer of Korea OSHA training centre and responsible for editorial member of 'Ships and Offshore Structure'. His previous experience includes the SIS engineering and the DCS engineering including design, implementation, manufacture review, site modification and the project engineering including schedule management and document management.

A METHOD TO CALCULATE LEVEL ICE LOADS ON MULTI LEGGED CONICAL STRUCTURES

J.M. Cabrera and **J.K. Paik***, Department of Naval Architecture and Ocean Engineering; The Korea Ship and Offshore Research Institute (Lloyd's Register Foundation Research Centre of Excellence), Pusan National University, Busan 46241, Republic of Korea
*Also: Department of Mechanical Engineering, University College London, London WC1E 7JE, UK

SUMMARY

Multi-legged structures are often used in arctic environments to withstand strong ice loads. Usually, these types of offshore structures have cones around their legs for the benefit of inducing level ice bending failure and thus reducing global ice loading. The present work introduces a method to predict global ice loads on multi-conical legged structures due to interaction with level ice by combining the *Croasdale method* and the *formula of Mellor*. The first one allows estimating forces of unbroken ice on front legs whilst the second one is applied to calculate broken ice forces on rear legs. Formation of ice circumferential cracks around the legs becomes an important assumption and it plays an important role in the calculations. Results are compared with available published data for multi-conical legged structures; nevertheless, further validation is suggested to prove applicability and to tune the proposed method.

Keywords: ice loads, multi legged structures, conical structures.

NOMENCLATURE

c	Cohesion ($N{\cdot}m^{-2}$)
E	Elasticity modulus of sea ice ($N{\cdot}m^{-2}$)
F_1	Ice action on one leg (N)
F_h	Level ice horizontal force (N)
F_S	Global limit stress action (N)
f_r	Perimeter reduction factor (-)
g	Acceleration of gravity ($m{\cdot}s^{-2}$)
H_B	Ice sheet breaking force (N)
H_L	Force to lift and shear the ice rubble on the top of the sheet (N)
H_P	Force to push the advancing ice sheet through the ice rubble (N)
H_R	Force to push the ice blocks up the slope through the ice rubble (N)
H_T	Force to turn the ice blocks at top of the cone (N)
h_i	Level ice thickness (m)
h_2	Rubble ice thickness (m)
k_j	Ice jamming factor (-)
k_n	Non-simultaneous failure factor (-)
k_s	Interference and sheltering factor (-)
l_b	Breaking of floating ice sheet against a sloping structure (m)
l_c	Length of circumferential crack (m)
L_c	Characteristic length (m)
p_G	Broken ice pressure ($N{\cdot}m^{-2}$)
R_f	Flexural strength of ice ($N{\cdot}m^{-2}$)
w	Width of sloping structure (m)
w_t	Width of broken ice track (m)
ρ_i	Density of sea ice ($kg{\cdot}m^{-3}$)
ρ_w	Density of seawater ($kg{\cdot}m^{-3}$)
ϕ	Friction angle of ice rubble (rad)
ν	Poisson ratio of ice (-)
η	Porosity of ice rubble (-)

1. INTRODUCTION

Multi-legged fixed offshore structures have been applied for the production of hydrocarbons in sub-arctic regions. They differ slightly from typical jacket structures since they have no brace connections at the sea level in order to avoid damage by sea ice features. Moreover, cones around the legs are often used in order to change ice failure mode from crushing to bending, which results in lower ice forces.

By not foretelling the right magnitude of global forces on offshore platforms during design stage, the structures can become endangered during extreme ice events. Therefore, design of multi-legged structures should include consideration of global actions that allow to assess global structural integrity and evaluate foundation sufficiency.

Calculation of global ice loads on multi-legged structures has been widely studied by different scholars. Saeki et al. [1] proposed a load calculation method for multi vertical legged structures which was implemented by Takeuchi et al. [2] for the cases of 4 legs structures and 9 legs structures. Toyama and Yashima [3] conducted model tests on a 9 legs structure and the results showed an agreement with their own predictions. Shkhinek et al. [4] used the Finite difference method to calculate global loads on 4 legs, considering 3 different scenarios and showed that the ISO/DIS 19906 recommendations are over conservative. Works on Multi Conical Legged Structures (MCLS) are addressed later.

The aim of this work is to present a simple yet useful method to calculate global level ice loads on Multi Conical Legged Structures (MCLS).

2. ICE BREAKING PROCESS

Experiments in ice model tanks, such as those in Kato et al. [5] and Huang et al. [6] remain as useful references to get a good insight into the MCLS-ice interaction problem. The first one employed a square arranged Upward Bending Multi Conical Legged Structure (UB-MCLS) and a square arranged Downward Bending Multi Conical Legged Structure (DB-MCLS) under continuous speed level ice loading. The second one conducted laboratory level ice experiments with a 4 legs vertical columns MCLS and a 4 legs UB-MCLS, with different structural natural frequencies and under different ice speeds.

Based on observations of the aforementioned authors, the level ice breaking process can be described as follows:

- The front conical leg (or the front legs, whichever is the case) interacts with the level ice sheet almost in an independent manner. First, radial cracks will develop and later a circumferential crack, which is associated with the maximum load on this leg, will occur just before bending failure of the ice sheet.
- Rear conical legs interact with damaged ice. The ice, which has already been ruined by the frontal legs, will be subsequently easily broken by the rear ones and thus, global load contribution from these must be less.
- Failed ice blocks accumulate between the legs. Kato et al. [5] observed deep jamming between the cones of their DB-MCLS experiment, so this extra force should be considered for predicting ice loads.

3. THE MCLS SIMPLE METHOD

3.1 ICE LOADS ON FRONT LEGS

Croasdale and Cammaert [7] developed a model to calculate ice actions on sloping structures (the so-called *Croasdale method* or *model based on elastic beam theory*). It involves the application of theory for the bending of a beam on an elastic foundation. Furthermore, it takes the formation of a circumferential crack into account, which plays an important role in the present proposed method.

Then, the estimation of the horizontal level ice force on conical front legs was estimated by means of the *Croasdale method* which requires substitution of the correct parameters in the following formula [7]:

$$F_h = \frac{H_B + H_P + H_R + H_L + H_T}{1 - \dfrac{H_B}{R_f l_c h_i}} \quad (1)$$

The characteristic length of a floating plate is defined as [7]:

$$L_c = \left(\frac{E h_i^3}{12 \rho_w g (1 - \nu^2)} \right)^{0.25} \quad (2)$$

The braking length of a floating ice sheet against an inclined structure can be expressed as:

$$l_b = \frac{\pi}{4} L_c \quad (3)$$

The perimeter of circumferential crack can be estimated by means of the following formula:

$$l_c = w + \pi \cdot l_b \quad (4)$$

Some of the previously mentioned geometrical parameters are illustrated in Fig. 1(a). For further details on the *Croasdale method*, the reader may refer to [7] and [8].

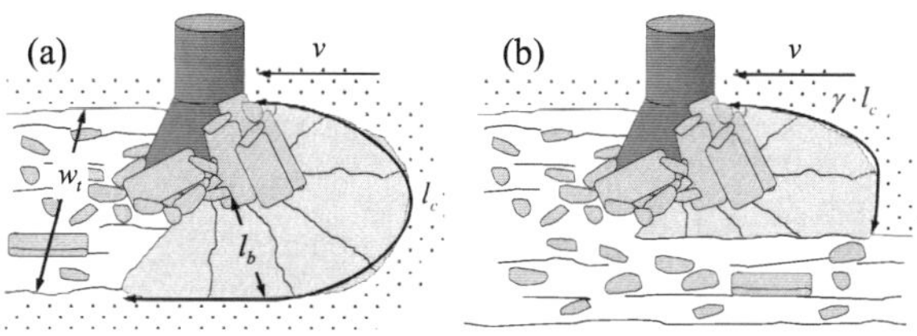

Figure 1: Ice sheet acting on a conical structure with fully developed circumferential crack (a) and not fully developed circumferential crack (b).

3.2 ICE LOADS ON REAR LEGS

Ice forces will generally be less on rear legs than in front legs. MCLSs with close distance between front legs can better protect rear legs; moreover, sheltering effects can be further influenced by mechanical properties of the ice sheet.

As the front leg breaks a floating ice sheet, a circumferential crack will form in front of it and at both sides of the conical structure. Then, a track of broken ice will be formed behind a front leg, which width depends on the leg width and the distance from the leg to its circumferential crack (see Fig. 1(a)):

$$w_t = w + 2l_b \tag{5}$$

Rear legs have the chance of being outside the broken ice tracks, in which case, the ice load would be that of a level ice sheet against a conical structure to be calculated by means of Eq. (1). Under the circumstance that the rear leg is too close to the broken ice track, its circumferential crack cannot be completely developed; in this situation, a perimeter reduction factor should be applied to Eq. (4), i.e., the intersection of the circumferential crack with the front legs developed tracks is subtracted from the total perimeter of circumferential crack for the calculation of the Croasdale's load; Fig. 1(b) shows the case of a not fully developed circumferential crack, where the perimeter reduction factor is denoted by γ.

Moreover, when a rear leg is inside a broken ice track, the ice blocks are considered to be a granular material and their action on a structure corresponds to the passive state which is the upper limit. Thus, the *formula of Mellor* [9] can be applied, which assumes that there is a pressure ridge formation in front of inclined walls that ensures the development of lateral earth pressure of broken ice as if they were vertical:

$$p_G = \frac{1}{2}\left(\frac{1+\sin\varphi}{1-\sin\varphi}\right)(1-\eta)\rho_i g(1-\rho_i/\rho_w)h_2 + 2c\left(\frac{1+\sin\varphi}{1-\sin\varphi}\right)^{1/2} \tag{6}$$

3.3 CALCULATION PROCEDURE

The simple method to calculate the global ice load on MCLS is now introduced. Consider a semi-infinite plate approaching to a MCLS at a certain intrusion angle. As the ice sheet touches the front cone, it will break under bending failure mode and the load can be calculated by means of the *Croasdale method* (Eq. (1)). A track of broken ice will form behind the front leg, which width is estimated by means of Eq. (5).

In the second stage, a rear leg will withstand a load that must be less than that for front legs, because the ice sheet has lost confinement. Then, the *Croasdale method* (Eq. (1)) can be applied with a perimeter reduction factor.

The farthest rear legs may not have to withstand level ice loads since the ice sheet is already broken; however, broken ice loads can be estimated trough the *formula of Mellor* (Eq. (6)). Combination of level ice loads and broken ice loads is also possible and this must be included in the computations by taking into account the position of broken ice tracks, perimeter reduction factors and granular material-leg contact surfaces.

The current procedure assumes that the ice sheet consists of a homogeneous material which travels at constant direction and with infinite field energy. Ice is continuously moving and hence add freeze may not occur. Moreover, influences of non-simultaneous failure of ice on the legs or strain rate dependence are out of the scope of this analytical model.

This method allows estimating overturning and twisting moments if the proper level arm is considered. The process diagram for the proposed calculation method is shown in Fig. 2, which can be easily solved by its implementation in a programming code.

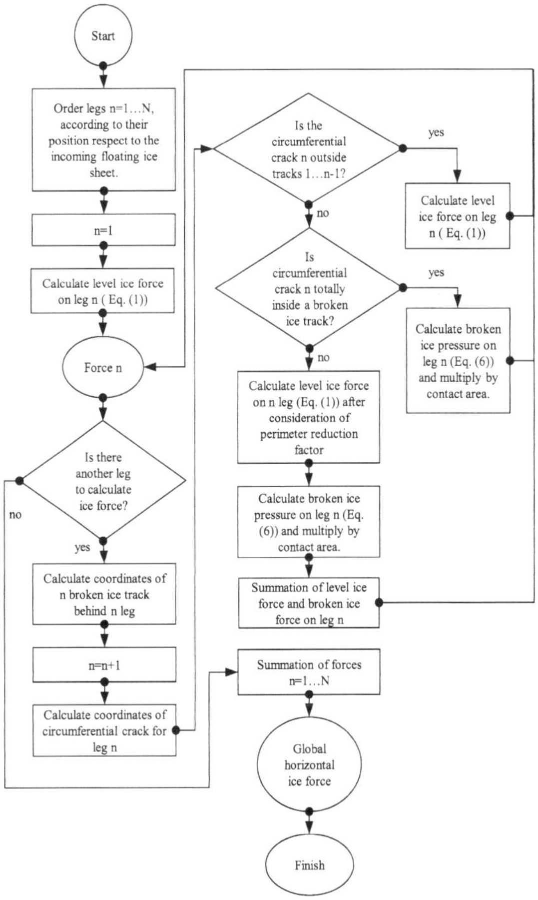

Figure 2: Procedure of the MCLS simple method.

4. PRELEMINARY RESULTS

Let the previously presented model be applied to the calculation of loads on some offshore structures. Four scenarios are studied, in which the number of legs and characteristics of sea ice are slightly varied, as indicated at the bottom of Figs. 3 and 4.

Scenarios 1 and 2 consist of a 4 legs platform with 13.716 m separation between axes, whilst scenarios 3 and 4 consider an 8 legs platform with same separation between axes. Scenarios 1 and 3 contemplate loading from thick strong ice, whilst characteristics for scenarios 2 and 4 are those for thinner softer ice. Upward breaking cones were utilized with 4.16 m diameter at the waterline and 60 degrees slope wall.

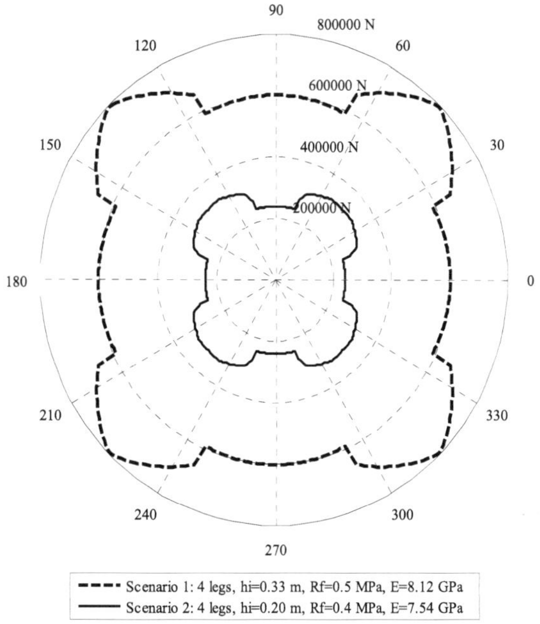

Figure 3: Level ice loads on a 4 conical legs structure.

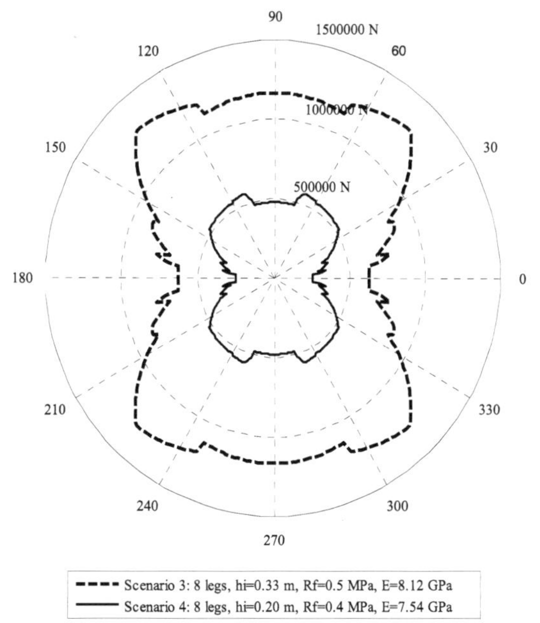

Figure 4: Level ice loads on an 8 conical legs structure.

The results were plotted on Figs. 3 and 4; it must be noted that the direction indicates the course to which drift ice "goes". They show strong dependency on the ice thickness which also enhances sheltering effects; let us observe the global action calculation method for multi legged structures as suggested in [8]:

$$F_S = k_s k_n k_j F_1 \quad (7)$$

The global limit stress action is expressed as the ice action on one leg which is multiplied by several factors. Assuming that the non-simultaneous failure factor is 0.9 and disregarding jamming effects, the calculated sheltering factors of the 4 legs platform are 2.85 for scenario 1 and 3.12 for scenario 2, which reveal that front legs perform better at reducing loads on rear legs when ice is thick and strong.

The reason behind this is that the size of circumferential cracks increases with ice thickness, elasticity modulus and flexural strength, and thus, larger tracks of damaged ice are formed behind front legs which reduce very much the load on rear legs. Similar results were obtained for the 8 legs structure, where the calculated sheltering factors were 4.63 and 5.38 for scenarios 3 and 4, respectively.

It was found that the load due to broken ice, has very low contribution to the global load. The broken ice load on one single leg is 2.5% of the unbroken ice load on one single leg for scenarios 1 and 3. It increases its importance for thin ice situation such as scenarios 2 and 4 where its relation respect to unbroken ice load is 3.43%. Therefore, the broken ice load could be neglected for practical purposes.

5. COMPARISON WITH AVAILABLE DATA

5.1 FIXED DB-MCLS AND UB-MCLS

It was decided to check whether the magnitude of our calculations match with those of publicly available data. Some results in the IHI Ice Model Tank were published in [5] for DB-MCLS and UB-MCLS. These numbers were tried to be reproduced by means of the MCLS simple method.

The physical model of Kato et al. [5] consisted of a 4 legs square arranged structure with 800 mm centre to centre separation; cones with 60 degrees slope angle and 300 mm diameter at the waterline were applied for both kind of tests. For the calculations, where no ice properties are reported, these were estimated as function of ice flexural strength and brine volume by means of the formula of Weeks and Assur in 1967 [10] for elasticity modulus and the first-year ice flexural strength equation of Timco and O'Brien in 1994 [11].

Calculation results for DB-MCLS were very close to the referred experimental results. Loads on one single downward bending cone were 35.5 N for [5] and 34.95 N for us, which show good agreement. Moreover, the comparison in Fig. 5 indicate that the MCLS method and experimental results have same tendency. If the proper non-simultaneous factor were applied to the calculated global force, practically same results would be obtained. Based on the investigations above, it is considered that the MCLS simple method can accurately predict ice loads on DB-MCLS.

Calculations were completed for the UB-MCLS as well. The MCLS simple method turned in conservative results. Load on one single upward bending cone was reported in [5] to be 74.4 N, whilst we estimated 102.74 N by employing the *Croasdale method* with 38% differences; then, the ice parameters used in the calculations for this kind of cones should be further studied. Moreover, the low value of experimental results, as shown in Fig. 6, suggest that non-simultaneous ice failure plays an important role in overall phenomenon.

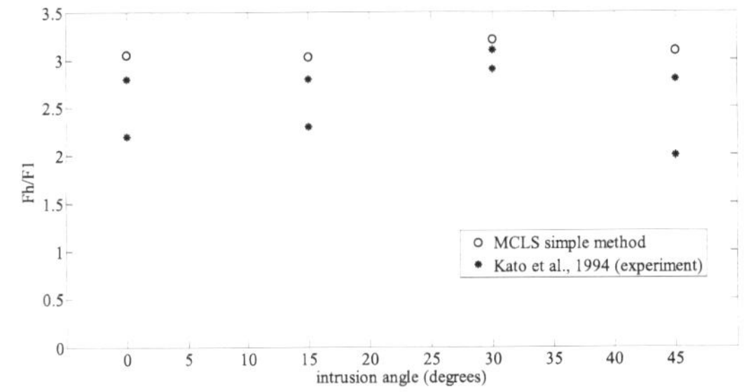

Figure 5: Comparison of ice forces on DB-MCLS.

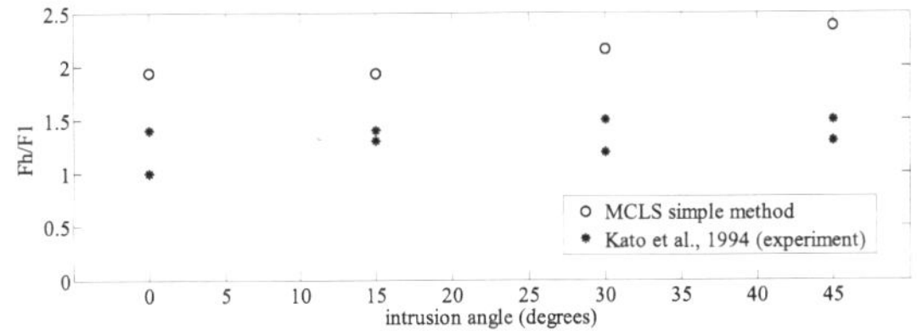

Figure 6: Comparison of ice forces on UB-MCLS.

5.2 COMPLIANT UB-MCLS

The experiments in [6] were studied for purposes of comparison. The tests were conducted in the ice tank at the Ice Engineering Laboratory of Tianjin University for 4 legs UB-MCLSs, where natural frequency of the structure and ice speed were varied trough the series. The characteristics of the tests are presented in Table 1.

The MCLS simple method was used to calculate the global ice forces for each of the scenarios in Table 1. Again, where input data was missing, the relations to calculate ice properties in [10] and [11] found useful application in this work.

Two comparison graphs in Figs. 7 and 8 show the results for 0 degree intrusion angle and 45 degree intrusion angle, respectively. It should be noted that only maximum values reported in [6] are plotted in the graph, whilst the complete results comprehend sets of time series with reported mean, maximum and minimum values.

For 0 degree intrusion angle, the global ice force as calculated was larger than experiment results about 94% to 350% increase, whilst for 45 degree intrusion angle the calculated results were larger between 0 and 122%.

It was observed that the calculations matched better the test results at 45 degree intrusion angle, high ice speed between 250 and 400 mm/s. Although no specific influence was observed of the natural frequency of the structure, dynamic response could be the reason for not matching results at 0 degree angle. Compliant structures are expected to have lower non-simultaneous ice failure on different legs which greatly reduce the level ice horizontal force.

Table 1: Scenarios for compliant UB-MCLSs comparison [6]

Scenario	Natural frequency x-direction (Hz)	Natural frequency y-direction (Hz)	Ice intrusion angle (°)	Ice speed (mm/s)	Average ice thickness (mm)	Average flexural strength (kPa)
1	5.67	5.69	0	10	27.9	39
2	5.67	5.69	0	20	27.9	39
3	5.67	5.69	0	50	27.8	39
4	5.67	5.69	0	100	27.5	40
5	5.67	5.69	0	150	27.6	40
6	5.67	5.69	0	200	27.9	40
7	5.67	5.69	0	250	28.3	40
8	5.67	5.69	0	300	33.6	45
9	5.67	5.69	0	350	33.9	45
10	5.67	5.69	0	400	32.3	45
11	9.78	9.86	0	10	34.1	45
12	9.78	9.86	0	20	34.3	41
13	9.78	9.86	0	50	34.2	41
14	9.78	9.86	0	100	33.4	41
15	9.78	9.86	0	150	33.8	41
16	9.78	9.86	0	200	32.5	41
17	9.78	9.86	0	250	32.1	44
18	9.78	9.86	0	300	32.4	44
19	9.78	9.86	0	350	32.3	44
20	9.78	9.86	0	400	33.2	44
21	5.67	5.69	45	10	28.1	38
22	5.67	5.69	45	20	28.3	38
23	5.67	5.69	45	50	28.6	38
24	5.67	5.69	45	100	29.1	38
25	5.67	5.69	45	150	29.9	39
26	5.67	5.69	45	200	30.8	39
27	5.67	5.69	45	250	31.1	39
28	5.67	5.69	45	300	31.4	39
29	5.67	5.69	45	350	31.7	37
30	5.67	5.69	45	400	32	37
31	9.78	9.86	45	10	26.2	42
32	9.78	9.86	45	20	26.7	42
33	9.78	9.86	45	50	26.8	42
34	9.78	9.86	45	100	27.3	42
35	9.78	9.86	45	150	27.6	42
36	9.78	9.86	45	200	28.2	42
37	9.78	9.86	45	250	28.7	45
38	9.78	9.86	45	300	27.1	45
39	9.78	9.86	45	350	28.1	45
40	9.78	9.86	45	400	29.2	45

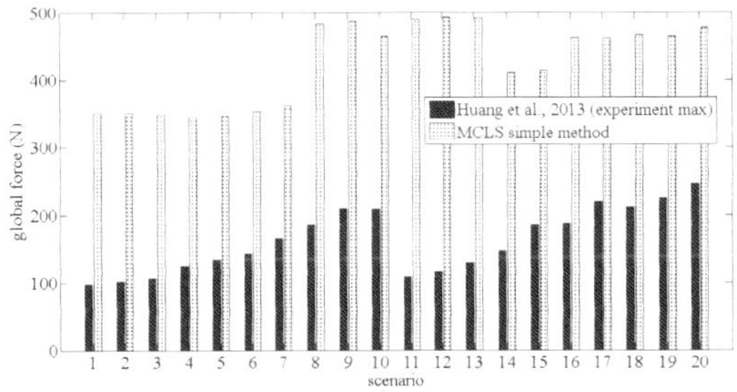

Figure 7: Comparison of ice forces at 0 degree intrusion angle.

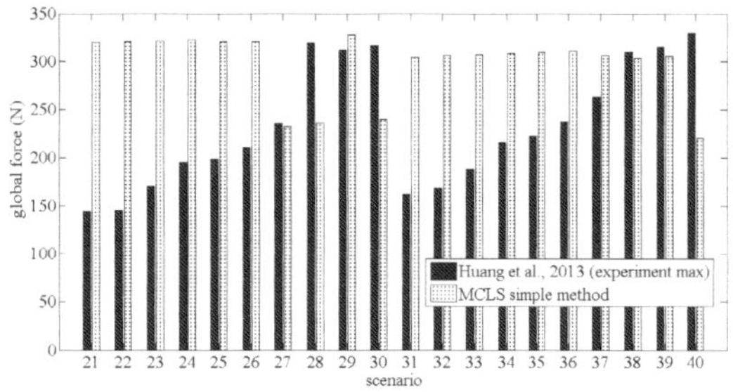

Figure 8: Comparison of ice forces at 45 degree intrusion angle.

6. CONCLUSIONS

The present paper has introduced a method to calculate the global ice force on Multi Conical Legged Structures (MCLS) which consists in calculating level ice forces on front legs and a combination of level ice and granular material loads on rear legs. This method can be a useful tool for calculating ice forces at early design stage of arctic offshore structures.

Preliminary results showed that sheltering effects are enhanced with ice thickness since larger broken ice tracks will be developed by front legs which will reduce ice loads on rear legs.

Comparison with publicly available data suggest that the method gives good agreement with test results for rigid structures and for downward bending cones; although predicted ice forces are conservative for upward bending cones. Moreover, ice forces calculated with the method become over conservative when the structure becomes compliant.

The authors suggest that future work in the calculation of ice loads on MCLS should include but be not limited to: sensibility to different level ice properties and other parameters such as water density, characterization of broken ice properties, clearance of ice blocks around downward and upward bending cones, jamming of ice on MCLS, non-simultaneous failure of ice on MCLS and application of non-linear computational techniques.

7. ACKNOWLEDGEMENTS

The study was undertaken at the Lloyd's Register Foundation (LRF) Research Centre of Excellence at Pusan National University, Busan, Korea.

8. REFERENCES

1. SAEKI, H., ONO, T., TAKEUCHI, T., SUENAGA, E.L., SAKAI, M., 'Total ice forces on the clusters of cylindrical piles', *Proceedings of the 5th OMAE, Tokyo, Japan, Vol. IV, pp. 461-466*, 1986.

2. TAKEUCHI, T., SAEKI, H., YAMASHITA, T., 'Total ice forces on multi-legged offshore structures', *Proceedings of the Third International Offshore and Polar Engineering Conference, Singapore, Vol. II, pp. 532-537*, 1993.

3. TOYAMA, Y., YASHIMA, N., 'Model tests of multi-legged structures in continuous and broken ice fields', *Proceedings of the 4th International Offshore and Polar Engineering Conference, Osaka, Japan, Vol. II, pp. 565-571*, 1994.

4. SHKHINEK, K.N. JILENKOV, A.G., BLANCHET, D., THOMAS, G.A.N., 'Ice loads on a four leg structure', *Proceedings of the 20th International Conference on Port and Ocean Engineering under Arctic Conditions, Luleå, Sweden, no. 43*, 2009.

5. KATO, K., ADACHI, M., KISHIMOTO, H., HAYASHIGUCHI, S., 'Model experiments for ice forces on multi conical legged structures', *Proceedings of the Fourth International Offshore and Polar Engineering Conference, Osaka, Japan, Vol. II, pp. 526-534*, 1994.

6. HUANG, Y., MA, J., TIAN, Y., 'Model tests for four-legged jacket platforms in ice: part 1. Model tests and results', *Cold Regions Science and Technology 95, 74-85*, 2013.

7. CROASDALE, K.R., CAMMAERT, A.B., 'An improved method for the calculation of ice loads on sloping structures in first-year ice', *Hydrotechnical Construction, Vol. 28, No. 3, 174-179*, 1994.

8. ISO 19906:2010, 'Petroleum and natural gas industries—Arctic offshore structures', *International Organization for Standarization*, 2010.

9. MELLOR, M., 'Ship resistance in thick brash ice', *Cold Regions Science and Technology, 3, 305-321*, 1980.

10. MOSLET, P.O., 'Field testing of uniaxial compression strength of columnar sea ice', *Cold Regions Science and Technology 48, 1-14*, 2007.

11. TIMCO, G.W., WEEKS, W.F., 'A review of the engineering properties of ice', *Cold Regions Science and Technology 60, 107-129*, 2010.

9. AUTHORS BIOGRAPHY

José Manuel Cabrera Miranda is a PhD candidate at Department of Naval Architecture and Ocean

Engineering of Pusan National University, Busan, Korea, under supervision of Prof. J.K. Paik. His research interests include offshore structures, fatigue, arctic and subsea engineering. His previous experience includes field development of offshore heavy oil fields and design of offshore fixed platforms.

Professor Jeom Kee Paik is a Professor of the Department of Naval Architecture and Ocean Engineering and serves as president of the Korea Ship and Offshore Research Institute (KOSORI) and as director of the Lloyd's Register Foundation Research Centre of Excellence at Pusan National University (PNU) in Korea. In 2015, he also joined University College London's Department of Mechanical Engineering as a professor and faculty member. He is a Fellow and Vice President of SNAME, USA and a Fellow and Council Member of RINA, UK. He was awarded William Froude Medal from RINA in 2015, the David W. Taylor Medal from SNAME in 2013, and a Doctor Honoris Causa from the University of Liege in Belgium in 2012. His research interests are safety engineering and design of ships and offshore structures in extreme and accidental conditions associated with risk assessment and management.

MECHANICAL PROPERTIES OF MILD AND HIGH-TENSILE STEEL UNDER IMPACT LOADS IN AN ARCTIC ENVIRONMENT

Jong Hwan Lee, Ki Jong Kim, Bo Gyeong Jung, Xu Han, Dae Kyeom Park, Bong Ju Kim, Jung Kwan Seo and Yeon Chul Ha, Korea Ship and Offshore Research Institute, Republic of South Korea
Toshiyuki Matsumoto, ClassNK, Japan
Su Hwan Byeon, STX Offshore and Shipbuilding Co., Ltd, Korea
JeomKeom Kee Paik, Korea Ship and Offshore Research Institute and Department of Mechanical Engineering, Korea and UK

SUMMARY

In the present study, tensile testing on material is undertaken with varying the temperature and strain rate. The coefficients of the Cowper-Symonds model as one of strain rate constitutive equations are suggested through the test database.

NOMENCLATURE

ε_{fd}	Dynamic fracture strain
ε_f	Fracture strain
ε_t	True strain
σ_{Yd}	Dynamic yield stress (MPa)
σ_Y	Yield stress (MPa)
$\dot{\varepsilon}$	Strain rate (/sec)
DHF	Dynamic hardening factor

1. INTRODUCTION

Ships and offshore structures in Arctic or severely low temperatures are exposed to various accident loads and ship collision, glacier collision, ship stranding and others and such loads can be led to calamities, loses and casualties. In order to understand the mechanism and movements of the structures occurring from sudden accidents and unexpected incidents, it is important to verify nonlinearity of materials used in them. Materials have vastly different material properties upon various temperatures and strain rate and they act as crucial parts in verifying actual structure response. This paper performed its test under various temperatures and strain rates for mild steel and high tensile steel and based on the results from the tests, it applied a Cowper-Symonds model as representative strain rate constitutive equation in naval and offshore field and suggested material coefficients for precise structure interpretations and responses on simulations.

In H.S.E., they performed strain rate dependency test up to 10 /sec under room temperature for marine structural steels. [8] They also performed quasi-static tensile test under room temperature and high temperature (RT to 500 °C). Brown and Vincent performed tensile tests on medium carbon steels for medium strain rate range of from 200 /sec to 900 /sec, using impact equipment under room temperature. [4] Manjoine performed tensile tests on low carbon steels (SAE 4140 and SAE 4340) for strain rate range of from 10 /sec to 30 /sec under room temperature.[9] It was verified that the steel plates used in shipbuilding industry have their dynamic yield stress increased than their static yield stress by from 100% to 150%. Campbell & Cooper performed their researches on low carbon steels in quasi-static areas up to 106 /sec strain rate.[5] As strain rate increased, yield stress also increased but amount increased proportion in tensile strength was not as big as in yield strength. Therefore, it was considered that strain hardening of materials were progressed as strain rates increased. Fujii et al. performed researches on changes in strength and fracture strain through tests on high-tensile steels for strain rate of from 0.000045 /sec to 480 /sec under the temperatures from -100 °C to 0 °C. Toyosada et al. performed tensile tests on EH36 rod shaped test specimens for intermediate strain rate range of from 0.0003 /sec to 30 /sec under the temperature from -145 °C to 0 °C.[14] Toyosada & Goto performed tensile tests on rod shaped test specimens made of high-tensile steel for from 0.0007 /sec to 50 /sec with various crosshead speeds under room temperature and measures temperature increase distributions along crack boundaries.[15] Itabashi & Kawata performed tests on various carbon steels for 0.001 /sec, 0.01 /sec, 0.1 /sec and 1000 /sec but limited to under room temperature.[10] Sia Nemat-Nasser &Guo performed various strain rate and temperature dependency tests for DH36, high-tensile steel renowned in shipbuilding industry.[12] They carried out researches with wide ranges of at 0.001 /sec, 0.1 /sec and 1000 /sec with quasi-static, low strain rate and high strain rate under temperatures from -196 °C to 727 °C. Using all those data from the tests mentioned above, we suggested a model applied with John-Cook model after applying dynamic strain aging then compared the results with the results from the tests mentioned above. Choung et al. performed dynamic tensile tests under temperatures at -40 °C, RT and 200 °C on shipbuilding steels of 2W50, EH36 for from 0.001 /sec to 200 /sec.[6] We indicated the DHFs as functions in each strain rate sections and proved a new formula deciding material coefficient D from Cowper-Symonds composition equation. The purpose of this research is to reflect a precise dynamic coefficient to structure designing by suggesting Cowper-Symonds material coefficient of each material under low temperatures, extremely low temperatures and room temperatures with various strain rate.

2. TEST SET-UP AND PROCEDURES

Traditionally, the types of dynamic effects in shipbuilding industry are classified into quasi-static and dynamic load as 0.1 /sec as criteria and especially under strain rate of a few hundred per second, a local plastic deformation causes temperature rise without having the time for the heat from it conducted to nearby area due to its fastness. In this paper, we performed test for strain rate of 0.001 /sec, 0.09 /sec, 1.8 /sec and 34 /sec and used UTM quasi-static tensile testing machine and MTS dynamic tensile testing machine. Figure 1 at 0.001 /sec of low strain rate shows environment chamber that contains UTM tensile testing machine and the readings from TDS data logger which receives temperature data at the condition. We used UTM tensile testers and 100 ton load cells for quasi-static tensile tests. In case of dynamic tensile tests with medium strain rate speed of 0.09 /sec and 1.8 /sec, we used MTS dynamic tensile testing machine attached with 50 ton load cells as shown in Figure 2 and we used MTS dynamic tensile testing machine attached with 10 ton load cells for intermediate strain rate of 34/sec as shown in Figure 3 The maximum speed of the UTM tensile testing machine used in this paper is less than 10mm/sec and a displacement control was applied. Displacement measurement was done by using an extensometer for low temperature in Figure 4, measuring 10 data per second along with temperature measurement using thermocouples. However, for the case of dynamic tests, the overall parts of the specimens cause inertial forces due to momentary deformation and generate vibrations and stress wave due to unbalanced forces. This occurs in the speed of a few times per second, which is a range of intermediate strain rate, and this is conducted to load cells. This is called 'Load Ringing Phenomenon'.

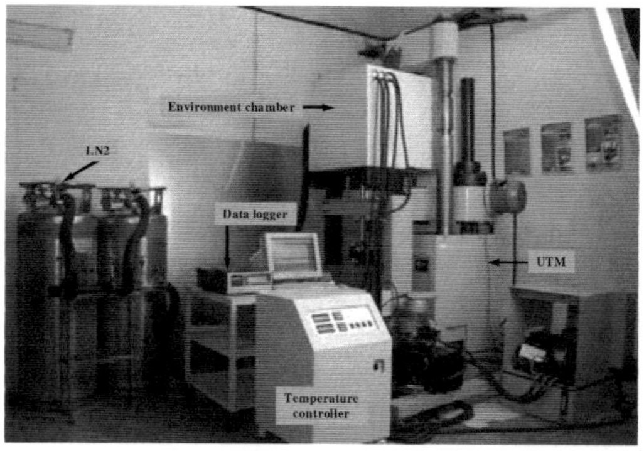

Figure 1: [Material testing machine for quasi-static with an environmental chamber]

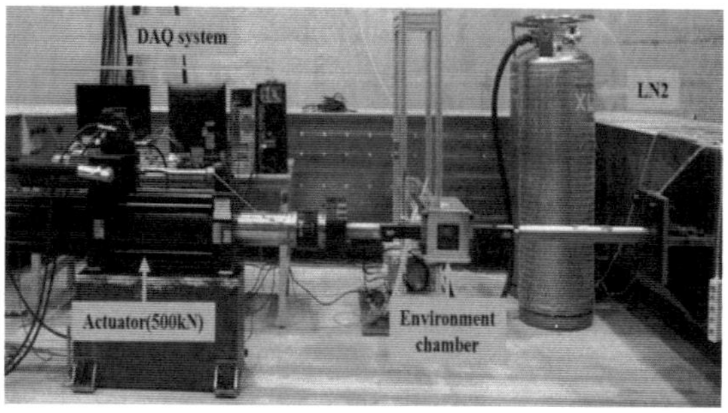

Figure 2: [Material testing machine for dynamic with an environmental chamber]

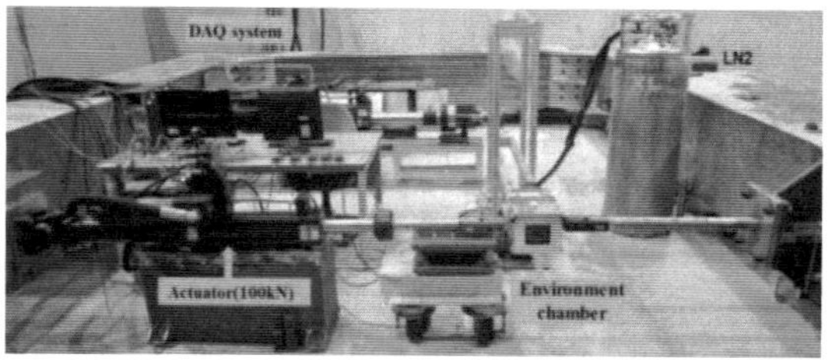

Figure 3: [Material testing machine for impact with an environmental chamber]

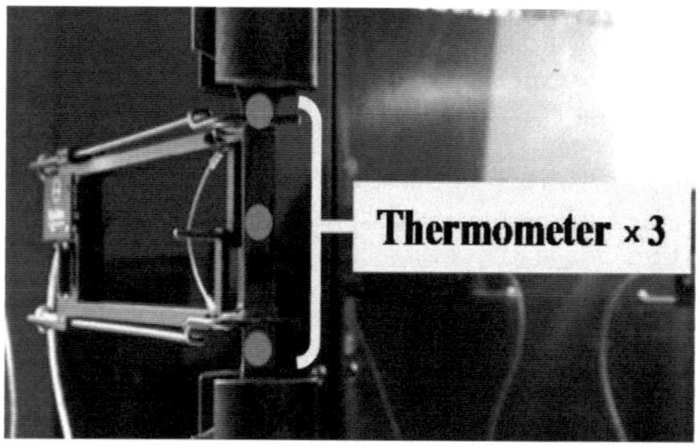

Figure 4: [Extensometer installation with temperature measurement]

In order to avoid that, we attached strain gauge onto grip part and horizontal part of the specimens, marked grids on the specimens using high speed camera as shown in Figure 5 and Figure 6 then calculated loads using each strain, cross-sectional area and young's modulus. On the other hand, we attached strain gauges used for RT and low temperatures in the center of gauge length then measured data from 1,000Hz/s, 20,000Hz/s and 50,000Hz/s to measure strain rate.

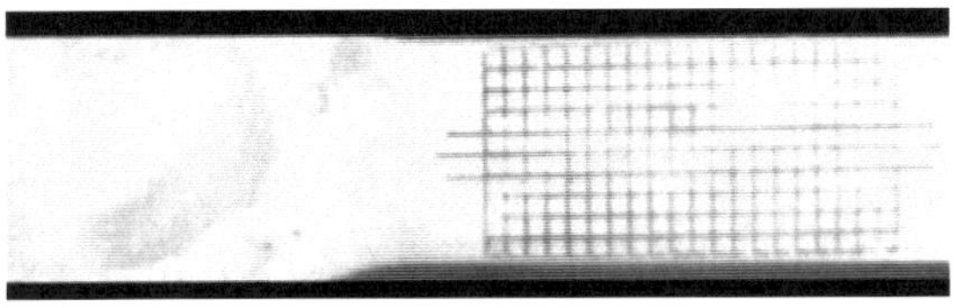

Figure 5: [Gage length with grid shape]

Figure 6: [High-speed camera with high resolution]

For quasi-static and dynamic tests, we stored signals and data in real time via DEWE measurement equipment.
In order to perform tests under low temperatures considering the temperatures in Arctic, we attached environment chambers onto each of tensile testers. The environment chambers for low temperatures were composed with LN2 (liquid nitrogen) as coolant. In order to measure air and test specimen temperatures, we attached T-type thermocouples inside of chambers and attached strain gauges onto the surface of test specimens to control the temperatures of the inside of the chambers and the specimens. We applied test methods based on ISO15579:2000 for low temperature tests.

The test specimens used for quasi-static tensile tests conformed to the regulations of ASTM E 8E/ 8M.[2] The following Figure. 7 and Figure 8 show the test specimens having gauge length of 50mm with conditions of strain rate of 0.09 /s and 1.8 /s respectably in their longitudinal directions and Figure 9 shows the test specimens that made 25mm of gauge length under impact loads. In case of quasi-static low temperature tensile tests, the test regulations and methods that are met to international standards are present. However, for medium strain rate speed tests, as there are no specific standards or suggestions, we manufactured the test specimens that are met with the maximum allowed loads for the load cells and the shapes of tester JIG. The 2 holes located at each side of grip part were made as pin-support type for the JIGs. The chemical proportions of each material used for the tests are as shown in Table 1 through Table 5 and the test ranges were shown in Table 6.

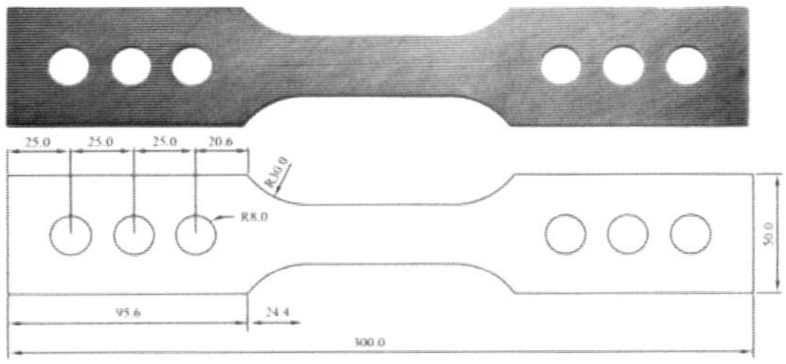

Figure 7: [Tension specimen for quasi-static]

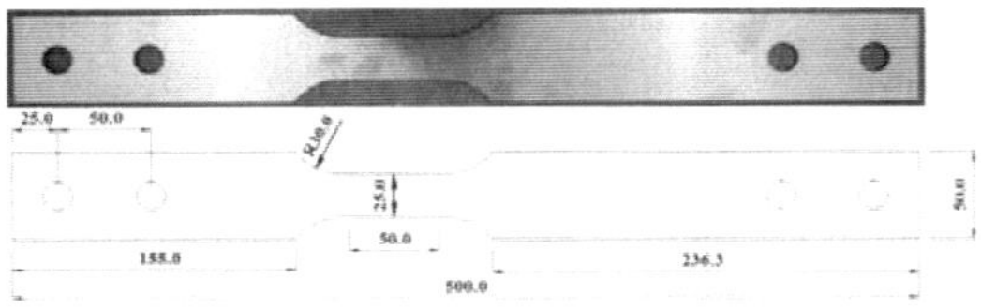

Figure 8: [Tension specimen for dynamic]

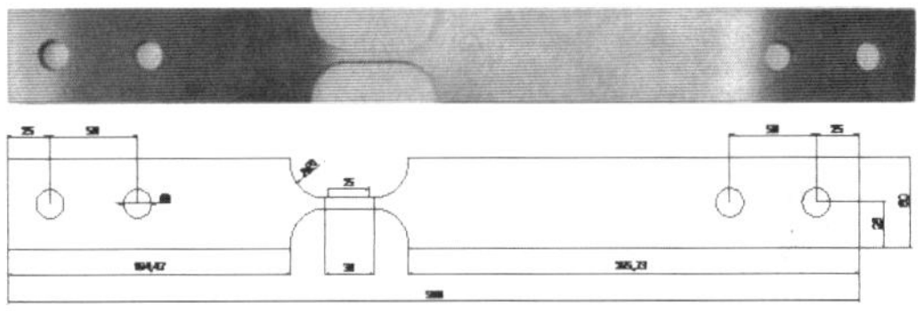

Figure 9: [Tension specimen for impact]

Table 1 Chemical composition of Grade A (%)

C	Si	Mn	P	S	Cr	Ni	Mo	Cu	Al
0.210	0.500	0.530	0.035	0.035	0.200	0.400	0.080	0.350	0.015

Table 2 Chemical composition of Grade D (%)

C	Si	Mn	P	S	Cr	Ni	Mo	Cu	Al
0.18 0	0.50 0	0.90 0	0.03 5	0.03 5	0.20 0	0.40 0	0.08 0	0.350 0	0.01 5

Table 3 Chemical composition of Grade AH32 (%)

C	Si	Mn	P	S	Cr	Ni	Mo	Cu	Al
0.180	0.10- 0.50	0.90- 1.60	0.035	0.035	0.200	0.400	0.080	0.350	0.015

Table 4 Chemical composition of Grade DH32 (%)

C	Si	Mn	P	S	Cr	Ni	Mo	Cu	Al
0.180	0.10- 0.50	0.90- 1.60	0.035	0.035	0.200	0.400	0.080	0.350	0.015

Table 5 Chemical composition of Grade DH36 (%)

C	Si	Mn	P	S	Cr	Ni	Mo	Cu	Al
0.131	0.425	1.175	0.015	0.006	0.04	0.02	0.017	0.013	0.023

Table 6 Experimental scope

Loading condition	Quasi-static	Dynamic		Impact
Velocity (mm/s)	0.05	4.5	90	850
Strain rate (/sec)	0.001	0.09	1.8	34
Temperature (℃)	-60, -20, RT			
Thickness(mm)	6			
Mild steel	Grade A, D			
High-tensile steel	Grade AH32, DH32, DH36			

3. RESULTS OF TENSILE TESTS FOR MILD STEELS AND HIGH-TENSILE STEELS

3.1 EFFECTS OF TEMPERATURES

Among those steels used for ships and offshore structures, we performed quasi-static and dynamic tensile tests for Grade A and D which are mild steels, Grade AH32, and DH36 which are high-tensile steels then indicated nominal stress - strain curve in Figure 10 ~ Figure 13. Table 7 contains mechanical properties from quasi-static tests.[1]

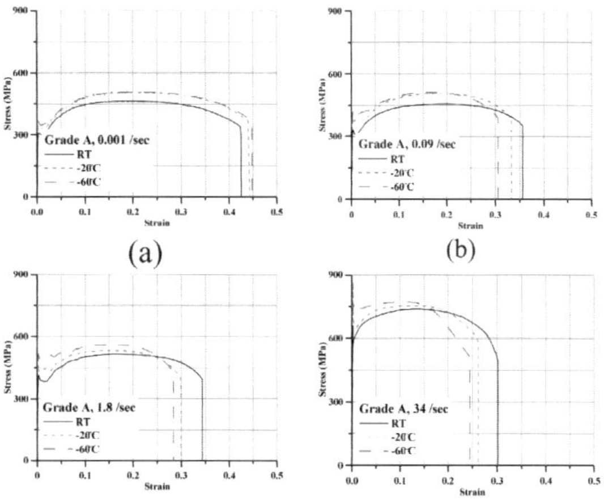

Figure 10: [Engineering stress-strain curve of Grade A as the mild steel with the variation of the strain rate : (a) 0.001 /sec, (b) 0.09 /sec, (c) 1.8 /sec, (d) 34 /sec]

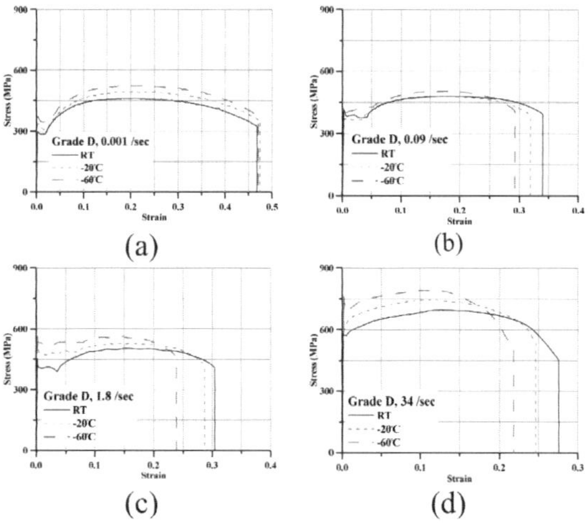

Figure 11: [Engineering stress-strain curve of Grade D as the mild steel with the variation of the strain rate : (a) 0.001 /sec, (b) 0.09 /sec, (c) 1.8 /sec, (d) 34 /sec]

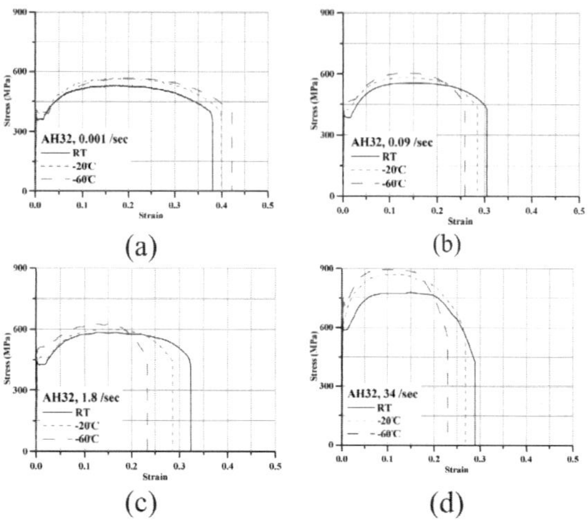

Figure 12: [Engineering stress-strain curve of Grade AH32 as the mild steel with the variation of the strain rate : (a) 0.001 /sec, (b) 0.09 /sec, (c) 1.8 /sec, (d) 34 /sec]

Table 6 mechanical properties from quasi-static tests

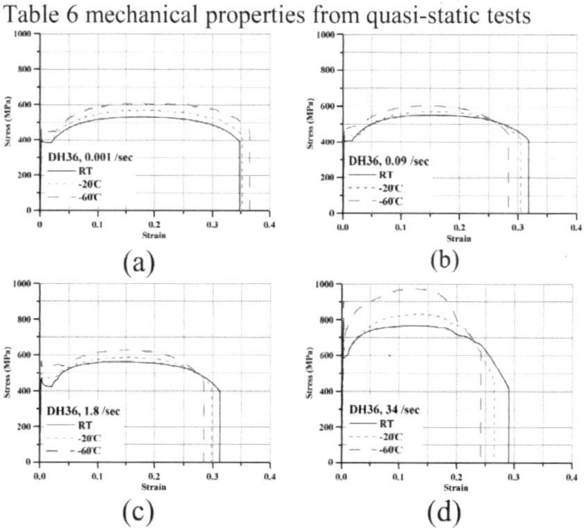

Figure 13: [Engineering stress-strain curve of Grade DH36 as the mild steel with the variation of the strain rate : (a) 0.001 /sec, (b) 0.09 /sec, (c) 1.8 /sec, (d) 34 /sec]

Table 7 Test results from quasi-static at 60, -20, RT

Temp. / Material		-60℃	-20℃	RT
Grade A	Δσ (MPa)	377.308	359.142	272.102
	m	0.070	0.071	0.061
Grade D	Δσ (MPa)	338.316	301.609	267.638
	m	0.065	0.070	0.063
Grade AH32	Δσ (MPa)	300.484	253.179	266.827
	m	0.054	0.048	0.054
Grade DH32	Δσ (MPa)	328.032	267.429	218.487
	m	0.060	0.054	0.046
Grade DH36	Δσ (MPa)	290.682	230.927	201.455
	m	0.048	0.043	0.040

As characteristics of general carbon steels, it was verified that both of mild and high-tensile steels showed both of their yield stress and tensile strength were increased as temperature decreased under the conditions with same strain rate.[7] In case of fracture strain, in strain rate of 0.001 /sec which is a quasi-static load velocity, all steels verify their tendencies that their fracture strains were increased as temperature goes down. However, after 0.09 /sec which is an intermediate strain rate, they showed their tendencies that their fracture strains were decreased as the temperature goes down. The reason to such decreases in their fracture strains in intermediate strain rate is considered as that fracture strain was decreased due to increase in strength caused by strain rate hardening as strain rate increases.

3.2 EFFECTS OF STRAIN RATE

Figure 14 to 17 shows stress - strain rate curve upon strain rate for each temperature on mild and high-tensile steels. For all of room temperature, -20 ℃ and -60 ℃, it indicated the tendency that higher strain rate caused increase in both yield stress and tensile stress and especially, stress was greatly increased with higher strain rate.[3] On the other hand, the tendency for fracture strain was to be decreased for all types of steels.

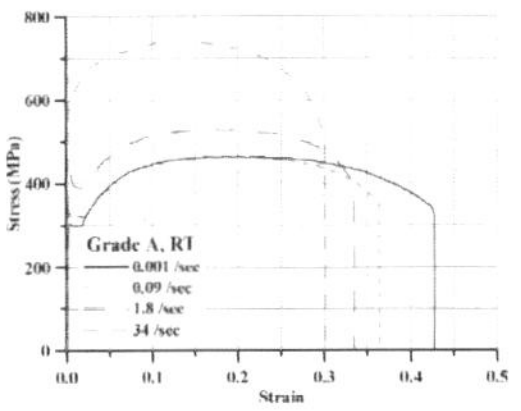

(a)

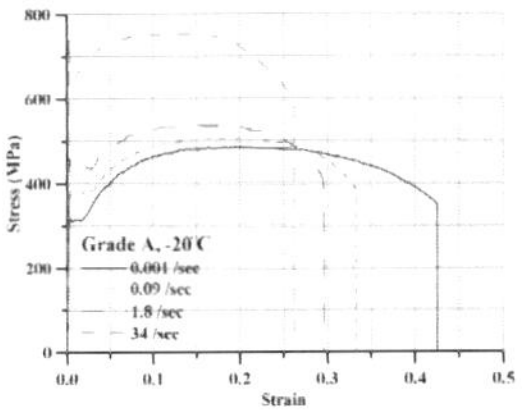

(b)

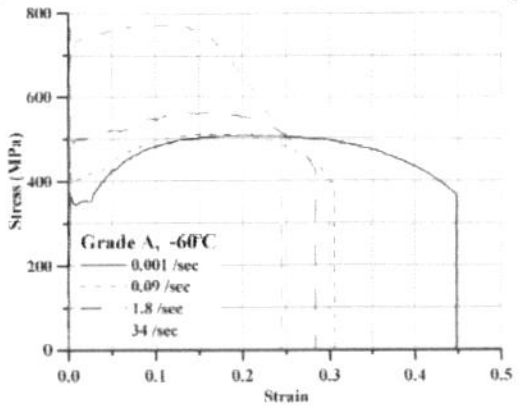

(c)

Figure 14: [Engineering stress-strain curve of Grade A as the mild steel with the variation of the temperature : (a) RT, (b) -20℃, (c) 60 ℃]

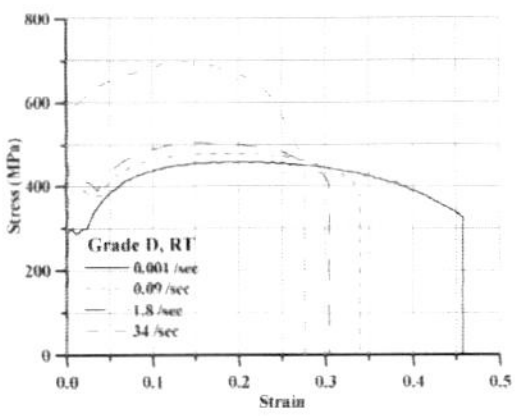

(a)

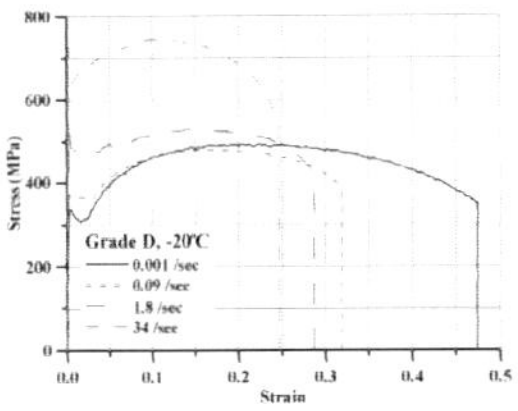

(b)

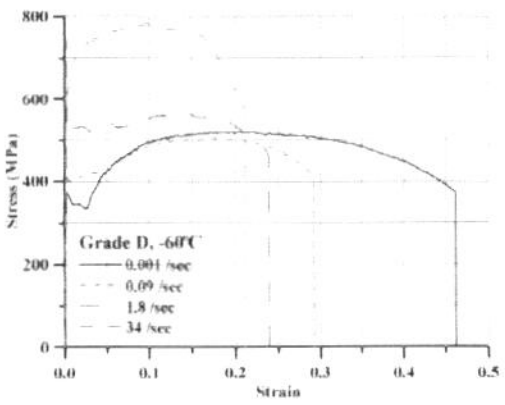

(c)

Figure 15: [Engineering stress-strain curve of Grade D as the mild steel with the variation of the temperature : (a) RT, (b) -20℃, (c) 60 ℃]

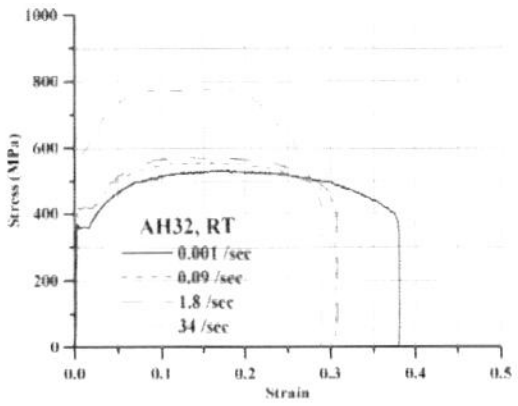

(a)

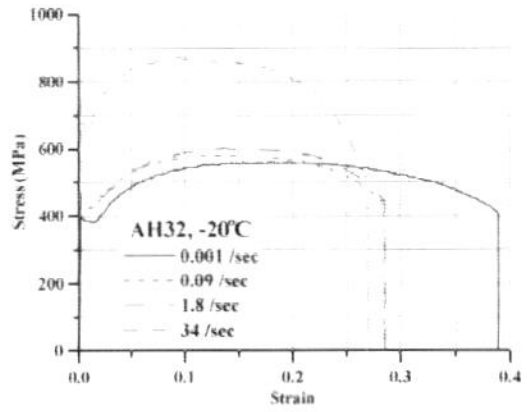

(b)

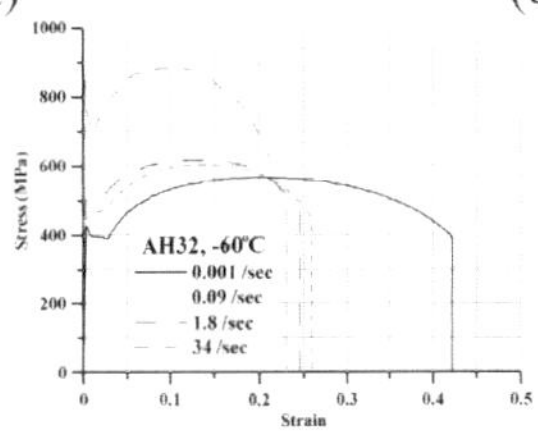

(c)

Figure 16: [Engineering stress-strain curve of Grade AH32 as the mild steel with the variation of the temperature : (a) RT, (b) -20℃, (c) 60 ℃]

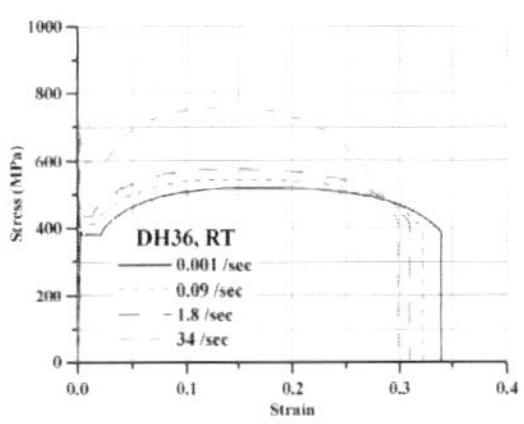

(a)

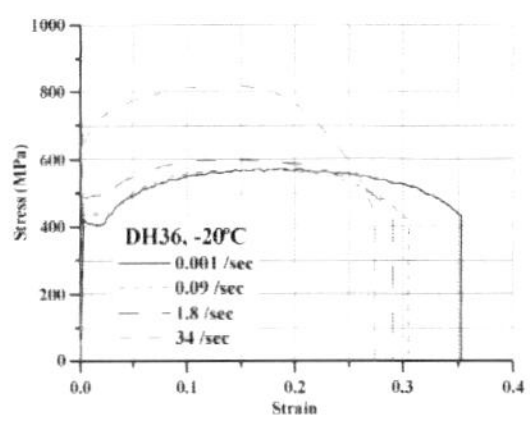

(b)

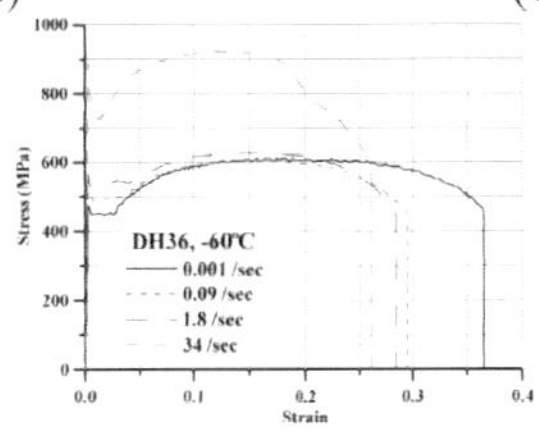

(c)

Figure 17: [Engineering stress-strain curve of Grade DH36 as the mild steel with the variation of the temperature : (a) RT, (b) -20℃, (c) 60 ℃]

4. DISCUSSION

4.1 STRAIN RATE CONSTITUTIVE EQUATION

Cowper-Symonds suggested the composition equation that consists of strain rate sensitivity as a function by displaying it as a dynamic yield stress compared to static yield stress as shown in Eq. (1).[11]

$$\frac{\sigma_{Yd}}{\sigma_Y} = 1.0 + \left(\frac{\dot{\varepsilon}}{C}\right)^{1/q} \quad (1)$$

In this paper, we compared and analyzed our results against material coefficients developed by other researchers previously by applying Cowper-Symonds model, a renowned and representative strain rate constitutive equation which did not consider thermal weakening.

Figure 18 shows a graph that displays the dynamic hardening coefficient against dynamic yield stress compared to static yield stress of mild steel fitted with the resultant values of other researches and the material coefficient suggested by Cowper-Symonds, C=40.4 and q=5 into purple lines.[5] As a result of displaying the values from our tests as dynamic hardening coefficients for each temperature of RT, -20 ℃ and -60 ℃, It was confirmed that all parts except the sections for 1.8 /sec of strain rate match with experimental results. Through this, we were able to verify the effects from low temperature are not significant and were able to verify that the existing coefficients C and q suggested by Cowper-Symonds can be equally applicable.

Figure 19 shows the values from our test data at RT for high-tensile steel fitted with dynamic hardening coefficient and fitted the values to the material coefficient suggested by Paik et al., C=3200 and q=5 then displayed them as purple dotted lines. The dynamic hardening coefficient through the values from the tests was similar to the curve which was substituted by the existing C and q values suggested by Pail et al. for the section up to the speed of 1.8 /sec. However, at the section for 34 /sec, the result showed higher values than the existing curve, resulting that it is more sensitive to increase in strain rate. However, as all structure designs must be strict and in conservative fashion in order to deal with unexpected issues, it was regarded that applying the existing C and q values can be ideal for the safety of designing structures for the industry. Furthermore, it was verified that those coefficients can be equally applicable for the Arctic.

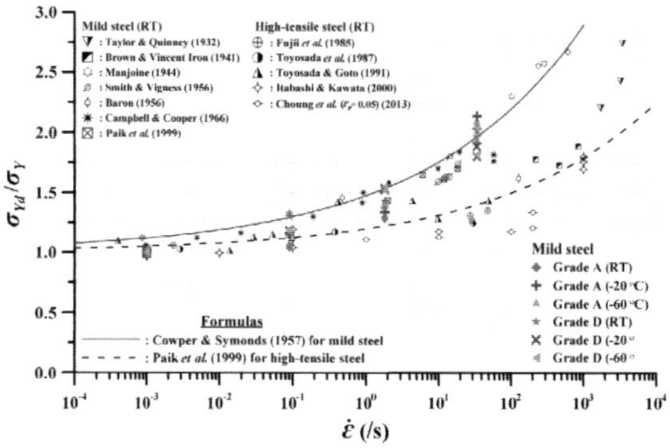

Figure 18: [Dynamic yield strength plotted versus strain rate for mild steel]

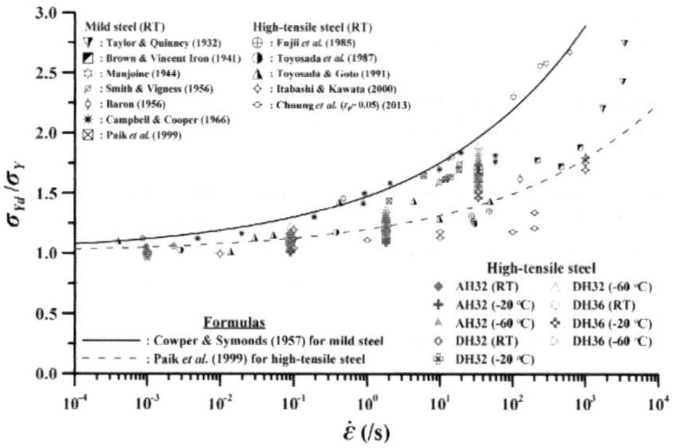

Figure 19: [Dynamic yield strength plotted versus strain rate for high-tensile steel]

Eq. (2) is Cowper-Symonds model that displays increase and decrease of dynamic fracture strain compared to static fracture strain which regards strain rate as its functions.

$$\frac{\varepsilon_{Fd}}{\varepsilon_F} = 1.0 + \left(\frac{\dot{\varepsilon}}{C}\right)^{1/q} \quad (1)$$

The results shown in Figure 20 were fitted with dynamic hardening coefficient to fracture strain of mild steel along with the material coefficients C and q values from Cowper-Symonds suggested by other researchers including C=40.4 and q=5. The tendency showed that the reciprocal substitution values acquired by substituting C=40.4 and q=5 that were suggested by Cowper-Symonds as the constitutive equation gave out either the same values or slightly higher values than the values from our tests. This was regarded as that by using the suggested coefficients by Cowper-Symonds will make structure designs safer with more conservative views.

Also, it was verified that those coefficients can be equally applicable for the Arctic.

Figure 21 shows a graph which displays the material coefficients C and q values suggested by Pail et al. and the ones derived by other researchers as the graph includes the test values on dynamic hardening coefficient to fracture strain of high-tensile steels. It was shown that the reciprocally substituted value of C=3200 and q=5 which was suggested by Paik et al. for the dynamic yield stress followed mostly to the tendency of the tested values.

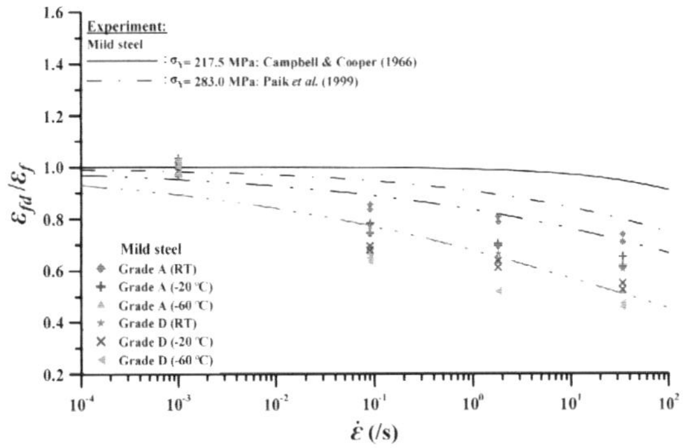

Figure 20: [Dynamic yield strength plotted versus strain rate for mild steel]

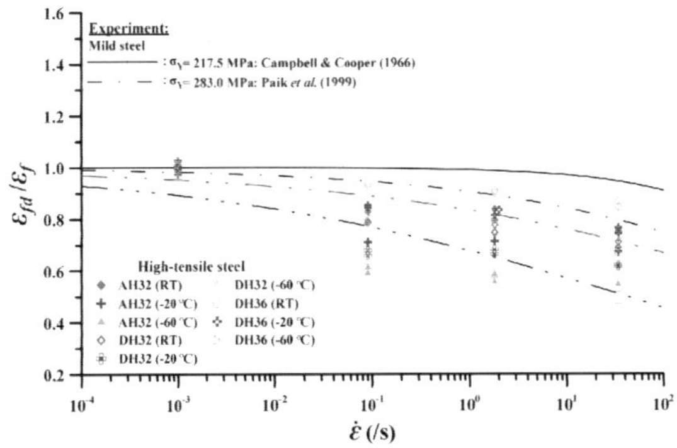

Figure 21: [Dynamic yield strength plotted versus strain rate for high-tensile steel]

5. CONCLUSION

In this paper, a series of material tensile tests were performed under various strain rate in extremely low, low and room temperature. In order to reflect the precise dynamic material properties upon temperatures and strain rate speeds to computer simulations, we newly suggested the material coefficient of Cowper-Symonds strain rate speed composition equation.

In case of mild and high-tensile steels, both of existing Cowper-Symonds and Paik et al. showed similar results to our test results hence it was verified that they are applicable as they are. Recently, the number of operations of ships and ocean plants in polar areas increases as well as increased amount of LNG carrier or FLNG are clear. In order to verify structure response occurring under various temperatures to accident loads, it is important to precisely understand nonlinearity of materials used in the structures and to reflect those into simulations. Therefore, by reflecting non-linear structure interpretation analysis results which are precisely reflecting material characteristics to the results from actual experiments and accidents, it is expected that such actions shall develop and improve required credibility.

6. ACKNOWLEDGEMENT

This research was supported by the Leading Foreign Research Institute Recruitment

Program through the National Research Foundation of Korea (NRF) funded by the Ministry of Science, Ict & future Planning (MSIP) (Grant no.: 2014040731). This work was part of the JDP titled 'Limit states design of type B independent LNG cargo tanks', which was funded by Class NK and STX Offshore & Shipbuilding Co., Ltd. The study was undertaken at the Lloyd's Register Foundation Research Centre of Excellence at Pusan National University.

Lloyd's Register Foundation (LRF), a UK registered charity and sole shareholder of Lloyd's Register Group Ltd, invests in science, engineering and technology for public benefit, worldwide.

7. REFERENCES

1. AMIN, H., 'Mechanical properties of very high strength steel at elevated temperature, Vol.64, pp27-35, 2014.

2. ASTM, 'Standard test methods for tension testing of metallic materials, E8, 2010.

3. BOYCE, B.Ll., 'The dynamic tensile behavior of tough, ultrahigh-strength steels at strain rate from $0.0002s^{-1}$ to 200 s^{-1} , Vol.36, pp263-271, 2009.

4. BROWN, A.F.C. and VINCENT, N.D.G., 'The relationship between stress and strain in the tensile impact test, 1941.

5. J. D. 'CAMPBELL and R. H. COOPER, 'yield and flow of low-carbon steel at medium strain rates, proc. Conf. physical basis of yield and fracture', Int. of physical soc., pp.77-87, 1966.

6. CHOUNG, J.M., 'Dynamic hardening behaviors of various marine structural steels considering dependencies on strain rate and temperature', Vol.32, pp.49-67, 2013.

7. HEIDARPOUR, A., 'Mechanical properties of partially damaged structural steel induced by high strain rate loading at elevated temperatures – An experimental investigation' Vol.76, pp.178-188, 2015.

8. H.S.E., 'Elevated temperature and high strain rate properties of offshore steels', Norwich NR3 1BQ, 2001.

9. MANJOINE, M. J., Influence of Rate of Strain and Temperature on Yield Stress of Mild Steel,. J. Appl. Mech., Vol. 11, pp. A.211- 218., 1944.

10. MASAKI and KAWATA., 'Carbon content effect on high strain rate tensile properties for carbon steels, Vol.24, pp.117-131, 2000.

11. PAIK, J.K. and THAYAMBALLI. A.K., Ultimate limit state design of steel-plated structures, Wiley, Chichester, UK, 2003.

12. SIA NEMAT-NASSER, 'Thermomechanical response of DH36 structural steel over a wide range of strain rates and temperatures', Vol35, pp.1023-1047, 2003.

13. SMITH and VIGNESS., 'The mechanical properties certain steels indicated by axial dynamic load tests', 1956.

14. TOYOSADA, M., 'The effect of strain rate on critical CTOD and J integral', 1987.

15. TOYOSADA, M. and KOTOH. K., 'Temperature rise distribution near a crack tip due to plastic work under high loading rate', 1991.

8. AUTHORS BIOGRAPHY

Jong Hwan Lee holds the current position of MSc in naval architecture and ocean engineering of Pusan national university and researcher at KMC Co., Ltd.

Ki Jong Kim holds the current position of MSc in naval architecture and ocean engineering of Pusan national university.

Bo Gyeong Jung holds the current position of MSc in naval architecture and ocean engineering of Pusan national university.

Xu Han holds the current position of MSc in naval architecture and ocean engineering of Pusan national university.

Dae Kyeom Park is responsible for researcher in Korea Ship and Offshore Research Institute

Bong Ju Kim is responsible for associate professor, Ph.D. at Pusan national university.

Jung Kwan Seo is responsible for associate professor, Ph.D. at Pusan national university.

Yeon Chul Ha is responsible for associate professor, Ph.D. at Pusan national university.

Toshiyuki Matsumoto is working at ClassNK.

Su Hwan Byeon is working at STX Offshore and Shipbuilding Co., Ltd.

Professor Jeom Kee Paik is a Professor of the Department of Naval Architecture and Ocean Engineering and serves as president of the Korea Ship and Offshore Research Institute (KOSORI) and as director of the Lloyd's Register Foundation Research Centre of Excellence at Pusan National University (PNU) in Korea. In 2015, he also joined University College London's Department of Mechanical Engineering as a professor and faculty member. He is a Fellow and Vice President of SNAME, USA and a Fellow and Council Member of RINA, UK. He was awarded William Froude Medal from RINA in 2015, the David W. Taylor Medal from SNAME in 2013, and a Doctor Honoris Causa from the University of Liege in Belgium in 2012. His research interests are safety engineering and design of ships and offshore structures in extreme and accidental conditions associated with risk assessment and management.

The Royal Institution of Naval Architects
8-9 Northumberland Street
London, WC2N 5DA
United Kingdom

ISBN 978-1-5108-9199-9

CURRAN ASSOCIATES INC.
proceedings.com